Wesley Chu, Tsau Young Lin (Eds.)

Foundations and Advances in Data Mining

Studies in Fuzziness and Soft Computing, Volume 180

Editor-in-chief
Prof. Janusz Kacprzyk
Systems Research Institute
Polish Academy of Sciences
ul. Newelska 6
01-447 Warsaw
Poland
E-mail: kacprzyk@ibspan.waw.pl

Wesley Chu
Tsau Young Lin
(Eds.)

Foundations and Advances in Data Mining

Springer

Professor Wesley Chu
University of California
at Los Angeles Department
of Computer Science
3731H Boelter Hall
Los Angeles, CA 90095-1596
USA
E-mail: wwc@cs.ucla.edu

Professor Tsau Young Lin
San Jose State University
Dept. Mathematics and Computer Science
San Jose, CA 95192-0103
USA
E-mail: tylin@mathcs.sjsu.edu

ISSN print edition: 1434-9922
ISSN electronic edition: 1860-0808
ISBN-10 3-642-42538-0 Springer Berlin Heidelberg New York
ISBN-13 978-3-642-42538-7 Springer Berlin Heidelberg New York

This work is subject to copyright. All rights are reserved, whether the whole or part of the material is concerned, specifically the rights of translation, reprinting, reuse of illustrations, recitation, broadcasting, reproduction on microfilm or in any other way, and storage in data banks. Duplication of this publication or parts thereof is permitted only under the provisions of the German Copyright Law of September 9, 1965, in its current version, and permission for use must always be obtained from Springer. Violations are liable for prosecution under the German Copyright Law.

Springer is a part of Springer Science+Business Media
springeronline.com
© Springer-Verlag Berlin Heidelberg 2005
Softcover re-print of the Hardcover 1st edition 2005

The use of general descriptive names, registered names, trademarks, etc. in this publication does not imply, even in the absence of a specific statement, that such names are exempt from the relevant protective laws and regulations and therefore free for general use.

Typesetting: by the authors and TechBooks using a Springer LaTeX macro package

Printed on acid-free paper   SPIN: 11362197   89/TechBooks   5 4 3 2 1 0

Preface

With the growing use of information technology and the recent advances in web systems, the amount of data available to users has increased exponentially. Thus, there is a critical need to understand the content of the data. As a result, data-mining has become a popular research topic in recent years for the treatment of the "data rich and information poor" syndrome. Currently, application oriented engineers are only concerned with their immediate problems, which results in an ad hoc method of problem solving. Researchers, on the other hand, lack an understanding of the practical issues of data-mining for real-world problems and often concentrate on issues (e.g. incremental performance improvements) that are of no significance to the practitioners.

In this volume, we hope to remedy these problems by (1) presenting a theoretical foundation of data-mining, and (2) providing important new directions for data-mining research. We have invited a set of well respected data mining theoreticians to present their views on the fundamental science of data mining. We have also called on researchers with practical data mining experiences to present new important data-mining topics.

This book is organized into two parts. The first part consists of four chapters presenting the foundations of data mining, which describe the theoretical point of view and the capabilities and limits of current available mining techniques. The second part consists of seven chapters which discuss the new data mining topics.

The first part of the book includes four chapters. The first chapter, authored by T. Poggio and S. Smale, is entitled "The Mathematics of Learning: Dealing with Data." The authors present the mathematical formula of learning theory. In particular, they present an algorithm for supervised learning by training sets and show that the algorithm performs well in a number of applications involving regression as well as binary classification. The second chapter, by H. Tsukimoto, is entitled "Logical Regression Analysis: From Mathematical Formulas to Linguistic Rules." He presents a solution for solving the accurate prediction and comprehensive rules in supervised learning. The author has developed a data mining technique called Logical Regression Analysis which consists of regression analysis, and the Approximation Method, that can provide comprehensive rules and also accurate prediction. The paper also shows how to apply the techniques for mining images. The third chapter, by T.Y. Lin, is entitled "A Feature/Attribute Theory for Association Mining and Constructing the Complete Feature Set" The author points out the importance of selecting correct attributes in data mining and develops a theory of features for association mining (AM). Based on the isomorphism theorem in AM, he concludes that it is sufficient to perform AM in canonical models, and constructs the complete feature set for every canonical model. Using the isomorphism theorem, the complete feature set can be derived for each relation. Though the number of possible features is enormous, it can be shown that the high frequency patterns features can be derived within polynomial time. The fourth chapter is entitled "A new theoretical framework for K-means-type clustering," and is authored by J. Peng and Y. Xia. The authors present generalized K-means type clustering method as the 0–1 semi-definite programming (SDP). The classical K-means algorithm, minimal sum of squares (MSSC), can be interpreted as a special heuristic. Moreover, the 0–1 SDP model can be further approximated by the relaxed and polynomially solvable linear and semi-definite programming. The 0–1 SDP model can be applied to MSCC and to other scenarios of clustering as well.

The second part of the book, from Chaps. 5 to 11, present seven topics covering recent advances in data mining. Chapter 5, entitled "Clustering via Decision Tree Construction," is authored by B. Liu, Y. Xia, and P. Yu. They propose a novel clustering technique based on supervised learning called decision tree construction. The key idea is to use a decision tree to partition the data space into cluster (or dense) regions and empty (or sparse) regions (which produce outliers and anomalies). This technique is able to find "natural" clusters in large high dimensional spaces efficiently. Experimental data shows that this technique is effective and scales well for large high dimensional datasets. Chapter 6, "Incremental Mining on Association Rules," is written by Wei-Guang Teng and Ming-Syan Chen. Due to the increasing use of the record-based databases where data is being constantly added, incremental mining is needed to keep the knowledge current. The authors propose incremental mining techniques to update the data mining on association rules. Chapter 7, is entitled "Mining Association Rules from Tabular Data Guided by Maximal Frequent Itemsets" and authored by Q. Zou, Y. Chen, W. Chu, and X. Lu. Since many scientific applications are in tabular format, the authors propose to use the maximum frequency itemset (MFI) as a road map to guide us towards generating association rules from tabular dataset. They propose to use information from previous searches to generate MFI and the experimental results show that such an approach to generating MFI yields significant improvements over conventional methods. Further, using tabular format rather than transaction data set to derive MFI can reduce the search space and the time needed for support-counting. The authors use spreadsheet to present rules and use spreadsheet operations to sort and select rules, which is a very convenient way to query and organize rules in a hierarchical fashion. An example was also given to illustrate the process of generating association rules from the tabular dataset using past medical surgery data to aid surgeons in their decision making. Chapter 8. entitled "Sequential Pattern Mining by Pattern-Growth: Principles and Extensions," presents the sequential pattern growth method and studies the principles and extensions of the method such as (1) mining constraint-based sequential patterns, (2) mining multi-level, multi dimensional sequential patters, and (3) mining top-k closed sequential patterns. They also discuss the applications in bio-sequence pattern analysis and clustering sequences. Chapter 9, entitled "Web Page Classification," is written by B. Choi and Z. Yao. It describes systems that automatically classify web pages into meaningful subject-based and genre-based categories. The authors describe tools for building automatic web page classification systems, which are essential for web mining and constructing semantic web. Chapter 10 is entitled "Web Mining – Concepts, Applications, and Research Directions." and was written by Jaideep Srivastava, Prasanna Desikan, and Vipin Kumar. The authors present the application of data mining techniques to extract knowledge from web content, structure, and usage. An overview of accomplishments in technology and applications in web mining is also included. Chapter 11. by Chris Clifton, Murat Kantarcioglu, and Jaideep Vaidya is entitled. "Privacy-Preserving Data Mining." The goal of privacy-preserving data mining is to develop data mining models that do not increase the risk of misuse of the data used to generate those models. The author presents two classes of privacy-preserving data-mining. The first is based on adding noise to the data before providing it to the data miner. Since real data values are not revealed. individual privacy is preserved. The second class is derived from the cryptographic community. The data sources collaborate to obtain data mining results without revealing anything except those results.

Finally, we would like to thank the authors for contributing their work in the volume and the reviewers for commenting on the readability and accuracy of the work. We hope the theories presented in this volume will give data mining practitioners a scientific perspective in data mining and thus provide more insight into their problems. We also hope that the new data mining topics will stimulate further research in these important directions.

California
March 2005

Wesley W. Chu
Tsau Young Lin

Contents

Part I

Foundations of Data Mining

The Mathematics of Learning: Dealing with Data[*]

T. Poggio[1] and S. Smale[2]

[1] CBCL, McGovern Institute, Artificial Intelligence Lab, BCS, MIT
tp@ai.mit.edu
[2] Toyota Technological Institute at Chicago and Professor in the Graduate School,
University of California, Berkeley
smale@math.berkeley.edu

Summary. Learning is key to developing systems tailored to a broad range of data analysis and information extraction tasks. We outline the mathematical foundations of learning theory and describe a key algorithm of it.

1 Introduction

The problem of understanding intelligence is said to be the greatest problem in science today and "the" problem for this century – as deciphering the genetic code was for the second half of the last one. Arguably, the problem of learning represents a gateway to understanding intelligence in brains and machines, to discovering how the human brain works and to making intelligent machines that learn from experience and improve their competences as children do. In engineering, learning techniques would make it possible to develop software that can be quickly customized to deal with the increasing amount of information and the flood of data around us. Examples abound. During the last decades, experiments in particle physics have produced a very large amount of data. Genome sequencing is doing the same in biology. The Internet is a vast repository of disparate information which changes rapidly and grows at an exponential rate: it is now significantly more than 100 Terabytes, while the Library of Congress is about 20 Terabytes. We believe that a set of techniques, based on a new area of science and engineering becoming known as "supervised learning" will become a key technology to extract information from the ocean of bits around us and make sense of it. Supervised learning, or learning-from-examples, refers to systems that are trained, instead of programmed, with a set of examples, that is a set of input-output pairs. Systems that could learn from example to perform a specific task would have many

---

[*]This paper is reprinted from Notices of the AMS, 50(5), 2003, pp. 537–544.

applications. A bank may use a program to screen loan applications and approve the "good" ones. Such a system would be trained with a set of data from previous loan applications and the experience with their defaults. In this example, a loan application is a point in a multidimensional space of variables characterizing its properties; its associated output is a binary "good" or "bad" label. In another example, a car manufacturer may want to have in its models, a system to detect pedestrians that may be about to cross the road to alert the driver of a possible danger while driving in downtown traffic. Such a system could be trained with positive and negative examples: images of pedestrians and images without people. In fact, software trained in this way with thousands of images has been recently tested in an experimental car of Daimler. It runs on a PC in the trunk and looks at the road in front of the car through a digital camera [26, 36, 43]. Algorithms have been developed that can produce a diagnosis of the type of cancer from a set of measurements of the expression level of many thousands human genes in a biopsy of the tumor measured with a cDNA microarray containing probes for a number of genes [46]. Again, the software learns the classification rule from a set of examples, that is from examples of expression patterns in a number of patients with known diagnoses. The challenge, in this case, is the high dimensionality of the input space – in the order of 20,000 genes – and the small number of examples available for training – around 50. In the future, similar learning techniques may be capable of some learning of a language and, in particular, to translate information from one language to another. What we assume in the above examples is a machine that is trained, instead of programmed, to perform a task, given data of the form $(x_i, y_i)_{i=1}^m$. Training means synthesizing a function that best represents the relation between the inputs x_i and the corresponding outputs y_i. The central question of learning theory is how well this function generalizes, that is how well it estimates the outputs for previously unseen inputs.

As we will see later more formally, learning techniques are similar to fitting a multivariate function to a certain number of measurement data. The key point, as we just mentioned, is that the fitting should be *predictive*, in the same way that fitting experimental data (see Fig. 1) from an experiment in physics can in principle uncover the underlying physical law, which is then used in a predictive way. In this sense, learning is also a principled method for distilling predictive and therefore scientific "theories" from the data. We begin by presenting a simple "regularization" algorithm which is important in learning theory and its applications. We then outline briefly some of its applications and its performance. Next we provide a compact derivation of it. We then provide general theoretical foundations of learning theory. In particular, we outline the key ideas of decomposing the generalization error of a solution of the learning problem into a sample and an approximation error component. Thus both probability theory and approximation theory play key roles in learning theory. We apply the two theoretical bounds to the algorithm and describe for it the tradeoff – which is key in learning theory and its

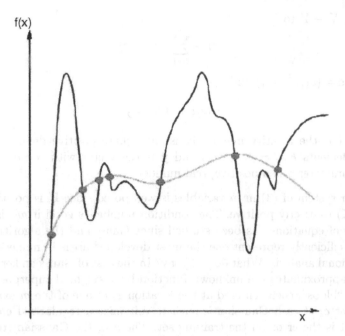

Fig. 1. How can we learn a function which is capable of generalization -- among the many functions which fit the examples equally well (here $m = 7$)?

applications -- between *number of examples* and *complexity of the hypothesis space*. We conclude with several remarks, both with an eye to history and to open problems for the future.

2 A Key Algorithm

2.1 The Algorithm

How can we fit the "training" set of data $S_m = (x_i, y_i)_{i=1}^m$ with a function $f : X \rightarrow Y$ -- with X a closed subset of $\mathbb{R}^n$ and $Y \subset \mathbb{R}$ -- that generalizes, eg is predictive? Here is an algorithm which does just that and which is almost magical for its simplicity and effectiveness:

1. Start with data $(x_i, y_i)_{i=1}^m$.
2. Choose a symmetric, positive definite function $K_x(x') = K(x, x')$, continuous on $X \times X$. A kernel $K(t, s)$ is *positive definite* if $\sum_{i,j=1}^n c_i c_j K(t_i, t_j) \geq 0$ for any $n \in \mathbb{N}$ and choice of $t_1, \ldots, t_n \in X$ and $c_1, \ldots, c_n \in \mathbb{R}$. An example of such a Mercer kernel is the Gaussian

$$K(x, x') = e^{-\frac{\|x - x'\|^2}{2\sigma^2}}. \qquad (1)$$

restricted to $X \times X$.

3. Set $f : X \to Y$ to

$$f(x) = \sum_{i=1}^{m} c_i K_{x_i}(x) .\tag{2}$$

where $\mathbf{c} = (c_1, \ldots, c_m)$ and

$$(m\gamma \mathbf{I} + \mathbf{K})\mathbf{c} = \mathbf{y} \tag{3}$$

where $\mathbf{I}$ is the identity matrix, $\mathbf{K}$ is the square positive definite matrix with elements $K_{i,j} = K(x_i, x_j)$ and $\mathbf{y}$ is the vector with coordinates y_i. The parameter γ is a positive, real number.

The linear system of (3) in m variables is well-posed since $\mathbf{K}$ is positive and $(m\gamma \mathbf{I} + \mathbf{K})$ is strictly positive. The condition number is good if $m\gamma$ is large. This type of equations has been studied since Gauss and the algorithms for solving it efficiently represent one the most developed areas in numerical and computational analysis. What does (2) say? In the case of Gaussian kernel, the equation approximates the unknown function by a weighted superposition of Gaussian "blobs", each centered at the location x_i of one of the m examples. The weight c_i of each Gaussian is such to minimize a regularized empirical error, that is the error on the training set. The σ of the Gaussian (together with γ, see later) controls the degree of smoothing, of noise tolerance and of generalization. Notice that for Gaussians with $\sigma \to 0$ the representation of (2) effectively becomes a "look-up" table that cannot generalize (it provides the correct $y = y_i$ only when $x = x_i$ and otherwise outputs 0).

2.2 Performance and Examples

The algorithm performs well in a number of applications involving regression as well as binary classification. In the latter case the y_i of the training set $(x_i, y_i)_{i=1}^{m}$ take the values $\{-1, +1\}$; the predicted label is then $\{-1, +1\}$, depending on the sign of the function f of (2). Regression applications are the oldest. Typically they involved fitting data in a small number of dimensions [44, 45, 53]. More recently, they also included typical learning applications, sometimes with a very high dimensionality. One example is the use of algorithms in computer graphics for synthesizing new images and videos [5,20,38]. The inverse problem of estimating facial expression and object pose from an image is another successful application [25]. Still another case is the control of mechanical arms. There are also applications in finance, as, for instance, the estimation of the price of derivative securities, such as stock options. In this case, the algorithm replaces the classical Black-Scholes equation (derived from first principles) by learning the map from an input space (volatility, underlying stock price, time to expiration of the option etc.) to the output space (the price of the option) from historical data [27]. Binary classification applications abound. The algorithm was used to perform binary classification on a number of problems [7,34]. It was also used to perform visual object recognition in a view-independent way and in particular face recognition and sex categorization from face images [8,39]. Other applications span bioinformatics for classification of human cancer from microarray data, text summarization, sound classification[1] Surprisingly, it has been realized quite recently that the same linear algorithm not only works well but is fully comparable in binary classification problems to the most popular classifiers of today (that turn out to be of the same family, see later).

2.3 Derivation

The algorithm described can be derived from Tikhonov regularization. To find the minimizer of the the error we may try to solve the problem – called Empirical Risk Minimization (ERM) – of finding the function in $\mathcal{H}$ which minimizes

$$\frac{1}{m}\sum_{i=1}^{m}(f(x_i)-y_i)^2$$

which is in general *ill-posed*, depending on the choice of the hypothesis space $\mathcal{H}$. Following Tikhonov (see for instance [19]) we minimize, instead, over the hypothesis space $\mathcal{H}_K$, for a fixed positive parameter γ, the regularized functional

$$\frac{1}{m}\sum_{i=1}^{m}(y_i-f(x_i))^2+\gamma\|f\|_K^2\ ,\tag{4}$$

where $\|f\|_K^2$ is the norm in $\mathcal{H}_K$ – the Reproducing Kernel Hilbert Space (RKHS), defined by the kernel K. The last term in (4) – called *regularizer* – forces, as we will see, smoothness and uniqueness of the solution. Let us first define the norm $\|f\|_K^2$. Consider the space of the linear span of $K_{\bar{x}_j}$. We use $\bar{x}_j$ to emphasize that the elements of X used in this construction do not have anything to do *in general* with the training set $(x_i)_{i=1}^m$. Define an inner product in this space by setting $\langle K_x, K_{\bar{x}_j}\rangle = K(x,\bar{x}_j)$ and extend linearly to $\sum_{j=1}^r a_j K_{\bar{x}_j}$. The completion of the space in the associated norm is the RKHS, that is a Hilbert space $\mathcal{H}_K$ with the norm $\|f\|_K^2$ (see [2, 10]). Note that $\langle f, K_x\rangle = f(x)$ for $f \in \mathcal{H}_K$ (just let $f = K_{\bar{x}_j}$ and extend linearly). To minimize the functional in (4) we take the functional derivative with respect to f, apply it to an element $\bar{f}$ of the RKHS and set it equal to 0. We obtain

$$\frac{1}{m}\sum_{i=1}^{m}(y_i-f(x_i))\bar{f}(x_i)-\gamma\langle f,\bar{f}\rangle=0\ .\tag{5}$$

---

[1]The very closely related Support Vector Machine (SVM) classifier was used for the same family of applications, and in particular for bioinformatics and for face recognition and car and pedestrian detection [25, 46].

Equation (5) must be valid for any $\bar{f}$. In particular, setting $\bar{f} = K_x$ gives

$$f(x) = \sum_{i=1}^{m} c_i K_{x_i}(x) \tag{6}$$

where

$$c_i = \frac{y_i - f(x_i)}{m\gamma} \tag{7}$$

since $\langle f, K_x \rangle = f(x)$. Equation (3) then follows, by substituting (6) into (7). Notice also that essentially the same derivation for a generic loss function $V(y, f(x))$, instead of $(f(x) - y)^2$, yields the same (6), but (3) is now different and, in general, nonlinear, depending on the form of V. In particular, the popular Support Vector Machine (SVM) regression and SVM classification algorithms correspond to special choices of non-quadratic V, one to provide a "robust" measure of error and the other to approximate the ideal loss function corresponding to binary (miss)classification. In both cases, the solution is still of the same form of (6) for any choice of the kernel K. The coefficients c_i are not given anymore by (7) but must be found solving a quadratic programming problem.

3 Theory

We give some further justification of the algorithm by sketching very briefly its foundations in some basic ideas of learning theory. Here the data $(x_i, y_i)_{i=1}^{m}$ is supposed random, so that there is an unknown probability measure ρ on the product space $X \times Y$ from which the data is drawn. This measure ρ defines a function

$$f_\rho : X \to Y \tag{8}$$

satisfying $f_\rho(x) = \int y d\rho_x$, where ρ_x is the conditional measure on $x \times Y$. From this construction f_ρ can be said to be the true input-output function reflecting the environment which produces the data. Thus a measurement of *the error* of f is

$$\int_X (f - f_\rho)^2 d\rho_X \tag{9}$$

where ρ_X is the measure on X induced by ρ (sometimes called the *marginal measure*). The goal of learning theory might be said to "find" f minimizing this error. Now to search for such an f, it is important to have a space $\mathcal{H}$ – the *hypothesis space* – in which to work ("learning does not take place in a vacuum"). Thus consider a convex space of continuous functions $f : X \to Y$, (remember $Y \subset \mathbb{R}$) which as a subset of $C(X)$ is *compact*, and where $C(X)$ is the Banach space of continuous functions with $\|f\| = \max_X |f(x)|$. A basic example is

$$\mathcal{H} = \overline{I_K(B_R)} \tag{10}$$

where $I_K : \mathcal{H}_K \to C(X)$ is the inclusion and B_R is the ball of radius R in $\mathcal{H}_K$. Starting from the data $(x_i, y_i)_{i=1}^m = z$ one may minimize $\frac{1}{m} \sum_{i=1}^m (f(x_i) - y_i)^2$ over $f \in \mathcal{H}$ to obtain a unique hypothesis $f_z : X \to Y$. This f_z is called the empirical optimum and we may focus on the problem of estimating

$$\int_X (f_z - f_\rho)^2 d\rho_X \qquad (11)$$

It is useful towards this end to break the problem into steps by defining a "true optimum" $f_{\mathcal{H}}$ relative to $\mathcal{H}$, by taking the minimum over $\mathcal{H}$ of $\int_X (f - f_\rho)^2$. Thus we may exhibit

$$\int_X (f_z - f_\rho)^2 = S(z, \mathcal{H}) + \int_X (f_{\mathcal{H}} - f_\rho)^2 = S(z, \mathcal{H}) + A(\mathcal{H}) \qquad (12)$$

where

$$S(z, \mathcal{H}) = \int_X (f_z - f_\rho)^2 - \int_X (f_{\mathcal{H}} - f_\rho)^2 \qquad (13)$$

The first term, (S) on the right in (12) must be estimated in probability over z and the estimate is called the *sample error* (sometime also the *estimation error*). It is naturally studied in the theory of probability and of empirical processes [16, 30, 31]. The second term (A) is dealt with via approximation theory (see [15] and [12- 14, 32, 33]) and is called the *approximation error*. The decomposition of (12) is related, but not equivalent, to the well known bias (A) and variance (S) decomposition in statistics.

3.1 Sample Error

First consider an estimate for the sample error, which will have the form:

$$S(z, \mathcal{H}) \leq \epsilon \qquad (14)$$

with high confidence, this confidence depending on ϵ and on the sample size m. Let us be more precise. Recall that the covering number or $\text{Cov}\#(\mathcal{H}, \eta)$ is the number of balls in $\mathcal{H}$ of radius η needed to cover $\mathcal{H}$.

Theorem 3.1 *Suppose* $|f(x) - y| \leq M$ *for all* $f \in \mathcal{H}$ *for almost all* $(x, y) \in X \times Y$. *Then*

$$Prob_{z \in (X \times Y)^m} \{S(z, \mathcal{H}) \leq \epsilon\} \leq 1 - \delta$$

where $\delta = \text{Cov}\#(\mathcal{H}, \frac{\epsilon}{24M}) e^{-\frac{m\epsilon}{288M^2}}$.

The result is Theorem C^* of [10], but earlier versions (usually without a topology on $\mathcal{H}$) have been proved by others, especially Vapnik, who formulated the notion of VC dimension to measure the complexity of the hypothesis space for the case of $\{0, 1\}$ functions. In a typical situation of Theorem 3.1 the hypothesis space $\mathcal{H}$ is taken to be as in (10), where B_R is the ball of radius R in a Reproducing Kernel Hilbert Space (RKHS) with a smooth K (or in a Sobolev space). In this context, R plays an analogous role to VC dimension [50]. Estimates for the covering numbers in these cases were provided by Cucker, Smale and Zhou [10, 54, 55]. The proof of Theorem 3.1 starts from Hoeffding inequality (which can be regarded as an exponential version of Chebyshev's inequality of probability theory). One applies this estimate to the function $X \times Y \to \mathbb{R}$ which takes (x, y) to $(f(x) - y)^2$. Then extending the estimate to the set of $f \in \mathcal{H}$ introduces the covering number into the picture. With a little more work, Theorem 3.1 is obtained.

3.2 Approximation Error

The approximation error $\int_X (f_{\mathcal{H}} - f_\rho)^2$ may be studied as follows. Suppose $B : L^2 \to L^2$ is a compact, strictly positive (selfadjoint) operator. Then let **E** be the Hilbert space

$$\{g \in L^2, \|B^{-s}g\| < \infty\}$$

with inner product $\langle g, h \rangle_{\mathbf{E}} = \langle B^{-s}g, B^{-s}h \rangle_{L^2}$. Suppose moreover that $\mathbf{E} \to L^2$ factors as $\mathbf{E} \to C(X) \to L^2$ with the inclusion $J_{\mathbf{E}} : \mathbf{E} \hookrightarrow C(X)$ well defined and compact. Let $\mathcal{H}$ be $\overline{J_{\mathbf{E}}(B_R)}$ when B_R is the ball of radius R in **E**. A theorem on the approximation error is

Theorem 3.2 *Let $0 < r < s$ and $\mathcal{H}$ be as above. Then*

$$\|f_\rho - f_{\mathcal{H}}\|^2 \le \left(\frac{1}{R}\right)^{\frac{2r}{s-r}} \|B^{-r}f_\rho\|^{\frac{2s}{s-r}}$$

We now use $\| \cdot \|$ for the norm in the space of square integrable functions on X, with measure ρ_X. For our main example of RKHS, take $B = L_K^{1/2}$, where K is a Mercer kernel and

$$L_K f(x) = \int_X f(x')K(x, x') \tag{15}$$

and we have taken the square root of the operator L_K. In this case **E** is $\mathcal{H}_K$ as above. Details and proofs may be found in [10] and in [48].

3.3 Sample and Approximation Error
for the Regularization Algorithm

The previous discussion depends upon a compact hypothesis space $\mathcal{H}$ from which the experimental optimum f_z and the true optimum $f_{\mathcal{H}}$ are taken. In the key algorithm of Sect. 2, the optimization is done over all $f \in \mathcal{H}_K$ with a regularized error function. The error analysis of Sects. 3.1 and 3.2 must therefore be extended. Thus let $f_{\gamma, z}$ be the empirical optimum for the regularized problem as exhibited in (4)

$$\frac{1}{m}\sum_{i=1}^{m}(y_i - f(x_i))^2 + \gamma\|f\|_K^2 .\tag{16}$$

Then

$$\int(f_{\gamma,z} - f_\rho)^2 \leq S(\gamma) + A(\gamma)\tag{17}$$

where $A(\gamma)$ (the approximation error in this context) is

$$A(\gamma) = \gamma^{1/2}\left\|L_K^{-\frac{1}{4}}f_\rho\right\|^2\tag{18}$$

and the sample error is

$$S(\gamma) = \frac{32M^2(\gamma + C)^2}{\gamma^2}v^*(m,\delta)\tag{19}$$

where $v^*(m,\delta)$ is the unique solution of

$$\frac{m}{4}v^3 - \ln\left(\frac{4m}{\delta}\right)v - c_1 = 0 .\tag{20}$$

Here $C, c_1 > 0$ depend only on X and K. For the proof one reduces to the case of compact $\mathcal{H}$ and applies Theorems 3.1 and 3.2. Thus finding the optimal solution is equivalent to finding the best tradeoff between A and S for a given m. In our case, this bias-variance problem is to minimize $S(\gamma) + A(\gamma)$ over $\gamma > 0$. There is a unique solution — a best γ – for the choice in (4). For this result and its consequences see [11].

4 Remarks

The Tradeoff between Sample Complexity and Hypothesis Space Complexity For a given, fixed hypothesis space $\mathcal{H}$ only the sample error component of the *error* of f_z can be be controlled (in (12) only $S(z,\mathcal{H})$ depends on the data). In this view, convergence of S to zero as the number of data increases (Theorem 3.1) is then the central problem in learning. Vapnik called consistency of ERM (eg convergence of the empirical error to the true error) the key problem in learning theory and in fact much modern work has focused on refining the necessary and sufficient conditions for consistency of ERM (the uniform Glivenko-Cantelli property of $\mathcal{H}$, finite V_γ dimension for $\gamma > 0$ etc., see [19]). More generally, however, there is a tradeoff between minimizing the sample error and minimizing the approximation error – what we referred to as the bias-variance problem. Increasing the number of data points m decreases the sample error. The effect of increasing the complexity of the hypothesis space is trickier. Usually the approximation error decreases but the sample error increases. This means that there is an optimal complexity of the hypothesis space for a given number of training data. In the case of the regularization algorithm described in this paper this tradeoff corresponds to an optimum value for γ as studied by [3, 11, 35]. In empirical work, the optimum value is often found through cross-validation techniques [53]. This tradeoff between approximation error and sample error is probably the most critical issue in determining good performance on a given problem. The class of regularization algorithms, such as (4), shows clearly that it is also a tradeoff – quoting Girosi – between the *curse of dimensionality* (not enough examples) and *the blessing of smoothness* (which decreases the effective "dimensionality" eg the complexity of the hypothesis space) through the parameter γ.

The Regularization Algorithm and Support Vector Machines. There is nothing to stop us from using the algorithm we described in this paper – that is square loss regularization – for *binary classification*. Whereas SVM classification arose from using – *with binary y* – the loss function

$$V(f(\mathbf{x}, y)) = (1 - yf(\mathbf{x}))_+ ,$$

we can perform least-squares regularized classification via the loss function

$$V(f(\mathbf{x}, y)) = (f(\mathbf{x}) - y)^2 .$$

This classification scheme was used at least as early as 1989 (for reviews see [7, 40] and then rediscovered again by many others (see [21, 49]), including Mangasarian (who refers to square loss regularization as "proximal vector machines") and Suykens (who uses the name "least square SVMs"). Rifkin [47] has confirmed the interesting empirical results by Mangasarian and Suykens: "classical" square loss regularization works well also for binary classification (examples are in Tables 1 and 2).

In references to supervised learning the Support Vector Machine method is often described (see for instance a recent issue of the Notices of the AMS [28]) according to the "traditional" approach, introduced by Vapnik and followed by almost everybody else. In this approach, one starts with the concepts of *separating hyperplanes* and *margin*. Given the data, one searches for the linear hyperplane that separates the positive and the negative examples, assumed to be linearly separable, with the largest margin (the margin is defined as

Table 1. A comparison of SVM and RLSC (Regularized Least Squares Classification) accuracy on a multiclass classification task (the 20newsgroups dataset with 20 classes and high dimensionality, around 50,000), performed using the standard "one vs. all" scheme based on the use of binary classifiers. The top row indicates the number of documents/class used for training. Entries in the table are the fraction of misclassified documents. From [47]

800		250		100		30	
SVM	RLSC	SVM	RLSC	SVM	RLSC	SVM	RLSC
0.131	0.129	0.167	0.165	0.214	0.211	0.311	0.309

Table 2. A comparison of SVM and RLSC accuracy on another multiclass classification task (the `sector105` dataset, consisting of 105 classes with dimensionality about 50,000). The top row indicates the number of documents/class used for training. Entries in the table are the fraction of misclassified documents. From [47]

52		20		10		3	
SVM	RLSC	SVM	RLSC	SVM	RLSC	SVM	RLSC
0.072	0.066	0.176	0.169	0.341	0.335	0.650	0.648

the distance from the hyperplane to the nearest example). Most articles and books follow this approach, go from the separable to the non-separable case and use a so-called "kernel trick" (!) to extend it to the nonlinear case. SVM for classification was introduced by Vapnik in the linear, separable case in terms of maximizing the margin. In the non-separable case, the margin motivation loses most of its meaning. A more general and simpler framework for deriving SVM algorithms for classification and regression is to regard them as special cases of regularization and follow the treatment of Sect. 2. In the case of linear functions $f(x) = w \cdot x$ and separable data, maximizing the margin is exactly equivalent to maximizing $\frac{1}{||w||}$, which is in turn equivalent to minimizing $||w||^2$, which corresponds to minimizing the RKHS norm.

The Regularization Algorithm and Learning Theory. The Mercer theorem was introduced in learning theory by Vapnik and RKHS by Girosi [22] and later by Vapnik [9, 50]. Poggio and Girosi [23, 40, 41] had introduced Tikhonov regularization in learning theory (the reformulation of Support Vector Machines as a special case of regularization can be found in [19]). Earlier, Gaussian Radial Basis Functions were proposed as an alternative to neural networks by Broomhead and Loewe. Of course, RKHS had been pioneered by Parzen and Wahba [37, 53] for applications closely related to learning, including data smoothing (for image processing and computer vision, see [4, 42]).

A Bayesian Interpretation. The learning algorithm (4) has an interesting Bayesian interpretation [52, 53]: the data term – that is the first term with the quadratic loss function – is a model of (Gaussian, additive) noise and the RKHS norm (the stabilizer) corresponds to a prior probability on the hypothesis space $\mathcal{H}$. Let us define $P[f|S_m]$ as the conditional probability of the function f given the training examples $S_m = (x_i, y_i)_{i=1}^m$, $P[S_m|f]$ as the conditional probability of S_m given f, i.e. a model of the noise, and $P[f]$ as the a priori probability of the random field f. Then Bayes theorem provides the posterior distribution as

$$P[f|S_m] = \frac{P[S_m|f] \, P[f]}{P(S_m)} \, .$$

If the noise is normally distributed with variance σ, then the probability $P[S_m|f]$ is

$$P[S_m|f] = \frac{1}{Z_L} e^{-\frac{1}{2\sigma^2} \sum_{i=1}^{m}(y_i - f(x_i))^2}$$

where Z_L is a normalization constant. If $P[f] = \frac{1}{Z_r} e^{-\|f\|_K^2}$ where Z_r is another normalization constant, then

$$P[f|S_m] = \frac{1}{Z_D Z_L Z_r} e^{-\left(\frac{1}{2\sigma^2} \sum_{i=1}^{m}(y_i - f(x_i))^2 + \|f\|_K^2\right)} .$$

One of the several possible estimates of f from $P[f|S_m]$ is the so called *Maximum A Posteriori* (MAP) estimate, that is

$$\max P[f|S_m] = \min \sum_{i=1}^{m}(y_i - f(\mathbf{x}_i))^2 + 2\sigma^2 \|f\|_K^2 .$$

which is the same as the regularization functional, if $\lambda = 2\sigma^2/m$ (for details and extensions to models of non Gaussian noise and different loss functions see [19]).

Necessary and Sufficient Conditions for Learnability. Compactness of the hypothesis space $\mathcal{H}$ is *sufficient* for consistency of ERM, that is for bounds of the type of Theorem 3.1 on the sample error. The *necessary and sufficient* condition is that $\mathcal{H}$ is a uniform Glivenko-Cantelli class of functions, in which case no specific topology is assumed for $\mathcal{H}^2$. There are several equivalent conditions on $\mathcal{H}$ such as finiteness of the V_γ dimension for all positive γ (which reduces to finiteness of the VC dimension for $\{0,1\}$ functions)[3]. We saw earlier that the regularization algorithm (4) ensures (through the resulting compactness of the "effective" hypothesis space) well-posedness of the problem. It also yields convergence of the empirical error to the true error (eg bounds such as Theorem 3.1). An open question is whether there is a connection between well-posedness and consistency. For well-posedness the critical condition is usually stability of the solution. In the learning problem, this condition refers to stability of the solution of ERM with respect to small changes of the training set S_m. In a similar way, the condition number (see [6] and especially [29]) characterizes the stability of the solution of (3). Is it possible that some specific

---

[2]*Definition: Let $\mathcal{F}$ be a class of functions f. $\mathcal{F}$ is a uniform Glivenko-Cantelli class if for every $\varepsilon > 0$*

$$\lim_{m \to \infty} \sup_{\rho} \mathbb{P} \left\{ \sup_{f \in \mathcal{F}} |E_{\rho_m} f - E_\rho f| > \varepsilon \right\} = 0 . \qquad (21)$$

where ρ_n is the empirical measure supported on a set $x_1, \ldots, x_n$.

[3]In [1] – following [17, 51] – a necessary and sufficient condition is proved for uniform convergence of $|I_{emp}[f] - I_{exp}[f]|$, in terms of the finiteness for all $\gamma > 0$ of a combinatorial quantity called V_γ dimension of $\mathcal{F}$ (which is the set $V(x), f(x), f \in \mathcal{H}$), under some assumptions on V. The result is based on a necessary and sufficient (distribution independent) condition proved by [51] which uses the metric entropy of $\mathcal{F}$ defined as $H_m(\epsilon, \mathcal{F}) = \sup_{x_m \in X^m} \log \mathcal{N}(\epsilon, \mathcal{F}, x_m)$, where $\mathcal{N}(\epsilon, \mathcal{F}, x_m)$ is the ϵ-covering of $\mathcal{F}$ wrt $l_{x_m}^\infty$ ($l_{x_m}^\infty$ is the l^∞ distance on the points x_m): *Theorem (Dudley, see [18]). $\mathcal{F}$ is a uniform Glivenko-Cantelli class iff* $\lim_{m \to \infty} \frac{H_m(\epsilon, \mathcal{F})}{m} = 0$ *for all* $\epsilon > 0$.

form of stability may be necessary and sufficient for consistency of ERM? *Such a result would be surprising* because, a priori, there is no reason why there should be a connection between well-posedness and consistency: they are both important requirements for ERM but they seem quite different and independent of each other.

Learning Theory, Sample Complexity and Brains. The theory of supervised learning outlined in this paper and in the references has achieved a remarkable degree of completeness and of practical success in many applications. Within it, many interesting problems remain open and are a fertile ground for interesting and useful mathematics. One may also take a broader view and ask: what next? One could argue that the most important aspect of intelligence and of the amazing performance of real brains is the ability to learn. How then do the learning machines we have described in the theory compare with brains? There are of course many aspects of biological learning that are not captured by the theory and several difficulties in making any comparison. One of the most obvious differences, however, is the ability of people and animals to learn from very few examples. The algorithms we have described can learn an object recognition task from a few thousand labeled images. This is a small number compared with the apparent dimensionality of the problem (thousands of pixels) but a child, or even a monkey, can learn the same task from just a few examples. Of course, evolution has probably done a part of the learning but so have we, when we choose for any given task an appropriate input representation for our learning machine. From this point of view, as Donald Geman has argued, the interesting limit is not "m goes to infinity," but rather "m goes to zero". Thus an important area for future theoretical and experimental work is learning from partially labeled examples (and the related area of active learning). In the first case there are only a small number ℓ of labeled pairs $(x_i, y_i)_{i=1}^{\ell}$ – for instance with y_i binary – and many unlabeled data $(x_i)_{\ell+1}^{m}$. $m \gg \ell$. Though interesting work has begun in this direction, a satisfactory theory that provides conditions under which unlabeled data can be used is still lacking. A comparison with real brains offers another, and probably related, challenge to learning theory. The "learning algorithms" we have described in this paper correspond to one-layer architectures. Are hierarchical architectures with more layers justifiable in terms of learning theory? It seems that the learning theory of the type we have outlined does not offer any general argument in favor of hierarchical learning machines for regression or classification. This is somewhat of a puzzle since the organization of cortex – for instance visual cortex – is strongly hierarchical. At the same time, hierarchical learning systems show superior performance in several engineering applications. For instance, a face categorization system in which a single SVM classifier combines the real-valued output of a few classifiers, each trained to a different component of faces – such as eye and nose –, outperforms a single classifier trained on full images of faces [25]. The theoretical issues surrounding hierarchical systems of this type are wide open, and likely to be of paramount importance for the next major development of efficient classifiers in several application domains. Why hierarchies? There may be reasons of *efficiency* – computational speed and use of computational resources. For instance, the lowest levels of the hierarchy may represent a dictionary of features that can be shared across multiple classification tasks (see [24]). Hierarchical system usually decompose a task in a series of simple computations at each level – often an advantage for fast implementations. There may also be the more fundamental issue of *sample complexity*. We mentioned that an obvious difference between our best classifiers and human learning is the number of examples required in tasks such as object detection. The theory described in this paper shows that the difficulty of a learning task depends on the size of the required hypothesis space. This complexity determines in turn how many training examples are needed to achieve a given level of generalization error. Thus the complexity of the hypothesis space sets the speed limit and the sample complexity for learning. If a task – like a visual recognition task – can be decomposed into low-complexity learning tasks, for each layer of a hierarchical learning machine, then each layer may require only a small number of training examples. Of course, not all classification tasks have a hierarchical representation. Roughly speaking, the issue is under which conditions a function of many variables can be approximated by a function of a small number of functions of subsets of the original variables. Neuroscience suggests that what humans can learn can be represented by hierarchies that are locally simple. Thus our ability of learning from just a few examples, and its limitations, may be related to the hierarchical architecture of cortex. This is just one of several possible connections, still to be characterized, between learning theory and the ultimate problem in natural science – the organization and the principles of higher brain functions.

Acknowledgments

Thanks to Felipe Cucker, Federico Girosi, Don Glaser, Sayan Mukherjee, Massimiliano Pontil, Martino Poggio and Ryan Rifkin.

References

1. N. Alon, S. Ben-David, N. Cesa-Bianchi, and D. Haussler. Scale-sensitive dimensions, uniform convergence, and learnability. *J. of the ACM*, 44(4):615–631, 1997.
2. N. Aronszajn. Theory of reproducing kernels. *Trans. Amer. Math. Soc.*, 686:337–404, 1950.
3. A.R. Barron. Approximation and estimation bounds for artificial neural networks. *Machine Learning*, 14:115–133, 1994.
4. M. Bertero, T. Poggio, and V. Torre. Ill-posed problems in early vision. *Proceedings of the IEEE*, 76:869–889, 1988.

5. D. Beymer and T. Poggio. Image representations for visual learning. *Science*, 272(5270):1905–1909, June 1996.
6. O. Bousquet and A. Elisseeff. Stability and generalization. *Journal of Machine Learning Research*, (2):499–526, 2002.
7. D.S. Broomhead and D. Lowe. Multivariable functional interpolation and adaptive networks. *Complex Systems*, 2:321–355, 1988.
8. R. Brunelli and T. Poggio. Hyberbf networks for real object recognition. In *Proceedings IJCAI*, Sydney, Australia, 1991.
9. C. Cortes and V. Vapnik. Support vector networks. *Machine Learning*, 20:1–25, 1995.
10. F. Cucker and S. Smale. On the mathematical foundations of learning. *Bulletin of AMS*, 39:1–49, 2001.
11. F. Cucker and S. Smale. Best choices for regularization parameters in learning theory: on the bias-variance problem. *Foundations of Computational Mathematics*, 2(4):413–428, 2002.
12. I. Daubechies. *Ten lectures on wavelets*. CBMS-NSF Regional Conferences Series in Applied Mathematics. SIAM, Philadelphia, PA, 1992.
13. R. DeVore, R. Howard, and C. Micchelli. Optimal nonlinear approximation. *Manuskripta Mathematika*, 1989.
14. R.A. DeVore, D. Donoho, M. Vetterli, and I. Daubechies. Data compression and harmonic analysis. *IEEE Transactions on Information Theory Numerica*, 44:2435–2476, 1998.
15. R.A. DeVore. Nonlinear approximation. *Acta Numerica*, 7:51–150, 1998.
16. L. Devroye, L. Györfi, and G. Lugosi. *A Probabilistic Theory of Pattern Recognition*. Number 31 in Applications of mathematics. Springer, New York, 1996.
17. R.M. Dudley. Universal Donsker classes and metric entropy. *Ann. Prob.*, 14(4):1306–1326. 1987.
18. R.M. Dudley, E. Gine. and J. Zinn. Uniform and universal glivenko-cantelli classes. *Journal of Theoretical Probability*, 4:485–510, 1991.
19. T. Evgeniou, M. Pontil, and T. Poggio. Regularization networks and support vector machines. *Advances in Computational Mathematics*, 13:1–50, 2000.
20. T. Ezzat. G. Geiger. and T. Poggio. Trainable videorealistic speech animation. In *Proceedings of ACM SIGGRAPH 2002, San Antonio, TX*, pp. 388–398, 2002.
21. G. Fung and O.L. Mangasarian. Proximal support vector machine classifiers. In *KDD 2001: Seventh ACM SIGKDD International Conference on Knowledge Discovery and Data Mining*, San Francisco. CA, 2001.
22. F. Girosi. An equivalence between sparse approximation and Support Vector Machines. *Neural Computation*, 10(6):1455–1480.
23. F. Girosi, M. Jones. and T. Poggio. Regularization theory and neural networks architectures. *Neural Computation*. 7:219–269, 1995.
24. T. Hastie, R. Tibshirani. and J. Friedman. *The Elements of Statistical Learning*. Springer Series in Statistics. Springer Verlag, Basel, 2001.
25. B. Heisele. T. Serre. M. Pontil. T. Vetter, and T. Poggio. Categorization by learning and combining object parts. In *Advances in Neural Information Processing Systems 14 (NIPS'01)*, volume 14, pp. 1239–1245. MIT Press, 2002.
26. B. Heisele, A. Verri. and T. Poggio. Learning and vision machines. *Proceedings of the IEEE*, 90:1164–1177. 2002.
27. J. Hutchinson, A. Lo. and T. Poggio. A nonparametric approach to pricing and hedging derivative securities via learning networks. *The Journal of Finance*, XLIX(3). July 1994.

28. R. Karp. Mathematical challenges from genomics and molecular biology. *Notices of the AMS*, 49:544–553, 2002.
29. S. Kutin and P. Niyogi. Almost-everywhere algorithmic stability and generalization error. Technical report TR-2002-03, University of Chicago, 2002.
30. S. Mendelson. Improving the sample complexity using global data. *IEEE Transactions on Information Theory*, 48(7):1977–1991, 2002.
31. S. Mendelson. Geometric parameters in learning theory. Submitted for publication, 2003.
32. C.A. Micchelli. Interpolation of scattered data: distance matrices and conditionally positive definite functions. *Constructive Approximation*, 2:11–22, 1986.
33. C.A. Micchelli and T.J. Rivlin. A survey of optimal recovery. In C.A. Micchelli and T.J. Rivlin, editors, *Optimal Estimation in Approximation Theory*, pp. 1–54. Plenum Press, New York, 1976.
34. J. Moody and C. Darken. Fast learning in networks of locally-tuned processing units. *Neural Computation*, 1(2):281–294, 1989.
35. P. Niyogi and F. Girosi. Generalization bounds for function approximation from scattered noisy data. *Advances in Computational Mathematics*, 10:51–80, 1999.
36. C. Papageorgiou, M. Oren, and T. Poggio. A general framework for object detection. In *Proceedings of the International Conference on Computer Vision*, Bombay, India, January 1998.
37. E. Parzen. An approach to time series analysis. *Ann. Math. Statist.*, 32:951–989, 1961.
38. T. Poggio and R. Brunelli. A novel approach to graphics. A.I. Memo No. 1354, Artificial Intelligence Laboratory, Massachusetts Institute of Technology, 1992.
39. T. Poggio and S. Edelman. A network that learns to recognize 3D objects. *Nature*, 343:263–266, 1990.
40. T. Poggio and F. Girosi. Networks for approximation and learning. *Proceedings of the IEEE*, 78(9), September 1990.
41. T. Poggio and F. Girosi. Regularization algorithms for learning that are equivalent to multilayer networks. *Science*, 247:978–982, 1990.
42. T. Poggio, V. Torre, and C. Koch. Computational vision and regularization theory. *Nature*, 317:314–319, 1985b.
43. T. Poggio and A. Verri. Introduction: Learning and vision at cbcl. *International Journal of Computer Vision*, 38–1, 2000.
44. M.J.D. Powell. Radial basis functions for multivariable interpolation: a review. In J.C. Mason and M.G. Cox, editors, *Algorithms for Approximation*. Clarendon Press, Oxford, 1987.
45. M.J.D. Powell. The theory of radial basis functions approximation in 1990. In W.A. Light, editor, *Advances in Numerical Analysis Volume II: Wavelets, Subdivision Algorithms and Radial Basis Functions*, pp. 105–210. Oxford University Press, 1992.
46. Ramaswamy, Tamayo, Rifkin, Mukherjee, Yeang, Angelo, Ladd, Reich, Latulippe, Mesirov, Poggio, Gerlad, Loda, Lander, and Golub. Multiclass cancer diagnosis using tumor gene expression signatures. *Proceedings of the National Academy of Science*, December 2001.
47. R.M. Rifkin. *Everything Old Is New Again: A Fresh Look at Historical Approaches to Machine Learning*. PhD thesis, Massachusetts Institute of Technology, 2002.
48. S. Smale and D. Zhou. Estimating the approximation error in learning theory. *Analysis and Applications*, 1:1–25, 2003.

49. J. Suykens, T. Van Gestel, J. De Brabanter, B. De Moor, and J. Vandewalle. *Least Squares Support Vector Machines.*

50. V.N. Vapnik. *Statistical Learning Theory.* Wiley, New York, 1998.

51. V.N. Vapnik and A.Y. Chervonenkis. On the uniform convergence of relative frequences of events to their probabilities. *Th. Prob. and its Applications,* 17(2):264–280, 1971.

52. G. Wahba. Smoothing and ill-posed problems. In M. Golberg, editor, *Solutions methods for integral equations and applications,* pp. 183–194. Plenum Press, New York, 1979.

53. G. Wahba. *Splines Models for Observational Data.* Series in Applied Mathematics, Vol. 59, SIAM, Philadelphia, 1990.

54. D. Zhou. The regularity of reproducing kernel hilbert spaces in learning theory. 2001. preprint.

55. D. Zhou. The covering number in learning theory. *J. Complexity,* 18:739–767, 2002.

Logical Regression Analysis: From Mathematical Formulas to Linguistic Rules

H. Tsukimoto

Tokyo Denki University
tsukimoto@c.dendai.ac.jp

1 Introduction

Data mining means the discovery of knowledge from (a large amount of) data, and so data mining should provide not only predictions but also knowledge such as rules that are comprehensible to humans. Data mining techniques should satisfy the two requirements, that is, **accurate predictions** and **comprehensible rules**.

Data mining consists of several processes such as preprocessing and learning. This paper deals with learning. The learning in data mining can be divided into supervised learning and unsupervised learning. This paper deals with supervised learning. that is, classification and regression.

The major data mining techniques are neural networks, statistics, decision trees, and association rules. When these techniques are applied to real data, which usually consist of discrete data and continuous data, they each have their own problems. In other words. there is no perfect technique, that is, there is no technique which can satisfy the two requirements of accurate predictions and comprehensible rules.

Neural networks are black boxes, that is, neural networks are incomprehensible. Multiple regression formulas, which are the typical statistical models, are black boxes too. Decision trees do not work well when classes are continuous [19]. that is. if accurate predictions are desired, comprehensibility has to be sacrificed. and if comprehensibility is desired, accurate predictions have to be sacrificed. Association rules, which are unsupervised learning techniques, do not work well when the right-hand sides of rules, which can be regarded as classes. are continuous [21]. The reason is almost the same as that for decision trees.

Neural networks and multiple regression formulas are mathematical formulas. Decision trees and association rules are linguistic rules. Mathematical formulas can provide predictions but cannot provide comprehensible rules. On the other hand, linguistic rules can provide comprehensible rules but cannot provide accurate predictions in continuous classes.

How can we solve the above problem? The solution is rule extraction from mathematical formulas. Rule extraction from mathematical formulas is needed for developing the perfect data mining techniques satisfying the two data mining requirements, accurate predictions and comprehensible rules.

Several researchers have been developing rule extraction techniques from neural networks, and therefore, neural networks can be used in data mining [14], whereas few researchers have studied rule extraction from linear formulas. We have developed a rule extraction technique from mathematical formulas such as neural networks and linear formulas. The technique is called the Approximation Method.

We have implemented the Approximation Method for neural networks in a data mining tool KINOsuite-PR [41].

Statistics have been developing many techniques. The outputs of several techniques are linear formulas. The Approximation Method can extract rules from the linear formulas, and so the combination of statistical techniques and the Approximation Method works well for data mining.

We have developed a data mining technique called Logical Regression Analysis(LRA). LRA consists of regression analysis[1] and the Approximation Method. Since there are several regression analyses, LRA has several versions. In other words, LRA can be applied to a variety of data, because a variety of regression analyses have been developed.

So far, in data mining, many researchers have been dealing with symbolic data or numerical data and few researchers have been dealing with image data.

One of LRA's merits is that LRA can deal with image data, that is, LRA can discover rules from image data. Therefore, this paper explains the data mining from images by LRA.

There are strong correlations between the pixels of images, and so the numbers of samples are small compared with the numbers of the pixels, that is, the numbers of attributes(variables).

A regression analysis for images must work well when the correlations among attributes are strong and the number of data is small. Nonparametric regression analysis works well when the correlations among attributes are strong and the number of data is small.

There are many kinds of images such as remote-sensing images, industrial images and medical images. Within medical images, there are many types such as brain images, lung images, and stomach images. Brain functions are the most complicated and there are a lot of unknown matters, and consequently the discovery of rules between brain areas and brain functions is a significant subject. Therefore, we have been dealing with functional brain images.

---

[1]The regression analysis includes the nonlinear regression analysis using neural networks.

We have been applying LRA using nonparametric regression analysis to fMRI images to discover the rules of brain functions. In this paper, for simplification, "LRA using nonparametric regression analysis" is abbreviated to "LRA".

LRA was applied to several experimental tasks. This paper reports the experimental results of finger tapping and calculation. The results of LRA include discoveries that have never been experimentally confirmed. Therefore, LRA has the possibility of providing new evidence in brain science.

Section 2 explains the problems in the major data mining techniques and how the problems can be solved by rule extraction from mathematical formulas. Section 3 surveys the rule extraction from neural networks. Section 4 explains the Approximation Method, which has been developed by the author. The Approximation Method is mathematically based on the multilinear function space, which is also explained. Section 5 explains the Approximation Method in the continuous domain. Section 6 discusses a few matters. Section 7 briefly explains the data mining from fMRI images. Section 8 describes the data mining from fMRI images by Logical Regression Analysis (LRA). Section 9 shows the experiments of calculations.

This paper is mainly based on [32, 33, 35], and [36].

2 The Problems of Major Data Mining Techniques

The major data mining techniques, that is, decision trees, neural networks, statistics and association rules, are reviewed in terms of the two requirements of accurate predictions and comprehensible rules.

2.1 Neural Networks

Neural networks can provide accurate predictions in the discrete domain and the continuous domain. The problem is that the training results of neural

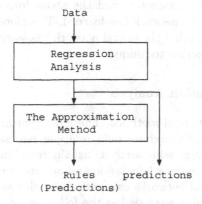

Fig. 1. Logical Regression Analysis

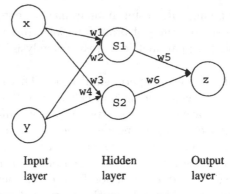

Fig. 2. Neural network

networks are sets of mathematical formulas, that is, neural networks are incomprehensible black boxes.

For example, Fig. 2 shows a trained neural network. In Fig. 2, x and y are inputs, z is an output, and t_i's are the outputs of the two hidden units. Each output is represented as follows:

$$t_1 = S(w_1 x + w_2 y + h_1),$$
$$t_2 = S(w_3 x + w_4 y + h_2),$$
$$z = S(w_5 t_1 + w_6 t_2 + h_3),$$

where w_i's stand for weight parameters, $S(x)$ stands for sigmoid function and h_i's stand for biases. The weight parameters are as follows:

$$w_1 = 2.51, \quad w_2 = -4.80, \quad w_3 = -4.90,$$
$$w_4 = 2.83, \quad w_5 = 2.52, \quad w_6 = -4.81,$$
$$h_1 = -0.83, \quad h_2 = -1.12, \quad h_3 = -0.82.$$

From the above weight parameters and the above formulas, we cannot understand what the neural network has learned. Therefore, the trained neural network is a black box. What the neural network has learned will be explained later using a rule extraction technique.

2.2 Multiple Regression Analysis

There are a lot of statistical methods. The most typical method is multiple regression analysis, and therefore, only multiple regression analysis is discussed here. Multiple regression analysis usually uses linear formulas, and so only linear regression analysis is possible and nonlinear regression analysis is impossible, while neural networks can perform nonlinear regression analysis. However, linear regression analysis has the following advantages.

1. The optimal solution can be calculated.
2. The linear regression analysis is the most widely used statistical method in the world.
3. Several regression analysis techniques such as nonparametric regression analysis and multivariate autoregression analysis have been developed based on the linear regression analysis.

Linear regression analysis can provide appropriate predictions in the continuous domain and discrete domain. The problem is that multiple regression formulas are mathematical formulas as showed below.

$$y = \sum_{i=1}^{n} a_i x_i + b .$$

where y is a dependent variable, x_is are independent variables, b is a constant, a_is are coefficients, and n is the number of the independent variables. The mathematical formulas are incomprehensible black boxes. A mathematical formula consisting of a few independent variables may be understandable. However, a mathematical formula consisting of a lot of independent variables cannot be understood. Moreover, in multivariate autoregression analysis, there are several mathematical formulas, and so the set of the mathematical formulas cannot be understood at all. Therefore, rule extraction from linear regression formulas is important.

2.3 Decision Trees

When a class is continuous, the class is discretized into several intervals. When the number of the intervals, that is, the number of the discretized classes, is small, comprehensible trees can be obtained, but the tree cannot provide accurate predictions. For example, Fig. 3 shows a tree where there are two

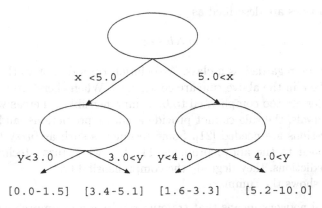

Fig. 3. Decision tree with a continuous class

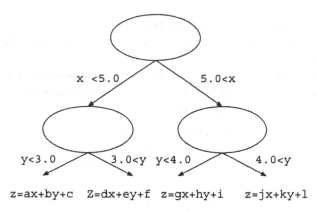

Fig. 4. Regression tree

continuous attributes and a continuous class. To improve the prediction accuracy, let the number of intervals be large. When the number of intervals is large, the trees obtained are too complicated to be comprehensible.

From the above simple discussion, we can conclude that it is impossible to obtain trees which can satisfy the two requirements for data mining techniques, accurate predictions and comprehensible rules at the same time. Therefore, decision trees cannot work well when classes are continuous [19].

As a solution for continuous classes, for example, Quinlan presented Cubist [40]. Cubist generates piecewise-linear models, which are a kind of regression trees. Figure 4 shows an example. As seen from this figure, the tree is an extension of linear formulas, and so the tree is a prediction model, that is, it is incomprehensible. As a result, this solution has solved the inaccurate prediction problem but has generated the incomprehensibility problem.

2.4 Association Rules

Association rules are described as

$$a \wedge b \rightarrow c \,,$$

where c can be regarded as a class. Association rules do not work well when "classes" like c in the above rule are continuous. When there are many intervals, the rules are too complicated to be comprehensible, whereas when there are few intervals, the rule cannot provide accurate predictions, and therefore some concessions are needed [21]. Some techniques such as fuzzy techniques can be applied to the above problem [11], but while fuzzy techniques can improve predictions, they degrade the comprehensibility.

Table 1 shows the summary.

$\triangle$ in neural network means that training results are incomprehensible.
$\triangle$ in linear regression means that training results are incomprehensible.

Table 1. Data mining techniques

Type	1	2	3	4
Class	Discrete	Discrete	Continuous	Continuous
Attribute	Discrete	Continuous	Discrete	Continuous
neural network	△	△	△	△
linear regression	△	△	△	△
regression tree	△	△	△	△
decision tree	∘	∘	△	△

△ in regression tree means that training results are incomprehensible.
△ means that decision trees cannot work well in continuous classes.

Association rules are omitted in Table 1, because association rules do not have classes. However, the evaluations for association rules are the same as those for decision trees.

Thus, we conclude that there is no technique which can satisfy the two requirements for data mining techniques, that is, the technique which can provide accurate predictions and comprehensible rules in the discrete domain and the continuous domain.

2.5 The Solution for the Problem

The solution for the above problem is extracting comprehensible rules from mathematical formulas such as neural networks and multiple regression formulas [38]. When rules are extracted from mathematical formulas, the rules are not used for predictions, but used only for human comprehension. The mathematical formulas are used to make the predictions. A set of a prediction model and a rule (or rules) extracted from the prediction model is the perfect data mining technique.

How a rule extracted from a prediction model is used is briefly explained. When a prediction model predicts, it only outputs a class or a figure. Humans cannot understand how or why the prediction model outputs the class or the figure. A rule extracted from the prediction model explains how or why the prediction model outputs the class or the figure.

For example, let a neural network be trained using process data consisting of three attributes (temperature (t), pressure (p), humidity (h)), and a class, the quality of a material (q). Let the rule extracted be as follows:

$$(200 \leq t \leq 300) \vee (p \leq 2.5) \vee (70 \leq h \leq 90) \rightarrow q \leq 0.2 .$$

Assume that the network is used for a prediction, and the inputs for the network are $t = 310$, $p = 2.0$, and $h = 60$ and the output from the network is 0.1, which means low quality. The above rule shows that the network outputs 0.1, because the pressure is 2.0, which is below 2.5, that is, $p \leq 2.5$ holds.

Without the rule, we cannot know how or why the network outputs 0.1 indicating the low quality. Note that the rule extracted from the neural network is not used for the predictions, which are made by the neural network.

An operator can understand that the neural network predicts low quality, because the pressure is low. Therefore, the operator can raise the pressure (for example, by manipulating a valve) to raise the quality. If the operator does not know why the neural network predicts the low quality, the operator cannot take an appropriate measure to raise the quality.

Because the operator understands how the neural network predicts, the operator can take an appropriate measure. That is, the extracted rule enables the operator to take the appropriate measure. If the neural network only predicts the low quality (without the extracted rule), the operator does not know what to do to raise the quality. The benefit of the extracted rule is very large.

2.6 KINOsuite-PR

The rule extraction technique for neural networks has been implemented in the data mining tool KINO- PR [41]. KINO stands for Knowledge INference by Observation and PR stands for Predictions and Rules. KINOsuite-PR is the first commercial data mining tool that has the rule extraction. The rule extraction technique is called the Approximation Method. Next, the Approximation Method is explained. Before the explanation, the rule extraction from neural networks is surveyed in Sect. 3.

3 The Survey of Rule Extraction from Neural Networks

This section briefly surveys algorithms for rule extraction from neural networks. Some rule extraction algorithms are based on the structurization of neural networks, but the rule extraction algorithms cannot be applied to neural networks trained by other training methods such as the back-propagation method. Some rule extraction algorithms are dedicated to hybrid models that include symbolic and subsymbolic knowledge.

The rule extraction algorithms mentioned above are outside the scope of this paper, because, in data mining, it is desired that rules can be extracted from any neural network trained by any training method.

There are several algorithms for rule extraction from neural networks [1,2]. The algorithms can be divided into decompositional algorithms and pedagogical algorithms. Decompositional algorithms extract rules from each unit in a neural network and aggregate them into a rule. For example, [5] is a decompositional algorithm. Pedagogical algorithms generate samples from a neural network and induce a rule from the samples. For example, [6] is a pedagogical algorithm. Decompositional algorithms can present training results of each unit in neural networks, and so we can understand the training results by the unit, while pedagogical algorithms can present only the results of neural networks, and so we cannot understand the training results by the unit. Therefore, decompositional algorithms are better than pedagogical algorithms in terms of understandability of the inner structures of neural networks.

Rule extraction algorithms are compared in several items, namely network structures, training methods, computational complexity, and values.

Network structure: This means the types of network structures the algorithm can be applied to. Several algorithms can be applied only to particular network structures. Most algorithms are applied to three-layer feedforward networks, while a few algorithms can be applied only to recurrent neural networks, where Deterministic Finite-state Automata(DFAs) are extracted [17].

Training method: This means the training methods the rule extraction algorithm can be applied to. Several rule extraction algorithms depend on training methods, that is, the rule extraction algorithms can extract rules only from the neural networks trained by a particular training method [5, 20]. The pedagogical algorithms basically do not depend on training methods.

Computational complexity: Most algorithms are exponential in computational complexity. For example, in pedagogical algorithms, the total number of samples generated from a neural network is 2^n, where n is the number of inputs to the neural network. It is very difficult to generate many samples and induce a rule from many samples. Therefore, it is necessary to reduce the computational complexity to a polynomial. Most decompositional algorithms are also exponential in computational complexity.

Values: Most algorithms can be applied only to discrete values and cannot be applied to continuous values.

The ideal algorithm can be applied to any neural network, can be applied to any training method, is polynomial in computational complexity, and can be applied to continuous values. There has been no algorithm which satisfies all of the items above. More detailed review can be found in [7].

4 The Approximation Method

4.1 The Outline

We have developed a rule extraction technique for mathematical formulas, which is called the Approximation Method. We presented an algorithm for extracting rules from linear formulas and extended it to the continuous domain [24]. Subsequently, we presented polynomial algorithms for extracting rules from linear formulas [26] and applied them to discover rules from numerical data [16]. We also developed an inductive learning algorithm based on rule extraction from linear formulas [29]. The data mining technique through regression analysis and the Approximation Method is called Logical regression Analysis(LRA). We have been applying LRA using nonparametric regression analysis to discover rules from functional brain images [9, 30, 34, 36, 37].

We extended the algorithms for linear formulas to the algorithms for neural networks [27], and modified them to improve the accuracies and simplicities [28, 33].

The Approximation Method for a linear formula is almost the same as the Approximation Method for a unit in a neural network. The Approximation Method for neural networks basically satisfies the four items listed in the preceding section, that is, the method can be applied to any neural network trained by any training method, is polynomial in computational complexity, and can be applied to continuous values. However, the method has a constraint, that is, the method can be applied only to neural networks whose units' output functions are monotone increasing. The Approximation Method is a decompositional method.

There are two kinds of domains, discrete domains and continuous domains. The continuous domain will be discussed later. The discrete domains can be reduced to $\{0, 1\}$ domains by dummy variables, so only these domains have to be discussed. In the $\{0, 1\}$ domain, the units of neural networks can be approximated to the nearest Boolean functions, which is the basic idea for the Approximation Method. The Approximation Method is based on the multilinear function space. The space, which is an extension of Boolean algebra of Boolean functions, can be made into a Euclidean space and includes linear functions and neural networks. The details can be found in [31] and [33].

4.2 The Basic Algorithm of the Approximation Method

A unit of a neural network or a linear function is a function. The basic algorithm approximates a function by a Boolean function. Note that "a function" in the following sentences means a unit of a neural network or a linear function.

Let $f(x_1, \ldots, x_n)$ stand for a function, and $(f_i)(i = 1, \ldots, 2^n)$ be the values of the function. Let the values of the function be the interval [0,1]. The values of a unit of a neural network are [0,1]. The values of a linear function can be normalized to [0,1] by some normalization method.

Let $g(x_1, \ldots, x_n)$ stand for a Boolean function, and $(g_i)(g_i = 0$ or $1, i = 1, \ldots, 2^n)$ be the values of the Boolean function. A Boolean function is represented by the following formula [3]:

$$g(x_1, \ldots, x_n) = \sum_{i=1}^{2^n} g_i a_i ,$$

where $\sum$ is disjunction, a_i is an atom, and g_i is the value at the domain corresponding to the atom. An atom is as follows:

$$a_i = \prod_{j=1}^{n} e(x_j) \ (i = 1, \ldots, 2^n) ,$$

where $\prod$ stands for conjunction, and $e(x_j) = x_j$ or $\overline{x_j}$, where $\overline{x}$ stands for the negation of x. The domain $(x_1, \ldots, x_n)$ corresponding to an atom is as follows:

When $e(x_j) = x_j$, $x_j = 1$, and when $e(x_j) = \overline{x_j}$, $x_j = 0$. The above formula can be easily verified.

The basic algorithm is as follows:

$$g_i = \begin{cases} 1 & (f_i \geq 0.5) , \\ 0 & (f_i < 0.5) . \end{cases}$$

This algorithm minimizes Euclidean distance. The Boolean function is represented as follows:

$$g(x_1, \ldots, x_n) = \sum_{i=1}^{2^n} g_i a_i ,$$

where g_i is calculated by the above algorithm.

Example 1. Figure 5 shows a case of two variables. Crosses stand for the values of a function and dots stand for the values of the Boolean function. $00, 01, 10$ and 11 stand for the domains, for example, 00 stands for $x = 0$, $y = 0$. In this case, there are four domains as follows:

$$(0,0), (0,1), (1,0), (1,1)$$

The atoms corresponding to the domains are as follows:

$$(0,0) \Leftrightarrow \overline{x}\overline{y}, (0,1) \Leftrightarrow \overline{x}y, (1,0) \Leftrightarrow x\overline{y}, (1,1) \Leftrightarrow xy .$$

The values of the Boolean function $g(x, y)$ are as follows:

$$g(0,0) = 0, \ g(0,1) = 1, \ g(1,0) = 0, \ g(1,1) = 1 .$$

Therefore, in the case of Fig. 5, the Boolean function is represented as follows:

$$g(x,y) = g(0,0)\overline{x}\overline{y} \vee g(0,1)\overline{x}y \vee g(1,0)x\overline{y} \vee g(1,1)xy .$$

The Boolean function is reduced as follows:

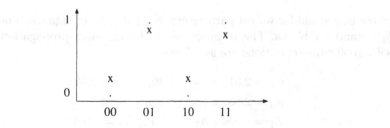

Fig. 5. Approximation

$$g(x,y) = g(0,0)\bar{x}\bar{y} \vee g(0,1)\bar{x}y \vee g(1,0)x\bar{y} \vee g(1,1)xy$$
$$g(x,y) = 0\bar{x}\bar{y} \vee 1\bar{x}y \vee 0x\bar{y} \vee 1xy$$
$$g(x,y) = \bar{x}y \vee xy$$
$$g(x,y) = y \, .$$

Example 2. **An example of a linear function**

Let a linear function be as follows:

$$z = 0.3x - 0.5y + 0.6$$

There are four domains as follows:

$$(0,0), (0,1), (1,0), (1,1)$$

The values of the function at the domains are as follows:

$$z(0,0) = 0.6, z(0,1) = 0.1, z(1,0) = 0.9, z(1,1) = 0.4 \, .$$

By approximating the function to a Boolean function $g(x,y)$,

$$g(0,0) = 1, g(0,1) = 0, g(1,0) = 1, g(1,1) = 0 \, .$$

are obtained. The Boolean function $g(x,y)$ is as follows:

$$g(x,y) = g(0,0)\bar{x}\bar{y} \vee g(0,1)\bar{x}y \vee g(1,0)x\bar{y} \vee g(1,1)xy$$
$$g(x,y) = 1\bar{x}\bar{y} \vee 0\bar{x}y \vee 1x\bar{y} \vee 0xy$$
$$g(x,y) = \bar{x}\bar{y} \vee x\bar{y}$$
$$g(x,y) = \bar{y} \, .$$

Example 3. **An example of a neural network**

We show what the neural network in Fig. 2 has learned by rule extraction. Two hidden units and the output unit are as follows:

$$t_1 = S(w_1x + w_2y + h_1) \, ,$$
$$t_2 = S(w_3x + w_4y + h_2) \, ,$$
$$z = S(w_5t_1 + w_6t_2 + h_3) \, ,$$

where w_i's stand for weight parameters, $S(x)$ stands for sigmoid function and h_i's stand for biases. The training results by the back-propagation method with 1000 time repetitions are as follows:

$$w_1 = 2.51, \ w_2 = -4.80, \ w_3 = -4.90 \, ,$$
$$w_4 = 2.83, \ w_5 = 2.52, \ w_6 = -4.81 \, ,$$
$$h_1 = -0.83, \ h_2 = -1.12, \ h_3 = -0.82 \, .$$

For example,
$$t_1 = S(2.51x - 4.80y - 0.83) ,$$

and the values of $t_1(1,1), t_1(1,0), t_1(0,1)$, and $t_1(0,0)$ are as follows:

$$t_1(1,1) = S(2.51 \cdot 1 - 4.80 \cdot 1 - 0.83) = S(-3.12) ,$$
$$t_1(1,0) = S(2.51 \cdot 1 - 4.80 \cdot 0 - 0.83) = S(1.68) ,$$
$$t_1(0,1) = S(2.51 \cdot 0 - 4.80 \cdot 1 - 0.83) = S(-5.63) ,$$
$$t_1(0,0) = S(2.51 \cdot 0 - 4.80 \cdot 0 - 0.83) = S(-0.83) .$$

$S(-3.12) \simeq 0$, $S(1.68) \simeq 1$, $S(-5.63) \simeq 0$, and $S(-0.83) \simeq 0$; therefore
$$t_1 \simeq x\bar{y} .$$

In a similar manner, t_2 is approximated by the following Boolean function:
$$t_2 = \bar{x}y$$

z is approximated as follows:
$$z = t_1 \bar{t_2} .$$

By substituting $t_1 = x\bar{y}$ and $t_2 = \bar{x}y$ in the above formula,
$$z = x\bar{y}\overline{\bar{x}y} = x\bar{y}(x \vee \bar{y}) = x\bar{y} \vee x\bar{y} = x\bar{y}$$

has been obtained. Thus, we understand that the neural network in Fig. 2 has learned $x\bar{y}$.

4.3 Multilinear Functions

The computational complexity of the basic algorithm is exponential, which is obvious from the discussion above. Therefore, a polynomial algorithm is needed. The multilinear function space is necessary for the polynomial algorithm [28, 33]. This subsection explains that the multilinear function space of the domain $\{0, 1\}$ is a Euclidean space. First, the multilinear functions are explained. Second, it is shown that the multilinear function space is the linear space spanned by the atoms of Boolean algebra of Boolean functions. Third, it is explained that the space can be made into a Euclidean space. Finally, it is explained that neural networks are multilinear functions.

What Multilinear Functions Are

Definition 1. *Multilinear functions of n variables are as follows:*

$$\sum_{i=1}^{2^n} a_i x_1^{e_{i1}} \cdots x_n^{e_{in}} ,$$

where a_i is real, x_i is a variable, and e_i is 0 or 1.

Example. Multilinear functions of 2 variables are as follows:

$$axy + bx + cy + d \, .$$

Multilinear functions do not contain any terms such as

$$x_1^{k_1} x_2^{k_2} \cdots x_n^{k_n}, \tag{1}$$

where $k_i \geq 2$. A function

$$f : \{0, 1\}^n \to \mathbf{R}$$

is a multilinear function, because $x_i^{k_i} = x_i$ holds in $\{0, 1\}$ and so there is no term like (1) in the functions. In other words, multilinear functions are functions which are linear when only one variable is considered and the other variables are regarded as parameters.

The Multilinear Function Space of the Domain $\{0, 1\}$ is the Linear Space Spanned by the Atoms of Boolean Algebra of Boolean Functions

Definition 2. *The atoms of Boolean algebra of Boolean functions of n variables are as follows:*

$$\phi_i = \prod_{j=1}^{n} e(x_j) \ (i = 1, \ldots, 2^n) \, ,$$

where $e(x_j) = \overline{x_j}$ *or* x_j.

Example. The atoms of Boolean algebra of Boolean functions of 2 variables are as follows:

$$x \wedge y, \quad x \wedge \overline{y}, \quad \overline{x} \wedge y, \quad \overline{x} \wedge \overline{y} \, .$$

Theorem 1. *The space of multilinear functions* $(\{0, 1\}^n \to \mathbf{R})$ *is the linear space spanned by the atoms of Boolean algebra of Boolean functions.*

The proof can be found in [31] and [33].

Example. A linear function of the atoms of 2 variables is

$$axy + bx\overline{y} + c\overline{x}y + d\overline{x}\overline{y} \, .$$

This function is transformed to the following:

$$pxy + qx + ry + s \, ,$$

where

$$p = a - b - c + d, \quad q = b - d, \quad r = c - d, \quad s = d \, .$$

A multilinear function

$$pxy + qx + ry + s$$

can be transformed into

$$axy + bx\bar{y} + c\bar{x}y + d\bar{x}\bar{y} ,$$

where
$$a = p + q + r + s, \quad b = q + s, \quad c = r + s, \quad d = s .$$

Now, it has been shown that the multilinear function space of the domain $\{0, 1\}$ is the linear space spanned by the atoms of Boolean algebra of Boolean functions. The dimension of the space is 2^n. Next, it is shown that the multilinear function space is made into a Euclidean space.

The Multilinear Function Space of the Domain $\{0, 1\}$ is a Euclidean Space

Definition 3. *The inner product is defined as follows:*

$$< f, g >= \sum_{\{0,1\}^n} fg .$$

The sum in the above formula is done over the whole domain.

Example. In the case of two variables,

$$< f, g > = \sum_{\{0,1\}^2} fg = f(0,0)g(0,0) + f(0,1)g(0,1)$$
$$+ f(1,0)g(1,0) + f(1,1)g(1,1) .$$

Theorem 2. *Atoms ϕ_is have unitarity and orthogonality.*

$$< \phi_i, \phi_i > = 1 \ (unitarity)$$
$$< \phi_i, \phi_j > = 0 (i \neq j) \ (orthogonality)$$

The proof can be found in [31] and [33].

Example. An example of unitarity and orthogonality of two variables is as follows:

$$< x\bar{y}, x\bar{y} > = \sum_{\{0,1\}^2} x\bar{y}x\bar{y} = \sum_{\{0,1\}^2} x^2\bar{y}^2$$
$$= \sum_{\{0,1\}^2} x\bar{y} = 1 \cdot \bar{1} + 1 \cdot \bar{0} + 0 \cdot \bar{1} + 0 \cdot \bar{0}$$
$$= 1 \cdot 0 + 1 \cdot 1 + 0 \cdot 0 + 0 \cdot 1 = 1 .$$

(The domain is $\{0,1\}$, and so $x^2 = x, \bar{y}^2 = \bar{y}$.)

$$< xy, x\bar{y} >= \sum_{\{0,1\}^2} xyx\bar{y} = \sum_{\{0,1\}^2} x^2 y\bar{y} = 0 .$$

Definition 4. *Norm is defined as follows:*

$$|f| = \sqrt{< f, f >} = \sqrt{\sum_{\{0,1\}^n} f^2} .$$

Example

$$|x\bar{y}| = \sqrt{< x\bar{y}, x\bar{y} >} = 1$$

From the above discussion, the space becomes a finite-dimensional inner product space, namely a Euclidean space.

Neural Networks are Multilinear Functions

Theorem 3. *When the domain is $\{0,1\}$, neural networks are multilinear functions.*

Proof. As described in this subsection, a function whose domain is $\{0,1\}$ is a multilinear function. Therefore, when the domain is $\{0,1\}$, neural networks, that is, the functions which neural networks learn are multilinear functions.

From the above discussions, in the domain of $\{0,1\}$, a neural network is a multilinear function, a Boolean function is a multilinear function, and the multilinear function space is a Euclidean space. Therefore, a neural network can be approximated to a Boolean function by Euclidean distance.

4.4 A Polynomial-Time Algorithm for Linear Formulas [29]

The Condition that $x_{i_1} \cdots x_{i_k} \bar{x}_{i_{k+1}} \cdots \bar{x}_{i_l}$ Exists in the Boolean Function After Approximation

The following theorem holds.

Theorem 4. *Let*

$$p_1 x_1 + \ldots + p_n x_n + p_{n+1}$$

stand for a linear function. Let

$$a_1 x_1 \cdots x_n + a_2 x_1 \cdots \overline{x_n} + \ldots + a_{2^n} \overline{x_1} \cdots \overline{x_n}$$

stand for the expansion by the atoms of Boolean algebra of Boolean functions. The a_i's are as follows:

$$a_1 = p_1 + \ldots + p_n + p_{n+1} ,$$

$$a_2 = p_1 + \ldots + p_{n-1} + p_{n+1} \,,$$

$$\ldots$$

$$a_{2^{n-1}} = p_1 + p_{n+1} \,,$$

$$a_{2^{n-1}+1} = p_2 + p_3 + \ldots + p_{n-1} + p_n + p_{n+1} \,,$$

$$\ldots$$

$$a_{2^n-1} = p_n + p_{n+1} \,,$$

$$a_{2^n} = p_{n+1} \,.$$

Proof. The following formula holds.

$$p_1 x_1 + \ldots + p_n x_n + p_{n+1} = a_1 x_1 \cdots x_n + a_2 x_1 \cdots \overline{x_n} + \ldots + a_{2^n} \overline{x_1} \cdots \overline{x_n} \,.$$

Each a_i can be easily calculated as follows. For example, let

$$x_1 = x_2 = \cdots = x_n = 1 \,.$$

Then the right-hand side of the above formula is

$$a_1$$

and the left-hand side is

$$p_1 + \ldots + p_n + p_{n+1} \,.$$

Thus,

$$a_1 = p_1 + \ldots + p_n + p_{n+1}$$

is obtained. Similarly, let

$$x_1 = x_2 = \cdots = x_{n-1} = 1, x_n = 0 \,,$$

then

$$a_2 = p_1 + \ldots + p_{n-1} + p_{n+1}$$

is obtained. The others can be calculated in the same manner. Thus, the theorem has been proved.

Theorem 5. *Let*

$$p_1 x_1 + \ldots + p_n x_n + p_{n+1}$$

stand for a linear function. The condition that $x_{i_1} \cdots x_{i_k} \overline{x}_{i_{k+1}} \cdots \overline{x}_{i_l}$ *exists in the Boolean function after approximation is as follows:*

$$\sum_{i_1}^{i_k} p_j + p_{n+1} + \sum_{1 \le j \le n, j \ne i_1, \ldots, i_l, p_j < 0} p_j \ge 0.5 \,.$$

Proof. Consider the existence condition of x_1 in the Boolean function after approximation. For simplification, this condition is called the existence condition. Because

$$x_1 = x_1 x_2 \cdots x_n \vee x_1 x_2 \cdots \overline{x_n} \vee \ldots \vee x_1 \overline{x_2} \cdots \overline{x_n} ,$$

the existence of x_1 equals the existence of the following terms:

$$x_1 x_2 \cdots x_n ,$$

$$x_1 x_2 \cdots \overline{x_n} ,$$

$$\cdots$$

$$x_1 \overline{x_2} \cdots \overline{x_n} .$$

The existence of the above terms means that all coefficients of these terms $a_1, a_2, \ldots, a_{2^{n-1}}$ are greater than or equal to 0.5 (See 4.2). That is,

$$MIN\{a_i\} \geq 0.5 (1 \leq i \leq 2^{n-1}) .$$

$MIN\{a_i\}$ will be denoted by $MINa_i$ for simplification. Because a_i's $(1 \leq i \leq 2^{n-1})$ are

$$a_1 = p_1 + \ldots + p_n + p_{n+1} ,$$

$$a_2 = p_1 + \ldots + p_{n-1} + p_{n+1} ,$$

$$\cdots$$

$$a_{2^{n-1}} = p_1 + p_{n+1} ,$$

each $a_i (1 \leq i \leq 2^{n-1})$ contains p_1. If each p_j is non-negative, $a_{2^{n-1}} (= p_1 + p_{n+1})$ is the minimum because the other a_i's contain other p_j's, and therefore the other a_i's are greater than or equal to $a_{2^{n-1}} (= p_1 + p_{n+1})$. Generally, since each p_j is not necessarily non-negative, the $MINa_i$ is a_i which contains all negative p_j. That is,

$$MINa_i = p_1 + p_{n+1} + \sum_{1 \leq j \leq n, j \neq 1, p_j < 0} p_j ,$$

which necessarily exists in $a_i (1 \leq i \leq 2^{n-1})$, because $a_i (1 \leq i \leq 2^{n-1})$ is

$$p_1 + p_{n+1} + \text{arbitrary sum of } p_j (2 \leq j \leq n)) .$$

From the above arguments, the existence condition of x_1, $MINa_i \geq 0.5$, is as follows:

$$p_1 + p_{n+1} + \sum_{1 \leq j \leq n, j \neq 1, p_j < 0} p_j \geq 0.5 .$$

Since

$$p_1 x_1 + \ldots + p_n x_n + p_{n+1}$$

is symmetric for x_i, the above formula holds for other variables; that is, the existence condition of x_i is

$$p_i + p_{n+1} + \sum_{1 \le j \le n, j \ne i, p_j < 0} p_j \ge 0.5 .$$

Similar discussions hold for $\overline{x}_i$, and so we have the following formula:

$$p_{n+1} + \sum_{1 \le j \le n, j \ne i, p_j < 0} p_j \ge 0.5 .$$

Similar discussions hold for higher order terms $x_{i_1} \cdots x_{i_k} \overline{x}_{i_{k+1}} \cdots \overline{x}_{i_l}$, and so we have the following formula:

$$\sum_{i_1}^{i_k} p_j + p_{n+1} + \sum_{1 \le j \le n, j \ne i_1, \dots, i_l, p_j < 0} p_j \ge 0.5 .$$

Generation of DNF Formulas

The algorithm generates terms using the above formula from the lowest order up to a certain order. A DNF formula can be generated by taking the disjunction of the terms generated by the above formula. A term whose existence has been confirmed does not need to be rechecked in higher order terms. For example, if the existence of x is confirmed, then it also implies the existence of $xy, xz. \dots$, because $x = x \vee xy \vee xz$; hence, it is unnecessary to check the existence of $xy, xz, \dots$. As can be seen from the above discussion, the generation method of DNF formulas includes reductions such as $xy \vee xz = x$.

Let

$$f = 0.65x_1 + 0.23x_2 + 0.15x_3 + 0.20x_4 + 0.02x_5$$

be the linear function. The existence condition of $x_{i_1} \cdots x_{i_k} \overline{x}_{i_{k+1}} \cdots \overline{x}_{i_l}$ is

$$\sum_{i_1}^{i_k} p_j + p_{n+1} + \sum_{1 \le j \le n, j \ne i_1, \dots, i_l, p_j < 0} p_j \ge 0.5 .$$

In this case, each p_i is positive and $p_{n+1} = 0$; therefore the above formula can be simplified to

$$\sum_{i_1}^{i_k} p_j \ge 0.5 .$$

For x_i, the existence condition is

$$p_i \ge 0.5 .$$

For $i = 1, 2, 3, 4, 5.$

$$p_1 \ge 0.5 ,$$

therefore

$$x_1$$

exists.

For $x_i x_j$, the existence condition is

$$p_i + p_j \geq 0.5 \ .$$

For $i, j = 2, 3, 4, 5$,

$$p_i + p_j < 0.5 \ ,$$

therefore no

$$x_i x_j$$

exists.

For $x_i x_j x_k$, the existence condition is

$$p_i + p_j + p_k \geq 0.5 \ .$$

For $i, j, k = 2, 3, 4, 5$,

$$p_2 + p_3 + p_4 \geq 0.5 \ ,$$

therefore

$$x_2 x_3 x_4$$

exists. Because higher order terms cannot be generated from x_5, the algorithm stops. Therefore, x_1 and $x_2 x_3 x_4$ exist and the DNF formula is the disjunction of these terms, that is,

$$x_1 \vee x_2 x_3 x_4 \ .$$

4.5 Polynomial-Time Algorithms for Neural Networks

The basic algorithm is exponential in computational complexity, and therefore, polynomial algorithms are needed. The authors have presented polynomial algorithms. The details can be found in [27, 28, 33]. The outline of the polynomial algorithms follows.

Let a unit of a neural network be as follows:

$$S(p_1 x_1 + \ldots + p_n x_n + p_{n+1}) \ ,$$

where $S(\cdot)$ is a sigmoid function. The Boolean function is obtained by the following algorithm.

1. Check if

$$x_{i_1} \cdots x_{i_k} \overline{x}_{i_{k+1}} \cdots \overline{x}_{i_l}$$

exists in the Boolean function after the approximation by the following formula:

$$S \left(p_{n+1} + \sum_{i_1}^{i_k} p_j + \sum_{1 \leq j \leq n, j \neq i_1, \ldots, i_l, p_j \leq 0} p_j \right) \geq 0.5$$

2. Connect the terms existing after the approximation by logical disjunction to make a DNF formula.
3. Execute the above procedures up to a certain (usually two or three) order.

After Boolean functions are extracted from all units, the Boolean functions are aggregated into a Boolean function (rule) for the network.

An example follows. Let

$$S(0.65x_1 + 0.23x_2 + 0.15x_3 + 0.20x_4 + 0.02x_5 - 0.5)$$

be a unit of a neural network.

For $x_i (i = 1, 2, 3, 4, 5)$,

$$S(p_1 + p_6) \geq 0.5 ,$$

and therefore

$$x_1$$

exists.

For $x_i x_j (i, j = 2, 3, 4, 5)$,

$$S(p_i + p_j + p_6) < 0.5 ,$$

and therefore no

$$x_i x_j$$

exists.

For $x_i x_j x_k (i, j, k = 2, 3, 4, 5)$,

$$S(p_2 + p_3 + p_4 + p_6) \geq 0.5 ,$$

and therefore

$$x_2 x_3 x_4$$

exists. Because higher order terms cannot be generated from x_5, the algorithm stops. Therefore, x_1 and $x_2 x_3 x_4$ exist and the DNF formula is the disjunction of these terms, that is,

$$x_1 \vee x_2 x_3 x_4 .$$

If accurate rules are obtained, the rules are complicated, and if simple rules are obtained, the rules are not accurate. It is difficult to obtain rules which are simple and accurate at the same time. We have presented a few techniques to obtain simple and accurate rules. For example, attribute selection works well for obtaining simple rules [28, 33].

4.6 Computational Complexity
of the Algorithm and Error Analysis

The computational complexity of generating the mth order terms is a polynomial of $\binom{n}{m}$, that is, a polynomial of n. Thus, the computational complexity of generating up to the kth order terms is $\sum_{m=1}^{m=k} \binom{n}{m}$, which is a polynomial of n. Therefore, the computational complexity of generating DNF formulas from trained neural networks is a polynomial of n. However, the sum of the number of up to the highest ($=n$th) order terms is $\sum_{m=1}^{m=n} \binom{n}{m} = 2^n$; therefore, the computational complexity of generating up to the highest order terms is exponential. If it is desired that the computational complexity be reduced to a polynomial, the terms are generated only up to a certain order. If it is desired that understandable rules are obtained from trained neural networks, higher order terms are unnecessary. Therefore, actual computational complexity is a low order polynomial.

In the case of the domain $\{0,1\}$, the following theorem holds [13].:

$$\sum_{|S|>k} \hat{f}(S)^2 \leq 2M2^{-k^{1/2}/20} ,$$

where f is a Boolean function, S is a term, $|S|$ is the order of S, k is any integer, $\hat{f}(S)$ denotes the Fourier Transform of f at S and M is the circuit's size of the function.

The above formula shows that the high order terms have very little power; that is, low order terms are informative. Therefore, a good approximation can be obtained by generating up to a certain order; that is, the computational complexity can be reduced to a polynomial by adding small errors.

5 The Approximation Method
in the Continuous Domain

When classes are continuous, extracting rules from neural networks is important. However, in the continuous domain, few algorithms have been proposed. For example, algorithms for extracting fuzzy rules have been presented [4], but fuzzy rules are described by linear functions, and so fuzzy rules are incomprehensible.

5.1 The Basic Idea

In this section, the algorithm is extended to continuous domains. Continuous domains can be normalized to [0,1] domains by some normalization method, so only these domains have to be discussed. First, we have to present a system

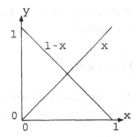

Fig. 6. Direct proportion and inverse proportion

of qualitative expressions corresponding to Boolean functions, in the [0,1] domain. The author presents the expression system generated by direct proportion, inverse proportion, conjunction and disjunction. Figure 6 shows the direct proportion and the inverse proportion. The inverse proportion ($y = 1 - x$) is a little different from the conventional one ($y = -x$), because $y = 1 - x$ is the natural extension of the negation in Boolean functions. The conjunction and disjunction will also be obtained by a natural extension. The functions generated by direct proportion, inverse proportion, conjunction and disjunction are called continuous Boolean functions, because they satisfy the axioms of Boolean algebra.

Since it is desired that a qualitative expression be obtained, some quantitative values should be ignored. For example, function "A" in Fig. 7 is different from direct proportion x but the function is a proportion. So the two functions should be identified as the same in the qualitative expression. That is, in [0,1], $x^k (k \geq 2)$ should be identified with x in the qualitative expression. Mathematically, a norm is necessary, by which the distance among the two functions is 0. The qualitative norm can be introduced.

In {0,1}, a unit of a neural network is a multilinear function. The multilinear function space is a Euclidean space. See Sect. 4. So the unit can be approximated to a Boolean function by Euclidean norm. In [0,1], similar facts hold, that is. a unit of a neural network is a multilinear function in the qualitative expression, that is. the qualitative norm, and the space of multilinear

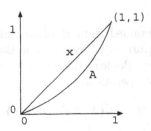

Fig. 7. Qualitative norm

functions is a Euclidean space in the qualitative norm. Thus the unit can be approximated to a continuous Boolean function by Euclidean norm.

5.2 Continuous Boolean Functions

This subsection briefly describes continuous Boolean functions [25, 31]. First, τ is defined, which is necessary for the definition of the qualitative norm.

Definition 5. τ_x *is defined as follows:*
Let $f(x)$ be a real polynomial function. Consider the following formula:

$$f(x) = p(x)(x - x^2) + q(x) \, ,$$

where $q(x) = ax + b$, where a and b are real, that is, $q(x)$ is the remainder. τ_x is defined as follows:

$$\tau_x : f(x) \rightarrow q(x) \, .$$

The above definition implies the following property:

$$\tau_x(x^k) = x \, ,$$

where $k \geq 2$.

Definition 6. *In the case of n variables,*
τ is defined as follows:

$$\tau = \prod_{i=1}^{n} \tau_{x_i} \, .$$

For example,

$$\tau(x^2 y^3 + y + 1) = xy + y + 1 \, .$$

Theorem 6. *The functions obtained from Boolean functions by extending the domain from $\{0,1\}$ to $[0,1]$ can satisfy all axioms of Boolean algebra with the logical operations defined below. Proof can be found in [25] and [31].*

AND: $\tau(fg)$,
OR: $\tau(f + g - fg)$,
NOT: $\tau(1 - f)$.

Therefore, the functions obtained from Boolean functions by extending the domain from $\{0,1\}$ to $[0,1]$ are called continuous Boolean functions. For example, xy and $1 - y(= \bar{y})$ are Boolean functions, where $x, y \in [0, 1]$. We show a simple example for logical operation.

$$(X \vee Y) \wedge (X \vee Y) = X \vee Y$$

is calculated as follows:

$$\tau\left((x + y - xy)(x + y - xy)\right)$$
$$= \tau(x^2 + y^2 + x^2y^2 + 2xy - 2x^2y - 2xy^2)$$
$$= x + y + xy + 2xy - 2xy - 2xy$$
$$= x + y - xy \ .$$

In the continuous domain, fuzzy rules can be obtained from trained neural networks by some algorithms [1] . The expression by continuous Boolean functions is more understandable than fuzzy rules, whereas continuous Boolean functions are worse than fuzzy rules in accuracy.

5.3 The Multilinear Function Space of the Domain [0,1]

Multilinear functions of the domain [0,1] are considered. In the domain [0,1], a qualitative norm has to be introduced.

Definition 7. *An inner product in the case of n variables is defined as follows:*

$$< f, g >= 2^n \int_0^1 \tau(fg)dx \ ,$$

where f and g are multilinear functions. The above definition can satisfy the properties of inner product [23, 24, 31].

Definition 8. *Norm $|f|$ is defined as follows:*

$$|f| = \sqrt{< f, f >}.$$

The distance between functions is roughly measured by the norm. For example, function A in Fig. 7, which stands for x^k, is different from x. However, by the norm, the distance between the two functions is 0, because τ in the norm

$$\sqrt{< f, g >} = \sqrt{2^n \int_0^1 \tau(fg)dx}$$

identifies $x^k(k \geq 2)$ with x. Therefore, the two functions are identified as being the same one in the norm. The norm can be regarded as a qualitative norm, because, roughly speaking, the norm identifies increasing functions as direct proportions, identifies decreasing functions as inverse proportions, and ignores the function values in the intermediate domain between 0 and 1.

Theorem 7. *The multilinear function space in the domain [0,1] is a Euclidean space with the above inner product: Proof can be found in [23, 24] and [31].*

The orthonormal system is as follows:

$$\phi_i = \prod_{j=1}^n e(x_j) \ (i = 1, \ldots, 2^n) \ ,$$

where $e(x_j) = 1 - x_j$ or x_j. It is easily understood that these orthonormal functions are the expansion of atoms in Boolean algebra of Boolean functions. In addition, it can easily be verified that the orthonormal system satisfies the following properties:

$$< \phi_i, \phi_j > = \begin{cases} 0 & (i \neq j), \\ 1 & (i = j), \end{cases}$$

$$f = \sum_{i=1}^{2^n} < f, \phi_i > \phi_i .$$

Example. In the case of 2 variables, the orthonormal functions are as follows:

$$xy, x(1-y), (1-x)y, (1-x)(1-y) .$$

and the representation by orthonormal functions of $x+y-xy$ of two variables (dimension 4) is as follows:

$$f = 1 \cdot xy + 1 \cdot x(1-y) + 1 \cdot (1-x)y + 0 \cdot (1-x)(1-y) .$$

When the domain is $[0,1]$, neural networks are approximated to multilinear functions with the following:

$$x^n = \begin{cases} x & (n \leq a) \\ 0 & (n > a), \end{cases}$$

where a is a natural number. When $a = 1$, the above approximation is the linear approximation.

5.4 The Polynomial Algorithm in the Continuous Domain

The polynomial algorithm for the continuous domain is the same as that for the discrete domain, because the multilinear function space of $[0,1]$ is the same as the multilinear function space of $\{0,1\}$. The rules obtained are continuous Boolean functions.

A theoretical analysis of the error in the case of the $[0,1]$ domain will be included in future work. However, experimental results show that the algorithm works well in the continuous domains, as explained in [28] and [33].

6 Discussions

6.1 On the Continuous Domain

When classes are continuous, other techniques such as decision tree do not work well as explained in Sect. 2. Therefore, rule extraction from mathematical formulas with continuous classes is important. Continuous Boolean functions work well for linear formulas, but cannot express the detailed information on neural networks, because the neural networks are nonlinear. The continuous Boolean functions are insufficient, and therefore, extracting rules described by intervals such as

$$(200 \leq t \leq 300) \vee (p \leq 2.5) \vee (70 \leq h \leq 90) \rightarrow q \leq 0.2$$

from neural networks is needed.

Usual real data consist of discrete data and continuous data. Therefore, rule extraction from a mixture of discrete attributes and continuous attributes is needed.

6.2 On the Prediction Domain

Training domains are much smaller than prediction domains. For example, in the case of *voting-records*, which consist of 16 binary attributes, the number of possible training data is $2^{16}(= 65536)$, while the number of training data is about 500. The outputs of a neural network are almost 100% accurate for about 500 training data, and are predicted values for the other approximately 65000 data. These predicted values are probabilistic, because the parameters for the neural network are initialized probabilistically. 65000 is much greater than 400, that is, the probabilistic part is much larger than the non-probabilistic part. Therefore, when a rule is extracted from a neural network, the predicted values of the neural network have to be dealt with probabilistically.

We have developed two types of algorithms. The first one deals with the whole domain equally [27]. The second one deals with only the training domain and basically ignores the prediction domain [28, 33]. Both algorithms can be regarded as opposite extremes. We also have developed an algorithm which deals with the prediction domains probabilistically [16]. Future work includes the development of algorithms dealing with the prediction domains appropriately.

6.3 Logical Regression Analysis

There are several mathematical formulas obtained by regression analyses. It is desired that rule extraction techniques from mathematical formulas be applied to nonparametric regression analysis, (multivariate) autoregression analysis, regression trees, differential equations (difference equations), and so on. See Fig. 8.

The data mining technique consisting of regression analysis and the Approximation Method is called Logical Regression Analysis (LRA). We have applied the Approximation Method to nonparametric regression analysis, that is, we have developed LRA using nonparametric regression analysis. We have been applying LRA using nonparametric regression analysis to fMRI images to discover the rules of brain functions.

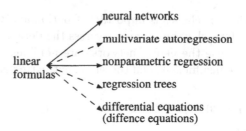

Fig. 8. The development of LRA

7 On the Data Mining from fMRI Images

The rest of the paper describes the data mining from fMRI images by LRA. This section explains the data mining from fMRI images.

The analysis of brain functions using functional magnetic resonance imaging (fMRI), positron emission tomography (PET), magnetoencephalography-(MEG) and so on is called non-invasive analysis of brain functions [18]. As a result of the ongoing development of non-invasive analysis of brain functions, detailed functional brain images can be obtained, from which the relations between brain areas and brain functions can be understood, for example, a relation between a few areas in the brain and an auditory function [10].

Several brain areas are responsible for a brain function. Some of them are connected in series, and others are connected in parallel. Brain areas connected in series are described by "AND" and brain areas connected in parallel are described by "OR". Therefore, the relations between brain areas and brain functions are described by rules.

Researchers are trying to heuristically discover the rules from functional brain images. Several statistical methods such as principal component analysis, have been developed. However, the statistical methods can only present some principal areas for a brain function. They cannot discover rules.

fMRI images can be dealt with by supervised inductive learning. However, the conventional inductive learning algorithms [19] do not work well for fMRI images, because there are strong correlations between attributes (pixels) and a small number of samples.

There are two solutions for the above two problems. The first one is the modification of the conventional inductive learning algorithms. The other one is nonparametric regression analysis. The modification of the conventional inductive learning algorithms would require a lot of effort. On the other hand, nonparametric regression analysis has been developed for the above two problems. Therefore, we have been using nonparametric regression analysis for the data mining from fMRI images.

The outputs of nonparametric regression analysis are linear formulas, which are not rules. However, we have already developed a rule extraction algorithm from regression formulas [16, 26, 29], that is, we have developed Logical Regression Analysis (LRA) as described above.

Since brains are three dimensional, three dimensional LRA is appropriate. However, the three dimensional LRA needs a huge computation time, for example, many years. Therefore, we applied two dimensional LRA to fMRI images as the first step.

We applied LRA to artificial data, and we confirmed that LRA works well for artificial data [30]. We applied LRA to several experimental tasks such as finger tappings and calculations. In the experiments of finger tapping, we compared the results of LRA with z-score [39], which is the typical conventional method. In the experiments, LRA could rediscover a little complicated relation, but z-score could not rediscover the relation. As the result, we confirmed that LRA works better than z-score. The details can be found in [37]. In the experiments of calculations, we confirmed that LRA worked well, that is, LRA rediscovered well-known facts regarding calculations, and discovered new facts regarding calculations. This paper reports the experiments of calculations.

8 The Data Mining from fMRI Images by Logical Regression Analysis

First, the data mining from fMRI images by Logical Regression Analysis is outlined. Second, nonparametric regression analysis is briefly explained. Finally, related techniques are described [34, 36, 37].

8.1 The Outline of Data Mining from fMRI Images by Logical Regression Analysis

The brain is 3-dimensional. In fMRI images, a set of 2-dimensional images(slices) represents a brain. See Fig. 9. 5 slices are obtained in Fig. 9. Figure 10 shows a real fMRI image. When an image consists of $64 \times 64 (= 4096)$ pixels, Fig. 10 can be represented as Fig. 11. In Fig. 11, white pixels mean activations and dot pixels mean inactivations. Each pixel has the value of the activation.

An experiment consists of several measurements. Figure 12 means that a subject repeats a task (for example, finger tapping) three times. "ON" in the upper part of the figure means that a subject executes the task and "OFF"

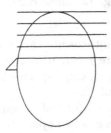

Fig. 9. fMRI images(3 dimension)

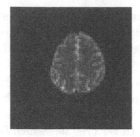

Fig. 10. A fMRI image

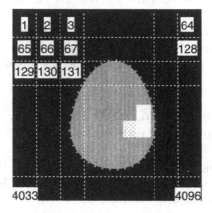

Fig. 11. An example of fMRI image

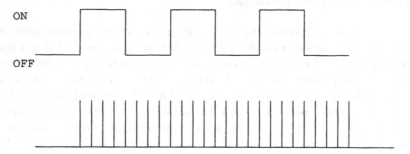

Fig. 12. Measurement

means that the subject does not execute the task, which is called rest. Bars in the lower part of the figure mean measurements. The figure means 24 measurements. When 24 images(samples) have been obtained, the data of a slice can be represented as Table 2.

Y(N) in the class stand for on (off) of an experimental task. From Table 2, machine learning algorithms can be applied to fMRI images. In the case of Table 2, the attributes are continuous and the class is discrete.

Table 2. fMRI Data

	1	2	⋯	4096	Class
S1	10	20	⋯	11	Y
S2	21	16	⋯	49	N
⋯	⋯	⋯	⋯	⋯	⋯
S24	16	39	⋯	98	N

Attributes (pixels) in image data have strong correlations between adjacent pixels. Moreover it is very difficult or impossible to obtain sufficient fMRI brain samples, and so there are few samples compared with the number of attributes (pixels). Therefore, the conventional supervised inductive learning algorithms such as C4.5 [19] do not work well, which was confirmed in [30].

Nonparametric regression analysis works well for strong correlations between attributes and a small number of samples. LRA using nonparametric regression analysis works well for the data mining from fMRI images.

Figure 13 shows the processing flow. First, the data is mapped to the standard brain [22]. Second, the brain parts of fMRI images are extracted using Standard Parametric Mapping:SPM (a software for brain images analysis [39]) Finally, LRA is applied to each slice.

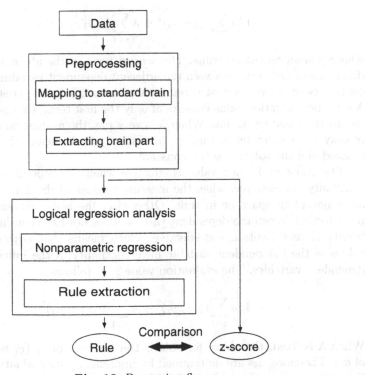

Fig. 13. Processing flow

8.2 Nonparametric Regression Analysis

First, for simplification, the 1-dimensional case is explained [8].

1-Dimensional Nonparametric Regression Analysis

Nonparametric regression analysis is as follows: Let y stand for a dependent variable and t stand for an independent variable and let $t_j (j = 1, \ldots, m)$ stand for measured values of t. Then, the regression formula is as follows:

$$y = \sum a_j t_j + e(j = 1, \ldots, m) ,$$

where a_j are real numbers and e is a zero-mean random variable. When there are n measured values of y,

$$y_i = \sum a_j t_{ij} + e_i (i = 1, \ldots, n) .$$

For example, in the case of Table 2, $m = 4096$, $n = 24$, $t_{11} = 10, t_{12} = 20, t_{1\ 4096} = 11$, and $y_1 = Y$.

In usual linear regression, error is minimized, while, in nonparametric regression analysis, error plus continuity or smoothness is minimized. When continuity is added to error, the evaluation value is as follows:

$$1/n \sum_{i=1}^{n}(y_i - \hat{y}_i)^2 + \lambda \sum_{i=1}^{n}(\hat{y}_{i+1} - \hat{y}_i)^2 ,$$

where $\hat{y}$ is an estimated value. The second term in the above formula is the difference of first order between the adjacent dependent variables, that is, the continuity of the dependent variable. λ is the coefficient of continuity. When λ is 0, the evaluation value consists of only the first term, that is, error, which means the usual regression. When λ is very big, the evaluation value consists of only the second term, that is, continuity, which means that the error is ignored and the solution $\hat{y}$ is a constant.

The above evaluation value is effective when the dependent variable has continuity, for example, when the measured values of the dependent variable are adjacent in space or in time. Otherwise, the above evaluation value is not effective. When the dependent variable does not have continuity, the continuity of coefficients a_js is effective, which means that adjacent measured values of the independent variable have continuity in the influence over the dependent variables. The evaluation value is as follows:

$$1/n \sum_{i=1}^{n}(y_i - \hat{y}_i)^2 + \lambda \sum_{j=1}^{m}(a_{j+1} - a_j)^2$$

When λ is fixed, the above formula is the function of a_i ($\hat{y}_i$ is the function of a_i). Therefore, a_is are determined by minimizing the evaluation value, and the optimal value of λ is determined by cross validation.

Calculation

Let $\mathbf{X}$ stand for $n \times m$ matrix. Let t_{ij} be an element of $\mathbf{X}$. Let $\mathbf{y}$ stand for a vector consisting of y_i. $m \times m$ matrix $\mathbf{C}$ is as follows:

$$\mathbf{C} = \begin{pmatrix} 1 & -1 & & & \\ -1 & 2 & -1 & & \\ & -1 & 2 & -1 & \\ & & & & \cdots \end{pmatrix}$$

Cross validation CV is as follows:

$$CV = n\tilde{\mathbf{y}}^{t}\tilde{\mathbf{y}}$$

$$\tilde{\mathbf{y}} = \mathbf{Diag}(\mathbf{I} - \mathbf{A})^{-1}(\mathbf{I} - \mathbf{A})\mathbf{y}$$
$$\mathbf{A} = \mathbf{X}(\mathbf{X}^{t}\mathbf{X} + (n-1)\lambda\mathbf{C})^{-1}\mathbf{X}^{t},$$

where $\mathbf{Diag}(\mathbf{A})$ is a diagonal matrix whose diagonal components are $\mathbf{A}$'s diagonal components. The coefficients $\hat{\mathbf{a}}$ are as follows:

$$\hat{\mathbf{a}} = (\mathbf{X}^{t}\mathbf{X} + n\lambda_{o}\mathbf{C})^{-1}\mathbf{X}^{t}\mathbf{y},$$

where λ_{o} is the optimal λ determined by cross validation.

2-Dimensional Nonparametric Regression Analysis

In 2-dimensional nonparametric regression analysis, the evaluation value for the continuity of coefficients a_{ij} is modified. In one dimension, there are two adjacent measured values, while, in two dimensions, there are four adjacent measured values. The evaluation value for the continuity of coefficients is not

$$(a_{i+1} - a_i)^2,$$

but the differences of first order between a pixel and the four adjacent pixels in the image. For example, in the case of pixel 66 in Fig. 11, the adjacent pixels are pixel 2, pixel 65, pixel 67. and pixel 130, and the evaluation value is as follows:

$$(a_2 - a_{66})^2 + (a_{65} - a_{66})^2 + (a_{67} - a_{66})^2 + (a_{130} - a_{66})^2.$$

Consequently, in the case of two dimensions, $\mathbf{C}$ is modified in the way described above. When continuity is evaluated, four adjacent pixels are considered, while smoothness is evaluated, eight adjacent pixels are considered, for example, in the case of pixel 66 in Fig. 11, the adjacent pixels are pixel 1, pixel 2, pixel 3, pixel 65. pixel 67, pixel 129, pixel 130 and pixel 131.

8.3 Related Techniques

This subsection briefly explains two popular techniques, z-score and independent component analysis, and compares them with LRA.

z-Score

z-score is widely used in fMRI images. z-score is calculated pixel-wise as follows:

$$\text{z-score} = \frac{|Mt - Mc|}{\sqrt{\sigma t^2 + \sigma c^2}} \,,$$

where

Mt: Average of task images
Mc: Average of rest images
σt: Standard deviation of task images
σc: Standard deviation of rest images

Task images mean the images in which a subject performs an experimental task. Rest images mean the images in which a subject does not perform an experimental task.

When z-score is 0, the average of task images equals the average of rest images. When z-score is 1 or 2, the difference between the average of task images and the average of rest images is big.

The areas whose z-scores are big are related to the experimental task. However, z-score does not tell which slices are related to the experimental task and does not tell the connections among the areas such as serial connection or parallel connection.

Independent Component Analysis

Independent component analysis (ICA) [12] can be roughly defined as follows. Let n independent source signals at a time t be denoted by

$$\mathbf{S}(t) = (S_1(t), \ldots, S_n(t)) \,.$$

Let m (mixed) observed signals at a time t be denoted by

$$\mathbf{X}(t) = (X_1(t), \ldots, X_m(t)) \,.$$

$\mathbf{X}$ (t) is assumed to be the linear mixture of $\mathbf{S}(t)$, that is,

$$\mathbf{X}(t) = A\mathbf{S}(t) \,,$$

where A is a matrix.

ICA obtains source signals $\mathbf{S}(t)$ from observed signals $\mathbf{X}$ under the assumption of the independence of source signals. Notice that the matrix A is unknown.

Independent Component Analysis (ICA) is applied to fMRI images. LRA is advantageous compared with ICA respecting the following points:

1. LRA uses classes. That is, LRA uses task/rest information, while ICA does not use task/rest information.
2. LRA conserves the spatial topologies in the images, while ICA cannot conserve the spatial topologies in the images.
3. LRA works well in the case of small samples, while it is not sure if ICA works well in the case of small samples.
4. LRA does not fall into a local minimum, while ICA falls into a local minimum.
5. LRA's outputs can represent the connections among areas, while ICA's outputs cannot represent the connections among the areas.

9 The Experiments of Calculations

In the experiments of calculations, we confirmed that LRA worked well, that is, we rediscovered well-known facts regarding calculations, and discovered new facts regarding calculations. In the experiment, a subject adds a number repeatedly in the brain. The experimental conditions follow:

Magnetic field:	1.5 tesla
Pixel number:	64 × 64
Subject number:	8
Task sample number:	34
Rest sample number:	36

Table 3 shows the errors of nonparametric regression analysis. Slice 0 is the image of the bottom of brain and slice 31 is the image of the top of brain. We focus on the slices whose errors are small, that is, the slices related to calculation. 133, . . . , 336 in the table are the ID numbers of subjects.

Table 4 summarizes the results of LRA. Numbers in parenthesis mean slice numbers. Figure 14, ... Fig. 19 show the main results.

White indicates high activity, dark gray indicates low activity, and black indicates non-brain parts. White and dark gray ares are connected by conjunction. For example, let A stand for the white area in Fig. 14 and B stand for the dark gray area in Fig. 14. Then, Fig. 14 is interpreted as $A \wedge \bar{B}$, which means area A is activated and area B is inactivated.

The extracted rules are represented by conjunctions, disjunctions and negations of areas. The conjunction of areas means the co-occurrent activation of the areas. The disjunction of areas means the parallel activation of the areas. The negation of an area means a negative correlation.

LRA can generate rules including disjunctions. However, the rules including disjunctions are too complicated to be interpreted by human experts in brain science, because they have paid little attention to the phenomena. Therefore, the rules including disjunctions are not generated in the experiments.

Researchers in brain science have paid attention to positive correlations, and have not paid attention to negative correlations. LRA can detect negative correlations. and so is expected to discover new facts.

Table 3. Results of nonparametric regression analysis

Slice	133	135	312	317	321	331	332	336
0	0.924	0.882	0.444	0.547	0.0039	0.870	0.455	0.306
1	0.418	0.030	0.546	0.587	0.298	0.814	0.028	0.946
2	0.375	0.538	0.337	0.435	0.278	0.723	0.381	0.798
3	0.016	0.510	0.585	0.430	0.282	0.743	0.402	0.798
4	0.456	0.437	0.519	0.446	0.157	0.636	0.419	0.058
5	0.120	0.469	0.473	0.376	0.265	0.698	0.385	0.366
6	0.965	0.434	0.602	0.138	0.380	0.475	0.420	0.541
7	1.001	0.230	0.430	0.309	0.119	0.175	0.482	0.547
8	1.001	0.388	0.434	0.222	0.478	0.246	0.387	0.704
9	0.968	0.473	0.362	0.281	0.390	0.409	0.193	0.913
10	1.001	0.008	0.447	0.357	0.341	0.358	0.227	0.908
11	1.001	0.066	0.383	0.380	0.167	0.275	0.115	0.914
12	1.001	0.736	0.302	0.312	0.397	0.021	0.181	0.909
13	0.828	0.793	0.525	0.222	0.455	0.845	0.204	0.733
14	0.550	0.822	0.349	0.523	0.023	0.229	0.130	0.474
15	0.528	0.805	0.298	0.569	0.107	0.439	0.338	0.374
16	0.571	0.778	0.494	0.509	0.008	0.354	0.377	0.493
17	0.009	0.007	0.159	0.615	0.238	0.159	0.561	0.774
18	0.089	0.060	0.663	0.010	0.011	0.033	0.519	0.711
19	0.642	0.238	0.573	0.405	0.185	0.426	0.470	0.689
20	0.887	0.514	0.383	0.376	0.149	0.177	0.214	0.430
21	0.282	0.532	0.256	0.028	0.018	0.219	0.303	0.548
22	0.281	0.415	0.613	0.167	0.045	0.213	0.352	0.528
23	0.521	0.422	0.229	0.227	0.048	0.306	0.050	0.450
24	0.814	0.270	0.401	0.439	0.013	0.212	0.350	0.570
25	0.336	0.394	0.411	0.195	0.469	0.148	0.414	0.689
26	0.603	0.008	0.390	0.180	0.477	0.107	0.358	0.541
27	0.535	0.062	0.324	0.191	0.308	0.279	0.455	0.413
28	0.719	0.010	0.371	0.271	0.167	0.436	0.237	0.649
29	0.942	0.310	0.400	0.257	0.169	0.353	0.023	0.775
30	0.898	0.360	0.547	0.283	0.209	0.467	0.464	0.157
31	0.746	0.026	0.023	0.445	0.187	0.197	0.084	0.195

Figures are taken from feet, and so the left side in the figures represents the right side of the brain, and the right side in the figures represents the left side of the brain. The upper side in the figures represents the front of the brain, and the lower side in the figures represents the rear of the brain.

Activation in the left angular gyrus and supramarginal gyrus was observed in 4 and 3 cases, respectively, and activation in the right angular gyrus and supramarginal gyrus was observed in 3 cases and 1 case, respectively. Clinical observations show that damage to the left angular and supramarginal gyrii causes acalculia which is defined as an impairment of the ability to calculate. Despite the strong association of acalculia and left posterior parietal lesions, there are certain characteristics of acalculia that have led to the suggestion

Table 4. Results of LRA

NO.		Left	Right

133	cingulate gyrus (17)	Cerebellum(3)
		superior frontal
		  gyrus(17)
		inferior frontal gyrus(17)
		superior temporal plane(17,18)
		middle frontal gyrus(18,21)
		angular gyrus(21,22)
			Cerebellum(0,5)
			middle frontal
			  gyrus(25)

135		inferior frontal
		  gyrus(17,18)
		superior temporal plane(17,18)
		precuneus(26,28)
			superior frontal
			  gyrus(26)
			superior parietal
			  gyrus(26,28)

312	cingulate gyrus (21)	inferior frontal gyrus(15) angular gyrus(21) supramarginal gyrus(18,21) middle frontal gyrus(23)	angular gyrus(17)
317	cingulate gyrus (27)	inferior frontal gyrus(18,21) cuneus(21,22,26,27)	angular gyrus(26)

321	cingulate gyrus (16,22,24)	inferior frontal gyrus(14)
		postcentral gyrus(16,18)
		cuneus(21)
		parieto-occipital sulcus(22)
		supramarginal gyrus(24)
			cuneus(16,18)
			parieto-occipital
			  sulcus(21)
			supramarginal
			  gyrus(22,24)

331	cingulate gyrus (25,26)	inferior frontal
		  gyrus(12,17,18)
		angular gyrus(17,18,26)
		supramarginal gyrus(17,18)
			angular
			  gyrus(17,18,26)
			middle temporal
			  gyrus(7,12,17)

332		inferior temporal
		  gyrus(1)
		Cerebellum(1)
		postcentral gyrus(33)
		middle temporal gyrus(12)
		pre-,post-central gyrus(14)
		angular gyrus(23)
		middle frontal gyrus(29)
			inferior temporal
			  gyrus(1)
			Cerebellum(1)
			middle frontal
			  gyrus(29)
			superior parietal
			  gyrus(29,43)

336		Cerebellum(0,5)
		middle temporal gyrus(4,5)
		middle frontal gyrus(30,31)
		precentral gyrus(31)
		superior frontal
		  gyrus(32)
			Cerebellum(0,5)
			superior parietal
			  gyrus(30)
			occipital
			  gyrus(11)

of a right-hemispheric contribution. Clinical observations also suggest that acalculia is caused by lesions not only in the left parietal region and frontal cortex but also in the right parietal region. Figure 14 shows slice 18 of subject 331 and Fig. 15 shows slice 26 of subject 331.

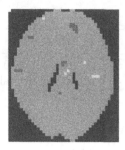 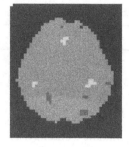 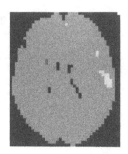

Fig. 14. Slice 18 331   **Fig. 15.** Slice 26 331   **Fig. 16.** Slice 17 135

Significant activation was observed in the left inferior frontal gyrus in 6 out of 8 cases. On the other hand, none was observed in the right inferior frontal gyrus. The result suggests that the left inferior frontal gyrus including Broca's area is activated in most subjects in connection with implicit verbal processes required for the present calculation task. Furthermore, significant activation in frontal region including middle and superior frontal regions was found in 8 cases (100%) in the left hemisphere and in 3 cases in the right hemisphere. The left dorsolateral prefrontal cortex may play an important role as a working memory for calculation. Figure 16 shows slice 17 of 135 and Fig. 17 shows slice 18 of subject 317.

In addition to these activated regions, activation in cingulate gyrus, cerebellum, central regions and occipital regions was found. The activated regions depended on individuals, suggesting different individual strategies. Occipital regions are related to spatial processing, and the cingulate gyrus is related to intensive attention. Central regions and the cerebellum are related to motor imagery. 5 out of 8 subjects used their cingulate gyrus, which means that they were intensively attentive. Figure 18 shows slice 17 of 133. 3 out of 8 subjects use cerebellum, which is thought not to be related to calculation. Figure 19 shows slice 1 of 332. The above two results are very interesting discoveries that have never been experimentally confirmed so far. The problem

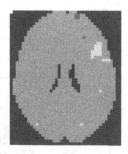 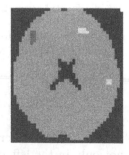 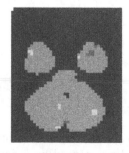

Fig. 17. Slice 18 317   **Fig. 18.** Slice 17 133   **Fig. 19.** Slice 1 332

of whether these regions are specifically related to mental calculation or not is to be investigated in further research with many subjects.

LRA has generated rules consisting of regions by conjunction and negation. As for conjunctions and negations, the results showed the inactivated regions simultaneously occurred with the activated regions. In the present experiment, inactivation in the brain region contralateral to the activated region was observed, suggesting inhibitory processes through corpus callosum. LRA has the possibility of providing new evidence in brain hemodynamics.

10 Conclusions

This paper has explained that rule extraction from mathematical formulas is needed for perfect data mining techniques. This paper has briefly reviewed the rule extraction techniques for neural networks and has briefly explained the Approximation Method developed by the author.

The author has developed a data mining technique called Logical Regression Analysis (LRA), which consists of regression analysis and the Approximation Method. One of LRA's merits is that LRA can deal with images. The author has been applying LRA using nonparametric regression analysis to fMRI images to discover the rules of brain functions.

The LRA works better than z-score and has discovered new "relations" respecting brain functions.

In the experiments, LRA was applied to slices, that is, 2-dimensional fMRI images. However, complicated tasks such as calculation are related to at least a few areas, and so the application of LRA to a set of a few slices is necessary for fruitful data mining from fMRI images. It is desired that LRA be applied to 3-dimensional fMRI images. However, the nonparametric regression analysis of 3-dimensional fMRI images needs a huge computational time. Therefore, the computational time should be reduced, which is included in future work.

There are a lot of open problems in the field of LRA, therefore the authors hope that researchers will tackle this field.

References

1. Andrews R, Diederich J, Tickle AB (1995) Survey and critique of techniques for extracting rules from trained artificial neural networks. *Knowledge-Based Systems* 8 (6):373–189.
2. Andrews R, Diederich J eds (1996) *Rules and Networks, Proceedings of the Rule Extraction from Trained Artificial Neural Networks Workshop, AISB'96.* Queensland University of Technology.
3. Birkhoff G, Bartee TC (1970) *Modern Applied Algebra.* McGraw–Hill.
4. Castro JL, Mantas CJ, Benitez JM (2002) Interpretation of Artificial Neural Networks by Means of Fuzzy Rules. *IEEE Transactions on Neural Networks* 13 (1):101–116.

5. Craven MW, Shavlik JW (1993) Learning symbolic rules using artificial neural networks. *Proceedings of the Tenth International Machine Learning Conference*: 73–80.

6. Craven MW, Shavlik JW (1994) Using sampling queries to extract rules from trained neural networks. *Proceedings of the Eleventh International Machine Learning Conference*: 37–45.

7. Etchells, TA (2003) *Rule Extraction from Neural Networks: A Practical and Efficient Approach*, Doctoral Dissertation, Liverpool John Moores University.

8. Eubank RL (1998) *Spline Smoothing and Nonparametric Regression*. Marcel Dekker, Newyork.

9. Kakimoto M, Morita C, Kikuchi Y, Tsukimoto H (2000) Data Mining from Functional Brain Images. *Proceedings of the First International Workshop on Multimedia Data Mining, MDM/KDD2000*: 91–97.

10. Kikuchi Y, Endo H, Yoshizawa S, Kita M, Nishimura C, Tanaka M, Kumagai T, Takeda T (1997) Human cortico-hippocampal activity related to auditory discrimination revealed by neuromagnetic field. *Neuro Report* 8:1657–1661.

11. Kuok CM, Fu A, Wong MH (1998) Mining fuzzy association rules in databases. *ACM SIGMOD record* 27 (1):1–12.

12. Lee TW (1998) *Independent Component Analysis*. Kluwer Academic Publishers.

13. Linial N, Mansour Y, Nisan N (1993) Constant Depth Circuits, Fourier Transform, and Learnability. *Journal of the ACM* 40 (3):607–620.

14. Lu HJ, Setiono R, Liu H (1998) Effective data mining using neural networks. *IEEE Trans. on Knowledge and Data Engineering* 8:957–961.

15. Miwa T et al. (1998) PLS regression and nonparametric regression in spectrum data analysis. Proc. 17th Japan SAS User Convention and Research Presentation, pp. 137–148.

16. Morita C, Tsukimoto H (1998) Knowledge discovery from numerical data. *Knowledge-based Systems* 10 (7):413–419.

17. Omlin CW, Giles CL (1996) Extraction of Rules from Discrete-time Recurrent Neural Networks. *Neural Networks* 9 (1):41–52.

18. Posner M.I, Raichle M.E (1997) *Images of Mind*. W H Freeman & Co.

19. Quinlan JR (1993) *C4.5: Programs for machine learning*. Morgan Kaufmann Pub.

20. Setiono R, Liu H (1995) Understanding Neural Networks via Rule Extraction. *Proceedings of The 14th International Joint Conference on Artificial Intelligence*: 480–485.

21. Srikant S, Agrawal R (1996) Mining quantitative association rules in large relational tables. *ACM SIGMOD conf. on management of Data*: 41–46.

22. Talairach J, Tournoux P (1988) *Coplanar Streoaxic atlas of the human brain*. Thieme Medica.

23. Tsukimoto H, Morita C (1994) The Discovery of Propositions in Noisy Data. *Machine Intelligence* 13:143–167. Oxford University Press.

24. Tsukimoto H (1994) The discovery of logical propositions in numerical data. *AAAI'94 Workshop on Knowledge Discovery in Databases*: 205–216.

25. Tsukimoto H (1994) On continuously valued logical functions satisfying all axioms of classical logic. *Systems and Computers in Japan* 25 (12):33–41. SCRIPTA TECHNICA, INC.

26. Tsukimoto H, Morita C (1995) Efficient algorithms for inductive learning-An application of multi-linear functions to inductive learning. *Machine Intelligence* 14:427–449. Oxford University Press.

27. Tsukimoto H (1996) An Algorithm for Extracting Propositions from Trained Neural Networks Using Multilinear Functions. *Rules and Networks, Proceedings of the Rule Extraction from Trained Artificial Neural Networks Workshop, AISB'96*: 103–114. Queensland University of Technology.
28. Tsukimoto H (1997) Extracting Propositions from Trained Neural Networks. *Proceedings of The 15th International Joint Conference on Artificial Intelligence*: 1098–1105.
29. Tsukimoto H, Morita C, Shimogori N (1997) An Inductive Learning Algorithm Based on Regression Analysis. *Systems and Computers in Japan* 28(3):62–70. SCRIPTA TECHNICA, INC.
30. Tsukimoto H, Morita C (1998) The Discovery of Rules from Brain Images. *Discovery Science, Proceedings of the First International Conference DS'98*: 198–209.
31. Tsukimoto H (1999) Symbol pattern integration using multilinear functions. In: Furuhashi T, Tano S. Jacobsen HA (eds) *Deep Fusion of Computational and Symbolic Processing*. Springer Verlag.
32. Tsukimoto H (1999) Rule extraction from prediction models. *The Third Pacific-Asia International Conference on Knowledge Discovery and Data Mining (PAKDD'99)*: 34–43.
33. Tsukimoto H (2000) Extracting Rules from Trained Neural Networks. *IEEE Transactions on Neural Networks* 11(2):377–389.
34. Tsukimoto H, Kakimoto M, Morita C, Kikuchi Y, Hatakeyama E, Miyazaki Y (2000) Knowledge Discovery from fMRI Brain Images by Logical Regression Analysis. LNAI 1967 *The Proceedings of The Third International Conference on Discovery Science*: 212–224.
35. Tsukimoto H, Morita C (2001) *Connectionism as Symbolicism:Artificial Neural Networks as Symbols*. Sanshusha.
36. Tsukimoto H, Kakimoto M, Morita C, Kikuchi Y (2002) Rule Discovery from fMRI Brain Images by Logical Regression Analysis. *Progress in Discovery Science*: 232–245. Springer-Verlag.
37. Tsukimoto H, Kakimoto M, Morita C, Kikuchi Y (2004) Nonparametric regression analysis of functional brain images. *Systems and Computers in Japan*. 35(1):67–78. John Wiley and Sons.
38. Zhou ZH, Jiang Y, Chen SF (2000) A general neural framework for classification rule mining. *International Journal of Computers, Systems and Signals* 1(2):154–168.
39. http://www.fil.ion.ucl.ac.uk/spm.
40. http://www.rulequest.com/cubist-info.html.
41. http://www2.toshiba.co.jp/datamining/index.htm.

A Feature/Attribute Theory
for Association Mining
and Constructing
the Complete Feature Set

Tsau Young Lin

Department of Computer Science
San Jose State University, San Jose, California 95192-0103
tylin@cs.sjsu.edu

Summary. A correct selection of features (attributes) is vital in data mining. For this aim, the complete set of features is constructed. Here are some important results: (1) Isomorphic relational tables have isomorphic patterns. Such an isomorphism classifies relational tables into isomorphic classes. (2) A unique canonical model for each isomorphic class is constructed; the canonical model is the bitmap indexes or its variants. (3) All possible features (attributes) is generated in the canonical model. (4) Through isomorphism theorem, all un-interpreted features of any table can be obtained.

Keywords: attributes, feature, data mining, granular, data model

1 Introduction

Traditional data mining algorithms search for patterns only in the given set of attributes. Unfortunately, in a typical database environment, the attributes are selected primarily for record-keeping, not for understanding of real world. Hence, it is highly possible that there are no visible patterns in the given set of attributes; see Sect. 2.2. The fundamental question is: Is there a suitable transformation of features/attributes so that

- The "invisible" patterns become visible in this new set?

Fortunately, the answer is yes. To answer this question, we critically analyze the essence of association mining. Based on it, we are able

- To construct the *complete* set of features for a given relational table.

Many applications will be in the forth coming volumes [10]. Here are some important results:(Continue the count from the abstract) (5) all high frequency patterns (generalized association rules) of the canonical model can be generated by a finite set of linear inequalities within polynomial time. (6) Through isomorphism theorem, all high frequency patterns of any relational table can be obtained.

1.1 Basics Terms in Association Mining (AM)

First, we recall (in fact, formalize) some basic terms. In traditional association rule mining, two measures, called the support and confidence, are the main criteria. Among the two, support is the essential measure. In this paper, we will consider **the support only**. In other words, we will be interested in the high frequency patterns that are not necessary in the form of rules. They could be viewed as *undirected association rules*, or just *associations*.

Association mining is originated from the market basket data [1]. However, in many software systems, the data mining tools are added to general DBMS. So we will be interested in data mining on *relational tables*. To be definitive, we have the following translations:

1. a relational table is a bag relation (i.e., repeated tuples are permitted [8])
2. an item is an attribute value,
3. a q-itemset is a subtuple of length q, or simply q-subtuple,
4. A q-subtuple is a q-association or (high frequency) q-pattern, if its occurrences are greater than or equal to a given threshold.

2 Background and Scope

2.1 Scope – A Feature Theory Based on the Finite Data

A feature is also called an attribute; the two terms have been used interchangeably. In the classical data model, an attribute is a representation of property, characteristic, and so forth [17]. It represents a human view of the universe (a slice of real world) – an intension view [5]. On the other hand, in modern data mining (DM), we are extracting information from the data. So in principle, the real world, including features (attributes), is encoded by and only by a finite set of data. This is an extension view or data view of the universe.

However, we should caution that each techniques of data mining often use some information (background knowledge) other than data [6]. So the encoding of the universe is different for different techniques. For examples association mining (AM) (Sect. 3) uses only the relational table, while clustering techniques utilize not only the table (of points), but also the geometry of the ambient space. So the respective feature theories will be different. In this paper, we will focus on Association Mining.

Next, we will show some peculiar phenomena of the finite encoding. Let Table 1θ and 2θ be the new tables derived from the tables in Sect. 2.2 by

Table 1. Ten point in (X,Y)-coordinate and Rotated coordinate

Segment#		Y	X'	Y'
S_1	2	0	1.99	0.17
S_2	$\sqrt{3} = 1.73$	1	1.64	1.15
S_3	$\sqrt{2} = 1.41$	$\sqrt{2}$	1.29	1.53
S_4	1	$\sqrt{3}$	0.85	1.81
S_5	0	2	−0.17	1.99
	X-Y coordinate Table 1A		Rotates -5 degree Table 1B	

rotating the coordinate systems θ degree. It should be easy to verify that (see Sect. 4.1 for the notion of isomorphism).

Proposition 2.1.1. *Table 1A, 1B and 1θ are isomorphic, so are the Table 2A, 2B and 2θ.*

This proposition says even though the rotations of the coordinate system generate infinitely many distinct features/attributes, they reduce to the same feature/attribute if the universe is encoded by a relational table. The main result of this paper is to determine all possible features of the encoded world.

2.2 Background – Mining Invisible Patterns

Let us consider a table of 5 points in X-Y-plane, as shown in Table 1A. The first column is the universe of the geometric objects. It has two attributes, which are the "X-Y coordinates." This table has no association rule of length 2. By transforming, the "X-Y coordinates" to "Polar coordinate system" (Table 2A). interestingly

Associations of length 2 appear .

The key question is how can we find such appropriate new features (polar coordinates).

Table 2. Ten points in polar coordinate and rotated coordinate

Segment#	Length	Direction	Length	Direction
S_1	2.0	0	2.0	5
S_2	2.0	30	2.0	35
S_3	2.0	45	2.0	45
S_4	2.0	60	2.0	65
S_5	2.0	90	2.0	95
	X-Y coordinate Table 2A		Rotates -5 degree Table 2B	

3 Formalizing Association Mining

In this section, we will critically analyze the association (rule) mining. Let us start with a general question: What is data mining? There is no universally accepted formal definition of data mining, however the following informal description (paraphrase from [7]) is rather universal:

- Deriving useful patterns from data.

This "definition" points out key ingredients: data, patterns, methodology of derivations and the real world meaning of patterns (useful-ness). We will analyze each of them.

3.1 Key Terms "Word" and "Symbol"

First we need to precisely define some key terms.

A *symbol* is a string of "bit and bytes." It has no *formal* real world meaning, more precisely, any real world interpretation (if there is one) does *not* participate in formal processing or computing. Mathematicians (in group theory, more specifically) use the term "word" for such purpose. However, in this paper, a "word" will be more than a symbol. A symbol is termed a *word*, if the intended real world meaning *does participate* in the formal processing or computing. In AI, there is a similar term, semantic primitive [2]; it is a symbol whose real world interpretation is not implemented. So in automated computing, a semantic primitive is a symbol.

3.2 What are Data? – A Table of Symbols

To understand the nature of the data, we will examine how the data is created: In traditional data processing, (1) we select a set of attributes, called relational schema. Then (2) we (knowledge) represent a set of real world entities by a table of words.

$$K_{map} : V \rightarrow K_{word} ; \quad v \longrightarrow k$$

where K_{word} is a table of words (this is actually the usual relational table). Each word, called an attribute value, represents a real world fact (to human); however the real world semantic is not implemented. Since K_{word} is a bag relation [8], it is more convenient to use the graph $K_{graph} = \{(v, K(v) \mid v \in V\}$. If the context is clear, we may drop the subscript, so K is a map, an image or a graph.

Next, how is the data processed? In traditional data processing environment, for example, the attribute name COLOR means exactly what a human thinks. Therefore its possible values are yellow, blue, and etc. More importantly,

- DBMS processes these data under *human commands*, and carries out the human perceived-semantics. Such processing is called Computing with Words.

However, in the system, COLOR, yellow, blue, and etc are "bits and bytes" without any meaning, they are pure symbols.

The same relational table is used by Association Mining (AM). But, the data are processed without human interventions, so the table of words K_{word} is processed as a table K_{symbol} of symbols.

$$DM : K_{word} \Rightarrow K_{symbol}$$

In summary,

- *The data (relational table) in AM is a table of symbols.*

3.3 What are Patterns? and Computing with Symbols

What are the possible patterns? The notion depends on the methodology. So we will examine the algorithms first. A typical AM algorithm treats words as symbols. It just *counts* and does not consult human for any possible real world meaning of any symbol. As we have observed in previous section no real world meaning of any symbols is stored in the system. So an AM algorithm is merely a computing of pure symbols. AM transforms a table K_{symbol} of symbols into a set A_{symbol} of association (rules)s of symbols. These associations are "expressions" of symbols. Therefore,

- All possible patterns of AM are expressions of the symbols of the relational table.

3.4 Interpretation and Realization of Patterns

The output of an AM algorithm is examined by human. So each symbol is alive again. Its interpretation (to human only) is assigned at the data creation time. So the patterns are interpreted by these interpretations of symbols.

1. Interpretation: A pattern, an expression of symbols, is an expression of words (to human). So a pattern is a mathematical expression of real world facts.
2. Realization: A mathematical expression of real world facts may or may not correspond to a real world phenomenon.

4 Understanding the Data – A Table of Symbols

In the previous section, we have concluded that the input data to AM is a table of symbols. In this section, we will explore the nature of such a table.

4.1 Isomorphism – Syntactic Nature of AM

We have explained how data is processed in (automated) data mining: The algorithms "forget" the real world meaning of each word, and regard the input data as pure symbols. Since no real world meaning of each symbol participates in the computing process if we replace the given set of symbols by a new set, then we can derive new patterns by simply replacing the symbols in "old" patterns. Formally, we have (Theorem 4.1. of [12])

Theorem 4.1.1. *Isomorphic relational tables have isomorphic patterns.*

Though this is a very important theorem, its proof does not increase the understanding. Its proof is in the appendix. Isomorphism is an equivalence relation defined on the family of all relational tables, so it classifies the tables into isomorphic classes.

Corollary 4.1.2. *A pattern is a property of an isomorphic class.*

The impacts of this simple theorem are rather far reaching. It essentially declares that patterns are syntactic in nature. They are patterns of the whole isomorphic class, even though many somorphic relations may have very different semantics.

Corollary 4.1.3. *The probability theory based on the item counting is a property of isomorphic class.*

We will illustrate the idea by an example. The following example is adopted from ([8], pp 702):

Example 4.1.4. In this example, we will illustrate the notion of isomorphism of tables and patterns. In Table 3, we present two "copies" of relational tables; they are obviously isomorphic (by adding prime' to one table you will get the other one). For patterns (support = 2), we have the following:

Isomorphic tables K and K' have isomorphic q-associations:

1. 1-association in K: 30, 40, *bar*, *baz*,
2. 1-association in K': 30', 40', *bar'*, *baz'*,
3. 2-association in K: (30, *bar*) and (40, *baz*),
4. 2-association in K': (30', *bar'*) and (40', *baz'*).

Two sets of q-association ($q = 1,2$) are obviously isomorphic in the sense that adding prime ' to associations in K become associations in K'.

Table 3. A Relational Table K and its Isomorphic Copy K′

V	$\rightarrow$	F	G	V	$\rightarrow$	F'	G'
e_1	$\rightarrow$	30	foo	e_1	$\rightarrow$	$30'$	foo'
e_2	$\rightarrow$	30	bar	e_2	$\rightarrow$	$30'$	bar'
e_3	$\rightarrow$	40	baz	e_3	$\rightarrow$	$40'$	baz'
e_4	$\rightarrow$	50	foo	e_4	$\rightarrow$	$50'$	foo'
e_5	$\rightarrow$	40	bar	e_5	$\rightarrow$	$40'$	bar'
e_6	$\rightarrow$	40	bar	e_6	$\rightarrow$	$40'$	bar'
e_7	$\rightarrow$	30	bar	e_7	$\rightarrow$	$30'$	bar'
e_8	$\rightarrow$	40	baz	e_8	$\rightarrow$	$40'$	baz'

4.2 Bitmaps and Granules – Intrinsic Representations

Due to the syntactic nature, as we have observed in last section, we can have
a more intrinsic representation, that is a representation in which only the
internal structure of the table is important, the real world meaning of each
attribute value can be ignored.

We will continue to use the same example. The following discussions essential excerpt from ([8], pp 702). Let us consider the bitmap indexes for K
(see Table 3) the first attributes, F, would have three bit-vectors. The first,
for value 30, is 11000110, because the first, second, sixth, and seventh tuple
have $F = 30$. The other two, for 40 and 50, respectively, are 00101001 and
00010000. A bitmap index for G would also have three bit-vectors: 10010000,
01001010, and 00100101. It should be obvious that we will have the exact
same bitmap table for K'.

Next, we note that a bit vector can be interpreted as a subset of V, called
an elementary granule. For example, the bit vector, 11000110, of $F = 30$ represents the subset $\{e_1, e_2, e_6, e_7\}$. Similarly, 00101001, of $F = 40$ represents the
subset $\{e_3, e_5, e_8\}$, and etc. Let us summarize the discussions in the following
proposition:

Proposition 4.2.1. *Using Table 4 as a translation table, we transform a table*
of symbols (Table 3) into its respective

1. a bitmap table, and Table 5.
2. a granular table, Table 6.

Conversely,

Proposition 4.2.2. *Using Table 4 as an interpretation table that interpret*

1. Table 5 and Table 6 into Table 3, where (to human) each symbol corresponds to a real world fact.
2. Note that F-granules (and G-granules too) are mutually disjoints and form
   a covering of V. So the granules of each attribute induces a partition on V
   (an equivalence relation).

Table 4. Translation Table

F-Value	Bit-Vectors	Granules
30	= 11000110	$(\{e1, e2, e6, e7\})$
40	= 00101001	$(\{e3, e5, e8\})$
50	= 00010000	$(\{e4\})$
G-Value	= Bit-Vectors	Granules
Foo	= 10010000	$(\{e1, e4\})$
Bar	= 01001010	$(\{e2, e5, e7\})$
Baz	= 00100101	$(\{e3, e6, e8\})$

Table 5. Contrasting Tables of Symbols and Bitmaps

Table K				Bitmap Table B_K	
V	$\rightarrow$	F	G	F-bit	G-bit
e_1	$\rightarrow$	30	foo	11000110	10010000
e_2	$\rightarrow$	30	bar	11000110	01001010
e_3	$\rightarrow$	40	baz	00101001	00100101
e_4	$\rightarrow$	50	foo	00010000	10010000
e_5	$\rightarrow$	40	bar	00101001	01001010
e_6	$\rightarrow$	30	baz	11000110	00100101
e_7	$\rightarrow$	30	bar	11000110	01001010
e_8	$\rightarrow$	40	baz	00101001	00100101

Table 6. Contrasting Tables of Symbols and Granules

Table K				Granular Table G_K	
U	$\rightarrow$	F	G	E_F	E_G
$v1$	$\rightarrow$	30	foo	$\{e1, e2, e6, e7\}$	$\{e1, e4\}$
$v2$	$\rightarrow$	30	bar	$\{e1, e2, e6, e7\}$	$\{e2, e5, e7\}$
$v3$	$\rightarrow$	40	baz	$\{e3, e5, e8\}$	$\{e3, e6, e8\}$
$v4$	$\rightarrow$	50	foo	$\{e4\}$	$\{e1, e4\}$
$v5$	$\rightarrow$	40	bar	$\{e3, e5, e8\}$	$\{e2, e5, e7\}$
$v6$	$\rightarrow$	30	baz	$\{e1, e2, e6, e7\}$	$\{e3, e6, e8\}$
e_7	$\rightarrow$	30	bar	$\{e1, e2, e6, e7\}$	$\{e2, e5, e7\}$
e_8	$\rightarrow$	40	baz	$\{e3, e5, e8\}$	$\{e3, e6, e8\}$

3. Each elementary granule, for example, the elementary granule $\{e_1, e_2, e_6, e_7\}$ of $F = 30$, consists of all entities that have (are mapped to) the same attribute value, in this case, F-value 30. In other words, F-granule $\{e_1, e_2, e_6, e_7\}$ is the inverse of the value $F = 30$.

It should be obvious that these discussions can be generalized: They are summarize in *Proposition* 5.1.1.

5 The Model and Language of High Frequency Patterns

As we have observed in Sect. 3.3, informally patterns are expressions (sub-tuples) of the symbols of the relational table. Traditional association mining considers only the "conjunction of symbols." Are there other possible expressions or formulas? A big Yes, if we look at a relational table as a logic system. There are many such logic views, for example, deductive database systems, Datalog [21], and Decision Logic [19] among others. For our purpose, such views are too "heavy", instead, we will take an algebraic approach. The idea is stated in [13] informally. There, the notion of "logic language" was introduced informally by considering the "logical formulas" of the names of elementary granules. Each "logical formula" (of names) corresponds to a set theoretical formula of elementary granules. In this section, we shall re-visit the idea more formally.

5.1 Granular Data Model (GDM) – Extending the Expressive Power

Based on example, we have discussed granular data model in Sect. 4.2. Now we will discuss the general case.

Let V be set of real world entities, $A = \{A^1, A^2, \ldots, A^n\}$ be a set of attributes. Let their (active) attribute domains be $C = \{C^1, C^2, \ldots, C^n\}$, where active is a database term to emphasize the fact that C^j is the set of distinct values that occur in the current representation. Each C^j, often denoted by $Dom(A^j)$, is a Cantor set.

A relational table K can be regarded as a map (knowledge representation)

$$K_{map} : V \longrightarrow Dom(A) = Dom(A^1) \times \ldots Dom(A^n)$$

Similarly, an attribute is also a map

$$A^j : V \longrightarrow Dom(A^j) ; \quad v \longrightarrow c .$$

The inverse of such an attribute map defines a partition on V (hence an equivalence relation); we will denote it by Q^j and list some of its properties in:

Proposition 5.1.1

1. The inverse image $S = (A^j)^{(-1)}(c)$ is an equivalence class of Q^j. We say S is elementary granule, and c is the name of it.
2. For a fixed order of V, S can be represented by a bit-vector. We also say c is the name of the bit vector.
3. By replacing each attribute value of the table K_{symbol} by its bit-vector or elementary granule (equivalence class), we have the bitmap table B_K or granular table G_K respectively.
4. The equivalence relations, $Q = \{Q^1, Q^2, \ldots, Q^n\}$, play the role of attributes in Table G_K and B_K.

5. For uniformity, we write $V/Q^j = Dom(Q^j)$, namely, we regard the quotient set as the attribute domain.
6. Theoretically, G_K and B_K conceptually represent the same granular data model; the difference is only in representations and is an internal matter.
7. We will regard the table K as an interpretation of G_K and B_K. The interpretation is an isomorphism (via a table similar to Table 4) By Theorem 4.1.1., the patterns in K, G_K, B_K are isomorphic and hence is the same (identified via interpretation).

• It is adequate to do the AM in G_K.

The canonical model G_K is uniquely determined by its universe V, and the family Q of equivalence relations. In other words, the pair (V, Q) determines and is determined by G_K.

Definition 5.1.1. *The pair (V, Q) is called granular data model (GDM).*

(V, Q) is a model of some rather simple kind of logic, where the only predicates are equivalence predicates (predicates that satisfy the reflexive, symmetric and transitive properties). It was considered by both Pawlak and Tony Lee and has been called knowledge base, relation lattice, granular strucutre [9, 13, 19].

Note that the set of all elementary granule in (V, Q) generate a sub-Boolean algebra of the power set of V. By abuse of notation, we will use (V, Q) to denote this algebra. Since G_K is a table format of (V, Q), we need to describe how G_K is "embedded" into the Boolean algebra. We will extend *Proposition* 5.1.1, Item 7 into

Proposition 5.1.2. *An attribute value of G_K, which is an elementary granule, is mapped to the same granule in (V, Q). A subtuple of G_K, consisting of a set of elementary granules is mapped into the granule that is the intersection of those elementary granules; note two subtuples may be mapped to the same granule.*

5.2 Algebraic Language and Granular Boolean Algebra

The attribute values in K are pure symbols. Now we will introduce a new Boolean algebra L_K as follows: We will use $\cup$ and $\cap$ as the join and meet of this Boolean algebra. L_K is a free Boolean algebra subject to the following conditions:

The $\cap$ between symbols in the same columns are

$$B_i^j \cap B_k^j = \quad \forall i \neq k \; \forall j$$

This condition reflects the fact that the elementary granules of the same column are mutually disjoint.

We can give a more algebraic description [3]. Let F be the free Boolean Algebra generated by the symbols in K. Let I be the ideal generated by

$$B_i^j \cap B_k^j \quad \forall \, i, k, j$$

Then the quotient algebra $F/I = L_K$.

We will regard this Boolean algebra as a language and call it

Granular algebraic language .

An attribute value in K can be regarded as the name of the corresponding elementary granule in G_K and the elementary granule is the meaning set of the name. Recall that GDM (V, Q) can be regarded as Boolean algebra of elementary granules, and G_K is "embedded" in (V, Q) (*Proposition* 5.1.2.) So the name-to-meaning set assignment, $K \to G_K$, can be extended to a homomorphism of Boolean algebras:

name-to-meaning: $L_K \longrightarrow (V, Q)$; formula $\longrightarrow$ meaning set .

- High frequency patterns of AM are formulas with large meaning set (the cardinality is large).

6 The Formal Theory of Features in AM

The theory developed here is heavily depended on the nature of association mining (AM) that are formalized in Sect. 3.

6.1 Feature Extractions and Constructions

Let us examine some informal assertions, e.g., [18]: "All new constructed features are defined in terms of original features,...." and "Feature extraction is a process that extracts a set of new features from the original features through some functional mapping." In summary the new feature is derived (by construction or extraction) from the given set of attributes. We will formalize the idea of features in association mining (AM). Perhaps, we should re-iterate that we are *not* formalizing the general notion of features that involves human view.

Let K be the given relational table that has attributes $A = \{A^1, \ldots A^n\}$. Next, let $A^{n+1} \ldots A^{n+m}$ be the *new* attributes that are constructed or extracted. As we remark in Sect. 5.1, an attribute is a mapping from the universe to a domain, so we have the following new mappings.

$$A^{n+k} : V \longrightarrow Dom(A^{n+k}) .$$

Now, let us consider the extended table, K_a, that includes both old and additional new attributes $\{A^1, \ldots A^n \ldots A^{n+m}\}$. In this extended table, by the meaning of feature construction, A^{n+k}, should be (extension) functionally dependent (EFD) on A. This fact implies, by definition of EFD, there is a mapping

$$f^{n+k} : Dom(A^1) \times \ldots \times Dom(A^n) \longrightarrow Dom(A^{n+k}) .$$

such that $A^{n+k} = f^{n+k} \circ (A^1 \times \ldots \times A^n$.

Those new extracted or constructed features, such as f^{n+k} is called derived feature.

6.2 Derived Features in GDM

Now we will consider the situation in G_K, the granular table of K. In this section, we will express EFD f^{n+k} in granular format, in other words, the granular form of f^{n+k} is:

$$V/(Q^1 \cap \ldots \cap Q^n) = V/Q^1 \times \ldots \times V/Q^n \longrightarrow V/Q^{n+k}$$

The first equality is a simple property of quotient sets. The second map is f^{n+k} in its granular form. The granular form of f^{n+k} implies that Q^{n+k} is a coarsening of $(Q^1 \cap \ldots \cap Q^k)$. So we have the following

Proposition 6.2. Q^{n+k} *is a derived feature of* G_K *if and only if* Q^{n+k} *is a coarsening of* $(Q^1 \cap \ldots \cap Q^n)$.

Let the original Table K have attributes $A = \{A^1, \ldots A^n\}$. Let $B \subseteq A$ and $Y \in A$ (e.g., $Y = A^{n+k}$ and $Y_E = Q^{n+k}$).

Proposition 6.3. Y *is a feature constructed from* B *if and only if the induced equivalence relation* Y_E *is a coarsening of the induced equivalence relation* $B_E = (Q^{j_1} \cap \ldots \cap Q^{j_m})$, *where* $Y \in A$ *and* $B \subseteq A$

The proposition says all the new constructed features are coarsening of the intersection of the original features.

7 Universal Model – Capture the Invisibles

Let $\Delta(V)$ be the set of all partitions on V (equivalence relations); $\Delta(V)$ forms a lattice, where meet is the intersection of equivalence relations and join is the "union," where the "union," denoted by $\cup_j Q^j$, is the smallest coarsening of all $Q^j, j = 1, 2, \ldots \Delta(V)$ is called the partition lattice.

Let $(V, Q = \{Q^1, \ldots, Q^n\})$ be a GDM. Let $L(Q)$ be the smallest sublattice of $\Delta(V)$ that contains Q, and $L^*(Q)$ be the set of all possible coarsenings of $(Q^1 \cap \ldots \cap Q^n)$. $L^*(Q)$ obviously forms a sublattice of $\Delta(V)$; the intersection and "union" of two coarsenings is a coarsening. From *Proposition* 6.2., we can easily establish

Theorem 7.1. *Let* G_K *be a granular table; its GDM is* (V, Q). *Then* $(V, L^*(Q))$ *is a GDM that consists of all possible features for* G_K.

The set of all possible features of G_K is the set D of all those derived features. By *Proposition* 6.2., D is the set of all those coarsenings of $(Q^1 \cap \ldots \cap Q^n)$. SO $(V, L^*(Q))$ is the desirable one.

Definition 7.2. *The $(V, L^*(Q))$ is the completion of (V, Q) and is called the universal model of K.*

We should point out that the cardinal number of $L^*(Q))$ is enormous; it is bounded by the Bell number B_n, where n is the cardinality of the smallest partition in $L^*(Q)$ [4].

8 Conclusions

1. A feature/attribute, from human view, is a characteristic or property of the universe (a set of entities). Traditional data processing takes such a view and use them to represent the universe (knowledge representation).
2. A feature/attribute, in data mining, is defined and encoded by data. So a feature in association mining is a partition of the universe. Under such a view, we have shown that a set of infinite many distinct human-view-features (rotations of coordinate systems) is reduced to a single data-encoded-feature (Sect. 2.1).
3. Such views are shared by those techniques, such as classification, that utilize only the relational table of symbols in their algorithms. The other techniques, such as clustering and neural network, that utilize additional background knowledge, do not share the same view.
4. In association mining, we have the following applications [10, 11]: All generalized associations can be generated by a finite set of integral linear inequalities within polynomial time.
5. Finally, we would like to note that by the isomorphism theorem, two isomorphic relations may have totally distinct semantics. So relations with additional structures that capture some semantics may be worthwhile to be explored; see [13, 15].

9 Appendix

9.1 General Isomorphism

Attributes A^i and A^j are isomorphic if and only if there is a one-to-one and onto map, $s : Dom(A^i) \longrightarrow Dom(A^j)$ such that $A^j(v) = s(A^i(v)) \; \forall \; v \; \in \; V$. The map s is called an isomorphism. Intuitively, two attributes (columns) are isomorphic if and only if one column turns into another one by properly renaming its attribute values.

Let $K = (V, A)$ and $H = (V, B)$ be two information tables, where $A = \{A^1, A^2, \ldots, A^n\}$ and $B = \{B^1, B^2, \ldots, B^m\}$. Then, K and H are said

76   T.Y. Lin

to be isomorphic if every A^i is isomorphic to some B^j, and vice versa. The isomorphism of relations is reflexive, symmetric, and transitive, so it classifies all relations into equivalence classes; we call them isomorphic classes.

Definition 9.1.1. *H is a simplified relational table of K, if H is isomorphic to K and only has non-isomorphic attributes.*

Theorem 9.1.2. *Let H be the simplified relational table of K. Then the patterns (large itemsets) of K can be obtained from those of H by elementary operations that will be defined below.*

To prove the Theorem, we will set up a lemma, in which we assume there are two isomorphic attributes B and B' in K, that is, degree K – degree $H = 1$. Let $s : Dom(B) \longrightarrow Dom(B')$ be the isomorphism and $b' = s(b)$. Let H be the new table in which B' has been removed.

Lemma 9.1.3. *The patterns of K can be generated from those of H by elementary operations, namely,*

1. If b is a large itemset in H, then b' and (b, b') are large in K.
2. If (a. ., b, c...) is a large itemset in H, then (a. ., b', c...) and (a. ., b, b', c,...) are large in K.
3. These are the only large itemsets in K.

The validity of this lemma is rather straightforward; and it provides the critical inductive step for Theorem; we ill skip the proof.

9.2 Semantics Issues

The two relations, Tables 7 and 8, are isomorphic, but their semantics are completely different. One table is about part, the other is about suppliers. These two relations have Isomorphic association rules;

1. Length one: TEN, TWENTY, March, SJ, LA in Table 7 and

Table 7. An Relational Table K

V	K	(S#	Business Amount (in m.)	Birth Day	CITY)
v_1	$\longrightarrow$	$(S_1$	TWENTY	MAR	NY
v_2	$\longrightarrow$	$(S_2$	TEN	MAR	SJ
v_3	$\longrightarrow$	$(S_3$	TEN	FEB	NY
v_4	$\longrightarrow$	$(S_4$	TEN	FEB	LA
v_5	$\longrightarrow$	$(S_5$	TWENTY	MAR	SJ
v_6	$\longrightarrow$	$(S_6$	TWENTY	MAR	SJ
v_7	$\longrightarrow$	$(S_7$	TWENTY	APR	SJ
v_8	$\longrightarrow$	$(S_8$	THIRTY	JAN	LA
v_9	$\longrightarrow$	$(S_9$	THIRTY	JAN	LA

Table 8. An Relational Table K′

V	K	(S#	Weight	Part Name	Material
v_1	⟶	(P_1	20	SCREW	STEEL
v_2	⟶	(P_2	10	SCREW	BRASS
v_3	⟶	(P_3	10	NAIL	STEEL
v_4	⟶	(P_4	10	NAIL	ALLOY
v_5	⟶	(P_5	20	SCREW	BRASS
v_6	⟶	(P_6	20	SCREW	BRASS
v_7	⟶	(P_7	20	PIN	BRASS
v_8	⟶	(P_8	30	HAMMER	ALLOY
v_9	⟶	(P_9	30	HAMMER	ALLOY

2. Length one: 10, 20, Screw, Brass, Alloy in Table 8
3. Length two: (TWENTY, MAR), (Mar, SJ), (TWENTY, SJ)in one Table 7,
4. Length two: (20, Screw), (screw, Brass), (20, Brass), Table 8

However, they have very non-isomorphic semantics:

1. Table 7: (TWENTY. SJ), that is, the business amount at San Jose is likely 20 millions; it is isomorphic to (20, Brass), which is not interesting.
2. Table 8: (SCREW. BRASS), that is, the screw is most likely made from Brass: it is isomorphic to (Mar, SJ), which is not interesting.

References

1. R. Agrawal, T. Imielinski, and A. Swami, "Mining Association Rules Between Sets of Items in Large Databases," in Proceeding of ACM-SIGMOD international Conference on Management of Data, pp. 207–216, Washington, DC, June, 1993
2. A. Barr and E. A. Feigenbaum, The handbook of Artificial Intelligence, Willam Kaufmann 1981.
3. G. Birkhoff and S. MacLane, A Survey of Modern Algebra, Macmillan, 1977
4. Richard A. Brualdi, Introductory Combinatorics, Prentice Hall, 1992.
5. C. J. Date, C. Date. An Introduction to Database Systems, 7th ed., Addison-Wesley. 2000.
6. Margaret H. Dunham. Data Mining Introduction and Advanced Topics Prentice Hall, 2003, ISBN 0-13-088892-3
7. U. M. Fayad, G. Piatetsky-Sjapiro, and P. Smyth, "From Data Mining to Knowledge Discovery: An overview." In Fayard, Piatetsky-Sjapiro, Smyth, and Uthurusamy eds., Knowledge Discovery in Databases, AAAI/MIT Press, 1996.
8. H Gracia-Molina, J. Ullman, and J. Windin, J, Database Systems The Complete Book, Prentice Hall. 2002.
9. T. T. Lee. "Algebraic Theory of Relational Databases," The Bell System Technical Journal Vol 62. No 10, December, 1983, pp. 3159–3204.

10. T. Y. Lin, "A mathematical Theory of Association Mining" In: Foundation and Novel Approach in Data Mining (Lin & et al), Spriner-Verlag, 2005, to appear.

11. T. Y. Lin, "Mining Associations by Solving Integral Linear Inequalities," in: the Proceedings of International Conference on Data Mining, Breighton, England, Nov 1–4, 2004.

12. T. Y. Lin "Attribute (Feature) Completion – The Theory of Attributes from Data Mining Prospect," in: the Proceedings of International Conference on Data Mining, Maebashi, Japan, Dec 9–12, 2002, pp. 282–289.

13. T. Y. Lin, "Data Mining and Machine Oriented Modeling: A Granular Computing Approach," Journal of Applied Intelligence, Kluwer, Vol. 13, No 2, September/October, 2000, pp. 113–124.

14. T. Y. Lin and M. Hadjimichael, "Non-Classificatory Generalization in Data Mining," in Proceedings of the 4th Workshop on Rough Sets, Fuzzy Sets, and Machine Discovery, November 6–8, Tokyo, Japan, 1996, 404–411.

15. E. Louie, T. Y. Lin, "Semantics Oriented Association Rules," In: 2002 World Congress of Computational Intelligence, Honolulu, Hawaii, May 12–17, 2002, 956–961 (paper # 5702)

16. E. Louie and T. Y. Lin, "Finding Association Rules using Fast Bit Computation: Machine-Oriented Modeling," in: Foundations of Intelligent Systems, Z. Ras and S. Ohsuga (eds), Lecture Notes in Artificial Intelligence 1932, Springer-Verlag, 2000, pp. 486–494. (ISMIS'00, Charlotte, NC, Oct 11–14, 2000)

17. H. Liu and H. Motoda, "Feature Transformation and Subset Selection," IEEE Intelligent Systems, Vol. 13, No. 2, March/April, pp. 26–28 (1998)

18. Hiroshi Motoda and Huan Liu "Feature Selection, Extraction and Construction," Communication of IICM (Institute of Information and Computing Machinery, Taiwan) Vol 5, No. 2, May 2002, pp. 67–72. (proceeding for the workshop "Toward the Foundation on Data Mining" in PAKDD2002, May 6, 2002.

19. Z. Pawlak, Rough sets. Theoretical Aspects of Reasoning about Data, Kluwer Academic Publishers, 1991

20. Z. Pawlak, Rough sets. International Journal of Information and Computer Science 11, 1982, pp. 341–356.

21. J. Ullman, Principles of Database and Knowledge-Base Systes, Vol 1, II, 1988, 1989, Computer Science Press.

A New Theoretical Framework
for K-Means-Type Clustering

J. Peng* and Y. Xia

Advanced optimization Lab, Department of Computing and Software McMaster University, Hamilton, Ontario L8S 4K1, Canada.
pengj@mcmaster.ca, xiay@optlab.mcmaster.ca

Summary. One of the fundamental clustering problems is to assign n points into k clusters based on the minimal sum-of-squares(MSSC), which is known to be NP-hard. In this paper, by using matrix arguments, we first model MSSC as a so-called 0–1 semidefinite programming (SDP). The classical K-means algorithm can be interpreted as a special heuristics for the underlying 0–1 SDP. Moreover, the 0–1 SDP model can be further approximated by the relaxed and polynomially solvable linear and semidefinite programming. This opens new avenues for solving MSSC. The 0–1 SDP model can be applied not only to MSSC, but also to other scenarios of clustering as well. In particular, we show that the recently proposed normalized k-cut and spectral clustering can also be embedded into the 0–1 SDP model in various kernel spaces.

1 Introduction

Clustering is one of major issues in data mining and machine learning with many applications arising from different disciplines including text retrieval, pattern recognition and web mining [12, 15]. Roughly speaking, clustering involves partition a given data set into subsets based on the closeness or similarity among the data. Typically, the similarities among entities in a data set are measured by a specific proximity function, which can be make precise in many ways. This results in many clustering problems and algorithms as well.

Most clustering algorithms belong to two classes: hierarchical clustering and partitioning. The hierarchical approach produces a nested series of partitions consisting of clusters either disjoint or included one into the other.

*The research of the first author was partially supported by the grant # RPG 249635-02 of the National Sciences and Engineering Research Council of Canada (NSERC) and a PREA award. This research was also Supported by the MITACS project "New Interior Point Methods and Software for Convex Conic-Linear Optimization and Their Application to Solve VLSI Circuit Layout Problems".

Those clustering algorithms are either agglomerative or divisive. An agglomerative clustering algorithm starts with every singleton entity as a cluster, and then proceeds by successively merging clusters until a stopping criterion is reached. A divisive approach starts with an initial cluster with all the entities in it, and then performs splitting until a stopping criterion is reached. In hierarchical clustering, an objective function is used locally as the merging or splitting criterion. In general, hierarchical algorithms can not provide optimal partitions for their criterion. In contrast, partitional methods assume given the number of clusters to be found and then look for the optimal partition based on the object function. Partitional methods produce only one partition. Most partitional methods can be further classified as deterministic or stochastic, depending on whether the traditional optimization technique or a random search of the state space is used in the process. There are several different ways to separate various clustering algorithms, for a comprehensive introduction to the topic, we refer to the book [12, 15], and for more recent results, see survey papers [4] and [13].

Among various criterion in clustering, the minimum sum of squared Euclidean distance from each entity to its assigned cluster center is the most intuitive and broadly used. Both hierarchical and partitional procedures for MSSC have been investigated. For example, Ward's [27] agglomerative approach for MSSC has a complexity of $O(n^2 \log n)$ where n is the number of entities. The divisive hierarchical approach is more difficult. In [9], the authors provided an algorithm running in $O(n^{d+1} \log n)$ time, where d is the dimension of the space to which the entities belong.

However, in many applications, assuming a hierarchical structure in partitioning based on MSSC is unpractical. In such a circumstance, the partitional approach directly minimizing the sum of squares distance is more applaudable. The traditional way to deal with this problem is to use some heuristics such as the well-known K-means [18]. To describe the algorithm, let us go into a bit more details.

Given a set S of n points in a d-dimensional Euclidean space, denoted by

$$S = \{\mathbf{s}_i = (s_{i1}, \ldots, s_{id})^T \in \mathbf{R}^d \quad i = 1, \ldots, n\}$$

the task of a partitional MSSC is to find an assignment of the n points into k disjoint clusters $\mathcal{S} = (S_1, \ldots, S_k)$ centered at cluster centers $\mathbf{c}_j$ $(j = 1, \ldots, k)$ based on the total sum-of-squared Euclidean distances from each point $\mathbf{s}_i$ to its assigned cluster centroid $\mathbf{c}_i$, i.e.,

$$f(S, \mathcal{S}) = \sum_{j=1}^{k} \sum_{i=1}^{|S_j|} \left\| \mathbf{s}_i^{(j)} - \mathbf{c}_j \right\|^2 ,$$

where $|S_j|$ is the number of points in S_j, and $\mathbf{s}_i^{(j)}$ is the ith point in S_j. Note that if the cluster centers are known, then the function $f(S, \mathcal{S})$ achieves its minimum when each point is assigned to its closest cluster center. Therefore,

MSSC can be described by the following bilevel programming problem (see for instance [2, 19]).

$$\min_{\mathbf{c}_1,\ldots,\mathbf{c}_k} \sum_{i=1}^{n} \min\{\|\mathbf{s}_i - \mathbf{c}_1\|^2, \ldots, \|\mathbf{s}_i - \mathbf{c}_k\|^2\}. \tag{1}$$

Geometrically speaking, assigning each point to the nearest center fits into a framework called *Voronoi Program*, and the resulting partition is named *Voronoi Partition*. On the other hand, if the points in cluster S_j are fixed, then the function

$$f(S_j, \mathcal{S}_j) = \sum_{i=1}^{|S_j|} \left\| \mathbf{s}_i^{(j)} - \mathbf{c}_j \right\|^2$$

is minimal when

$$\mathbf{c}_j = \frac{1}{|S_j|} \sum_{i=1}^{|S_j|} \mathbf{s}_i^{(j)}.$$

The classical K-means algorithm [18], based on the above two observations, is described as follows:

K-means clustering algorithm

(1) Choose k cluster centers randomly generated in a domain containing all the points,
(2) Assign each point to the closest cluster center,
(3) Recompute the cluster centers using the current cluster memberships,
(4) If a convergence criterion is met, stop; Otherwise go to step 2.

Another way to model MSSC is based on the assignment. Let $X = [x_{ij}] \in \Re^{n \times k}$ be the assignment matrix defined by

$$x_{ij} = \begin{cases} 1 & \text{If } \mathbf{s}_i \text{ is assigned to } S_j; \\ 0 & \text{Otherwise}. \end{cases}$$

As a consequence, the cluster center of the cluster S_j, as the mean of all the points in the cluster, is defined by

$$\mathbf{c}_j = \frac{\sum_{l=1}^{n} x_{lj} \mathbf{s}_l}{\sum_{l=1}^{n} x_{lj}}.$$

Using this fact, we can represent (1) as

$$\min_{x_{ij}} \sum_{j=1}^{k} \sum_{i=1}^{n} x_{ij} \left\| \mathbf{s}_i - \frac{\sum_{l=1}^{n} x_{lj} \mathbf{s}_l}{\sum_{l=1}^{n} x_{lj}} \right\|^2 \tag{2}$$

$$S.T. \ \sum_{j=1}^{k} x_{ij} = 1 \ (i = 1, \dots, n) \tag{3}$$

$$\sum_{i=1}^{n} x_{ij} \geq 1 \ (j = 1, \dots, k) \tag{4}$$

$$x_{ij} \in \{0, 1\} \ (i = 1, \dots, n; \ j = 1, \dots, k) \tag{5}$$

The constraint (3) ensures that each point $\mathbf{s}_i$ is assigned to one and only one cluster, and (4) ensures that there are exactly k clusters. This is a mixed integer programming with nonlinear objective [8], which is NP-hard. The difficulty of the problem consists of two parts. First, the constraints are discrete. Secondly the objective is nonlinear and nonconvex. Both the difficulties in the objective as well as in the constraints make MSSC extremely hard to solve.

Many different approaches have been proposed for attacking (2) both in the communities of machine learning and optimization [1,3,8]. Most methods for (2) are heuristics that can locate only a good local solution, not the exact global solution for (2). Only a few works are dedicated to the exact algorithm for (2) as listed in the references of [3].

Approximation methods provide a useful approach for (2). There are several different ways to approximate (2). For example, by solving the so-called K-medians problem we can obtain a 2-approximately optimal solution for (2) in $O(n^{d+1})$ time [10]. In [22], Mutousek proposed a geometric approximation method that can find an $(1 + \epsilon)$ approximately optimal solution for (2) in $O(n \log^k n)$ time, where the constant hidden in the big-O notation depends polynomially on ϵ. Another efficient way of approximation is to attack the original problem (typically NP-hard) by solving a relaxed polynomially solvable problem. This has been well studied in the field of optimization, in particular, in the areas of combinatorial optimization and semidefinite programming [5]. We noted that recently, Xing and Jordan [29] considered the SDP relaxation for the so-called normalized k-cut spectral clustering.

In the present paper, we focus on developing approximation methods for (2) based on linear and semidefinite programming (LP/SDP) relaxation. A crucial step in relaxing (2) is to rewrite the objective in (2) as a simple convex function of matrix argument that can be tackled easily, while the constraint set still enjoy certain geometric properties. This was possibly first suggested in [6] where the authors owed the idea to an anonymous referee. However, the authors of [6] did not explore the idea in depth to design any usable algorithm. A similar effort was made in [30] where the authors rewrote the objective in (2) as a convex quadratic function in which the argument is a $n \times k$ orthonormal matrix.

Our model follows the same stream as in [6,30]. However, different from the approach [30] where the authors used only a quadratic objective and simple spectral relaxation, we elaborate more on how to characterize (2) exactly by means of matrix arguments. In particular, we show that MSSC can be modelled as the so-called 0–1 semidefinite programming (SDP), which can be further relaxed to polynomially solvable linear programming (LP) and SDP. Several different relaxation forms are discussed. We also show that variants of K-means can be viewed as heuristics for the underlying 0–1 SDP.

Our model provides novel avenues not only for solving MSSC, but also for solving clustering problems based on some other criterions. For example, the clustering based on normalized cuts can also be embedded into our model. Moreover, our investigation reveals some interesting links between the well-known K-means and some recently proposed algorithms like spectral clustering.

The paper is organized as follows. In Sect. 2, we show that MSSC can be modelled as 0–1 SDP, which allows convex relaxation such as SDP and LP. In Sect. 3, we discuss algorithms and challenges for solving our 0–1 SDP model. Section 4 devotes to the discussion on the links between our model and some other recent models for clustering. Finally we close the paper by few concluding remarks.

2 Equivalence of MSSC to 0–1 SDP

In this section, we establish the equivalence between MSSC and 0–1 SDP. We start with a brief introduction to SDP and 0–1 SDP.

In general, SDP refers to the problem of minimizing (or maximizing) a linear function over the intersection of a polyhedron and the cone of symmetric and positive semidefinite matrices. The canonical SDP takes the following form

$$
\textbf{(SDP)} \quad \begin{cases} \min \ \mathrm{Tr}(WZ) \\ S.T. \ \mathrm{Tr}(B_i Z) = b_i \quad \text{for } i = 1, \dots, m \\ \quad Z \succeq 0 \end{cases}
$$

Here $\mathrm{Tr}(.)$ denotes the trace of the matrix, and $Z \succeq 0$ means that Z is positive semidefinite. If we replace the constraint $Z \succeq 0$ by the requirement that $Z^2 = Z$, then we end up with the following problem

$$
\textbf{(0–1 SDP)} \quad \begin{cases} \min \ \mathrm{Tr}(WZ) \\ S.T. \ \mathrm{Tr}(B_i Z) = b_i \quad \text{for } i = 1, \dots, m \\ \quad Z^2 = Z, Z = Z^T \end{cases}
$$

We call it 0–1 SDP owing to the similarity of the constraint $Z^2 = Z$ to the classical 0–1 requirement in integer programming.

We next show that MSSC can be modelled as 0–1 SDP. By rearranging the items in the objective of (2), we have

$$f(S,\mathcal{S}) = \sum_{i=1}^{n} \|\mathbf{s}_i\|^2 \left(\sum_{j=1}^{k} x_{ij} \right) - \sum_{j=1}^{k} \frac{\|\sum_{i=1}^{n} x_{ij}\mathbf{s}_i\|^2}{\sum_{i=1}^{n} x_{ij}} \qquad (6)$$

$$= \mathrm{Tr}(W_S^T W_S) - \sum_{j=1}^{k} \frac{\|\sum_{i=1}^{n} x_{ij}\mathbf{s}_i\|^2}{\sum_{i=1}^{n} x_{ij}},$$

where $W_S \in \Re^{n \times d}$ denotes the matrix whose ith row is the vector $\mathbf{s}_i$. Since X is an assignment matrix, we have

$$X^T X = \mathrm{diag}\left(\sum_{i=1}^{n} x_{i1}^2, \ldots, \sum_{i=1}^{n} x_{ik}^2 \right) = \mathrm{diag}\left(\sum_{i=1}^{n} x_{i1}, \ldots, \sum_{i=1}^{n} x_{ik} \right).$$

Let

$$Z := [z_{ij}] = X(X^T X)^{-1} X^T,$$

we can write (6) as $\mathrm{Tr}(W_S W_S^T(I - Z)) = \mathrm{Tr}(W_S^T W_S) - \mathrm{Tr}(W_S^T W_S Z)$. Obviously Z is a projection matrix satisfying $Z^2 = Z$ with nonnegative elements. For any integer m, let e_m be the all one vector in $\Re^m$. We can write the constraint (3) as

$$X e^k = e^n.$$

It follows immediately

$$Z e^n = Z X e^k = X e^k = e^n.$$

Moreover, the trace of Z should equal to k, the number of clusters, i.e.,

$$\mathrm{Tr}(Z) = k.$$

Therefore, we have the following 0–1 SDP model for MSSC

$$\min \ \mathrm{Tr}(W_S W_S^T(I - Z)) \qquad (7)$$
$$Z e = e, \mathrm{Tr}(Z) = k,$$
$$Z \geq 0, Z = Z^T, Z^2 = Z.$$

We first give a technical result about positive semidefinite matrix that will be used in our later analysis.

Lemma 1. *For any symmetric positive semidefinite matrix $Z \in \Re^{n \times n}$, there exists an index $i_0 \in \{1, \ldots, n\}$ such that*

$$Z_{i_0 i_0} = \max_{i,j} Z_{ij}.$$

Proof. For any positive semidefinite matrix Z, it is easy to see that

$$Z_{ii} \geq 0, \quad i = 1, \ldots, n.$$

Suppose the statement of the lemma does not hold, i.e., there exists $i_0 \neq j_0$ such that

$$Z_{i_0 j_0} = \max_{i,j} Z_{ij} > 0 .$$

Then the submatrix

$$\begin{pmatrix} Z_{i_0 i_0} & Z_{i_0 j_0} \\ Z_{j_0 i_0} & Z_{j_0 j_0} \end{pmatrix}$$

is not positive semidefinite. This contradicts to the assumptuion in the lemma.

Now we are ready to establish the equivalence between the models (7) and (2).

Theorem 2.1. *Solving the 0–1 SDP problem (7) is equivalent to finding a global solution of the integer programming problem (2).*

Proof. From the construction of the 0–1 SDP model (7), we know that one can easily construct a feasible solution for (7) from a feasible solution of (2). Therefore, it remains to show that from a global solution of (7), we can obtain a feasible solution of (2).

Suppose that Z is a global minimum of (7). Obviously Z is positive semidefinite. From Lemma 1 we conclude that there exists an index i_1 such that

$$Z_{i_1 i_1} = \max\{Z_{ij} : 1 \le i, j \le n\} > 0 .$$

Let us define the index set

$$\mathcal{I}_1 = \{j : Z_{i_1 j} > 0\} .$$

Since $Z^2 = Z$. we have

$$\sum_{j \in \mathcal{I}_1} (Z_{i_1 j})^2 = Z_{i_1 i_1} ,$$

which implies

$$\sum_{j \in \mathcal{I}_1} \frac{Z_{i_1 j}}{Z_{i_1 i_1}} Z_{i_1 j} = 1 .$$

From the choice of i_1 and the constraint

$$\sum_{j=1}^{n} Z_{i_1 j} = \sum_{j \in \mathcal{I}_1} Z_{i_1 j} = 1 ,$$

we can conclude that

$$Z_{i_1 j} = Z_{i_1 i_1}, \quad \forall j \in \mathcal{I}_1 .$$

This further implies that the submatrix $Z_{\mathcal{I}_1 \mathcal{I}_1}$ is a matrix whose elements are all equivalent, and we can decompose the matrix Z into a bock matrix with the following structure

$$Z = \begin{pmatrix} Z_{\mathcal{J}_1 \mathcal{J}_1} & 0 \\ 0 & Z_{\bar{\mathcal{I}}_1 \bar{\mathcal{I}}_1} \end{pmatrix}, \tag{8}$$

where $\bar{\mathcal{I}}_1 = \{i : i \notin \mathcal{I}_1\}$. Since $\sum_{i \in \mathcal{I}_1} Z_{ii} = 1$ and $\left(Z_{\mathcal{I}_1 \mathcal{I}_1}\right)^2 = Z_{\mathcal{I}_1 \mathcal{I}_1}$, we can consider the reduced 0–1 SDP as follows

$$\min \ \mathrm{Tr}\left((\mathrm{W_S W_S^T})\right)_{\bar{\mathcal{I}}_1 \bar{\mathcal{I}}_1} (I - Z)_{\bar{\mathcal{I}}_1 \bar{\mathcal{I}}_1} \tag{9}$$

$$Z_{\bar{\mathcal{I}}_1 \bar{\mathcal{I}}_1} e = e, \mathrm{Tr}(Z_{\bar{\mathcal{I}}_1 \bar{\mathcal{I}}_1}) = k - 1 \,,$$

$$Z_{\bar{\mathcal{I}}_1 \bar{\mathcal{I}}_1} \geq 0, Z_{\bar{\mathcal{I}}_1 \bar{\mathcal{I}}_1}^2 = Z_{\bar{\mathcal{I}}_1 \bar{\mathcal{I}}_1} \,.$$

Repeating the above process, we can show that if Z is a global minimum of the 0–1 SDP, then it can be decomposed into a diagonal block matrix as

$$Z = \mathrm{diag}\left(Z_{\mathcal{I}_1 \mathcal{I}_1}, \ldots, Z_{\mathcal{I}_k \mathcal{I}_k}\right),$$

where each block matrix $Z_{\mathcal{I}_l \mathcal{I}_l}$ is a nonnegative projection matrix whose elements are equal, and the sum of each column or each row equals to 1.

Now let us define the assignment matrix $X \in \Re^{n \times k}$

$$X_{ij} = \begin{cases} 1 & \text{if } i \in \mathcal{I}_j \\ 0 & \text{otherwise} \end{cases}$$

One can easily verify that $Z = X(X^T X)^{-1} X^T$. Our above discussion illustrates that from a feasible solution of (7), we can obtain an assignment matrix that satisfies the condition in (2). This finishes the proof of the theorem.

By comparing (7) with (2), we find that the objective in (7) is linear, while the constraint in (7) is still nonlinear, even more complex than the 0–1 constraint in (2). The most difficult part in the constraint of (7) is the requirement that $Z^2 = Z$. Several different ways for solving (7) will be discussed in the next section.

3 Algorithms for Solving 0–1 SDP

In this section, we focus on various algorithms for solving the 0–1 SDP model (7). From a viewpoint of the algorithm design, we can separate these algorithms into two groups. The first group consists of the so-called feasible iterative algorithms, while the second group contains approximation algorithms (might be infeasible at some stage in the process) based on relaxation. It is worthwhile pointing out that our discussion will focus on the design of the algorithm as well as the links among various techniques, not on the implementation details of the algorithm and numerical testing.

3.1 Feasible Iterative Algorithms

We first discuss the so-called feasible iterative algorithms in which all the iterates are feasible regarding the constraints in (7), while the objective is reduced step by step until some termination criterion is reached. A general procedure for feasible iterative algorithms can be described as follows:

Feasible Iterative Algorithm

Step 1: Choose a starting matrix Z^0 satisfying all the constraints in (7),

Step 2: Use a heuristics to update the matrix Z^k such that the value of the objective function in (7) is decreased,

Step 3: Check the termination criterion. If the criterion is reached, then stop; Otherwise go to Step 2.

We point out that the classical K-means algorithm described in the introduction can be interpreted as a special feasible iterative scheme for attacking (7). To see this, let us recall our discussion on the equivalence between MSSC and (7), one can verify that, at each iterate, all the constraints in (7) are satisfied by the matrix transformed from the K-means algorithm. It is also easy to see that, many variants of the K-means algorithm such as the variants proposed in [11. 14], can also be interpreted as specific iterative schemes for (7).

3.2 Approximation Algorithms Based on LP/SDP Relaxations

In the section we discuss the algorithms in the second group that are based on LP/SDP relaxation. We starts with a general procedure for those algorithm.

Approximation Algorithm Based on Relaxation

Step 1: Choose a relaxation model for (7),

Step 2: Solve the relaxed problem for an approximate solution,

Step 3: Use a rounding procedure to extract a feasible solution to (7) from the approximate solution.

The relaxation step has an important role in the whole algorithm. For example, if the approximation solution obtained from Step 2 is feasible for (7), then it is exactly an optimal solution of (7). On the other hand, when the approximation solution is not feasible regarding (7), we have to use a rounding procedure to extract a feasible solution. In what follows we discuss how to design a rounding procedure.

First, we note that when Z^* is a solution of (7), it can be shown that the matrix $Z^* W_S$ contains k different rows, and each of these k different rows represents one center in the final clusters. A good approximate solution, although it might not be feasible for (7), should give us some indications on how to locate a feasible solution. Motivated by the above-mentioned observation, we can cast the rows of the matrix $Z W_S$ as a candidate set for the potential approximate centers in the final clustering. This leads to the following rounding procedure.

A Rounding Procedure

Step 0: Input: an approximate solution Z and the matrix W_S,
Step 1: Select k rows from the rows of the matrix $ZW_S{}^2$ as the initial centers,
Step 2: Apply the classical K-means to the original MSSC using the selected initial centers.

We mention that in [29], Xing and Jordan proposed a rounding procedure based on the singular value decomposition $Z = U^T U$ of Z. In their approach, Xing and Jordan first cast the rows of U^T as points in the space, and then they employed the classical K-means to cluster those points.

Another way for extracting a feasible solution is to utilize branch and cut. In order to use branch and cut, we recall the fact that any feasible solution Z of (7) satisfies the following condition

$$Z_{ij}(Z_{ij} - Z_{ii}) = 0, \quad i, j = 1, \ldots, n.$$

If an approximate solution meets the above requirement, then it is a feasible solution and thus an optimal solution to (7). Otherwise, suppose that there exist indices i, j such that

$$Z_{ij}(Z_{ii} - Z_{ij}) \neq 0,$$

then we can add cut $Z_{ii} = Z_{ij}$ or $Z_{ij} = 0$ to get two subproblems. By combining such a branch-cut procedure with our linear relaxation model, we can find the exact solution to (7) in finite time, as the number of different branches is at most 2^{n^2}.

To summarize, as shown in our above discussion, finding a good approximation (or a nice relaxation) is essential for the success of approximation algorithms. This will be the main focus in the following subsections.

Relaxations Based on SDP

In this subsection, we describe few SDP-based relaxations for (7). First we recall that in (7), the argument Z is stipulated to be a projection matrix, i.e., $Z^2 = Z$, which implies that the matrix Z is a positive semidefinite matrix whose eigenvalues are either 0 or 1. A straightforward relaxation to (7) is replacing the requirement $Z^2 = Z$ by the relaxed condition

$$I \succeq Z \succeq 0.$$

Note that in (7), we further stipulate that all the entries of Z are nonnegative, and the sum of each row(or each column) of Z equals to 1. This means the eigenvalues of Z is always less than 1. In this circumstance, the constraint

---

[2]For example, we can select k rows from ZW_S based on the frequency of the row in the matrix, or arbitrarily select k centers.

$Z \preceq I$ becomes superfluous and can be waived. Therefore, we obtain the following SDP relaxation for MSSC

$$\min \ \mathrm{Tr}\big(W_S W_S^T (I - Z)\big) \tag{10}$$
$$Ze = e, \mathrm{Tr}(Z) = k \ ,$$
$$Z \geq 0, Z \succeq 0.$$

The above problem is feasible and bounded below. We can apply many existing optimization solvers such as interior-point methods to solve (10). It is known that an approximate solution to (10) can be found in polynomial time.

We noted that in [29], the model (10) with a slightly different linear constraint[3] was used as a relaxation to the so-called normalized k-cut clustering. As we shall show in Sect. 4, the model (10) can always provides better approximation to (7) than the spectral clustering. This was also observed and pointed out by Xing and Jordan [29].

However, we would like to point out here that although there exist theoretically polynomial algorithm for solving (10), most of the present optimization solvers are unable to handle the problem in large size efficiently.

Another interesting relaxation to (7) is to further relax (10) by dropping some constraints. For example, if we remove the nonnegative requirement on the elements of Z, then we obtain the following simple SDP problem

$$\min \ \mathrm{Tr}\big(W_S W_S^T (I - Z)\big) \tag{11}$$
$$Ze = e, \mathrm{Tr}(Z) = k \ ,$$
$$I \succeq Z \succeq 0 \ .$$

The above problem can be equivalently stated as

$$\max \ \mathrm{Tr}\big(W_S W_S^T Z\big) \tag{12}$$
$$Ze = e, \mathrm{Tr}(Z) = k \ ,$$
$$I \succeq Z \succeq 0 \ .$$

In the sequel we discuss how to solve (12). Note that if Z is a feasible solution for (12), then we have

$$\frac{1}{\sqrt{n}} Ze = \frac{1}{\sqrt{n}} e \ ,$$

which implies $\frac{1}{\sqrt{n}} e$ is an eigenvector of Z with eigenvalue 1. Therefore, we can write any feasible solution of (12) Z as

$$Z = Q\Gamma Q^T + \frac{1}{n} ee^T \ ,$$

where $Q \in \Re^{n \times (n-1)}$ is a matrix satisfying the condition:

---

[3]In [29], the constraint $Ze = e$ in (7) is replaced by $Zd = d$ where d is a positive scaling vector associated with the affinity matrix, and the constraint $Z \preceq I$ can not be waived.

C.1 The matrix $[Q : \frac{1}{\sqrt{n}}e]$ is orthogonal,

and $\Gamma = \mathrm{diag}\,(\gamma_1,\ldots,\gamma_{n-1})$ is a nonnegative diagonal matrix. It follows

$$k - 1 = \mathrm{Tr}(Z) - 1 = \mathrm{Tr}(Q\Gamma Q^T) = \mathrm{Tr}(Q^T Q \Gamma) = \mathrm{Tr}(\Gamma)\,.$$

Therefore, we can reduce (12) to

$$\begin{aligned} \max\ & \mathrm{Tr}(W_S W_S^T Q \Gamma Q^T) = \mathrm{Tr}(Q^T W_S W_S^T Q \Gamma) && (13)\\ & \mathrm{Tr}(\Gamma) = k - 1\,,\\ & I_{n-1} \succeq \Gamma \succeq 0\,. \end{aligned}$$

Let $\lambda_1(Q^T W_S W_S^T Q),\ldots,\lambda_{n-1}(Q^T W_S W_S^T Q)$ be the eigenvalues of the matrix $Q^T W_S W_S^T Q$ listed in the order of decreasing values. The optimal solution of (13) can be achieved if and only if

$$\mathrm{Tr}(Q^T W_S W_S^T Q \Gamma) = \sum_{i=1}^{k-1} \lambda_i(Q^T W_S W_S^T Q)\,.$$

Note that for any matrix Q satisfying Condition C.1, the summation of the first $k-1$ largest eigenvalues of the matrix $Q^T W_S W_S^T Q$ are independent of the choice of Q. This gives us an easy way to solve (13) and correspondingly (12). The algorithmic scheme for solving (12) can be described as follows:

Algorithm

Step 1: Choose a matrix Q satisfying C.1,
Step 2: Use singular value decomposition method to compute the first $k-1$ largest eigenvalues of the matrix $Q^T W_S W_S^T Q$ and their corresponding eigenvectors $v_1,\ldots,v_{k-1}$,
Step 3: Set

$$Z = \frac{1}{n}e^T e + \sum_{i=1}^{k-1} (Qv_i)^T Qv_i\,.$$

It should be mentioned that if $k = 2$, then Step 2 in the above algorithm uses the eigenvector corresponding to the largest eigenvalue of $Q^T W_S W_S^T Q$. This eigenvector has an important role in Shi and Malik' work [25] (See also [28]) for image segmentation where the clustering problem with $k = 2$ was discussed.

LP Relaxation

In this subsection, we propose an LP relaxation for (7). First we observe that if s_i and s_j, s_j and s_k belong to the same clusters, then s_i and s_k belong to the same cluster. In such a circumstance, from the definition of the matrix Z we can conclude that

$$Z_{ij} = Z_{jk} = Z_{ik} = Z_{ii} = Z_{jj} = Z_{kk} \ .$$

Such a relationship can be partially characterized by the following inequality

$$Z_{ij} + Z_{ik} \leq Z_{ii} + Z_{jk} \ .$$

Correspondingly, we can define a metric polyhedron MET[4] by

$$\mathrm{MET} = \{ Z = [z_{ij}] : z_{ij} \leq z_{ii}, \quad z_{ij} + z_{ik} \leq z_{ii} + z_{jk} \} \ .$$

Therefore, we have the following new model

$$\min \ \mathrm{Tr}(W_S W_S^T (I - Z)) \tag{14}$$
$$Ze = e, \mathrm{Tr}(Z) = k \ ,$$
$$Z \geq 0 \ ,$$
$$Z \in \mathrm{MET} \ .$$

If the optimal solution of (14) is not a feasible solution of (7), then we need to refer to the rounding procedure that we described earlier to extract a feasible solution for (7).

Solving (14) directly for large-size data set is clearly unpractical due to the huge amount $(O(n^3))$ of constraints. In what follows we report some preliminary numerical results for small-size data set. Our implementation is done on an IBM RS-6000 workstation and the package CPLEX 7.1 with AMPL interface is used to solve the LP model (14).

The first data set we use to test our algorithm is the Soybean data (small) from the UCI Machine Learning Repository[5], see also [21]. This data set has 47 instances and each instance has 35 normalized attributes. It is known this data set has 4 clusters. As shown by the following table, for k from 2 to 4, we found the exact clusters by solving (14).

The Soybean data

K	Objective	CPU Time(s)
2	404.4593	4.26
3	215.2593	1.51
4	205.9637	1.68

The second test set is the Ruspini data set from [24]. This data set, consisting of 75 points in $\Re^2$ with four groups, is popular for illustrating clustering techniques [15]. The numerical result is listed as follows:

---

[4] A similar polyhedron MET had been used by Karisch and Rendl, Leisser and Rendl in their works [16,17] on graph partitioning. We changed slightly the definition of MET in [17] to adapt to our problem.

[5] http://www.ics.uci.edu/ mlearn/MLRepository.html

The Ruspini's data

K	Objective	CPU Time(s)
2	893380	27.81
3	510630	66.58
4	12881	7.22
5	10127	9.47

We observed that in our experiments, for all cases $k = 2, \ldots, 5$, the solution of (14) is not feasible for (7). However, the resulting matrix is quite close to a feasible solution of (7). Therefore, we use the classical K-means to get the final clusters. After a few iterations, the algorithm terminated and reported the numerical results that match the best known results in the literature for the same problem.

The third test set is the Späth's postal zones data [26]. This data set contains 89 entities and each entity has 3 features. Correspondingly, we transform all the entities into points in $\Re^3$. It is known that the data set has 7 groups. In all cases that k runs from 2 to 9, we were able to find the exact solution of (7) via solving (14).

The Spath's Postal Zone data

K	Objective	CPU Time(s)
2	$6.0255 * 10^{11}$	283.26
3	$2.9451 * 10^{11}$	418.07
4	$1.0447 * 10^{11}$	99.54
5	$5.9761 * 10^{10}$	60.67
6	$3.5908 * 10^{10}$	52.55
7	$2.1983 * 10^{10}$	61.78
8	$1.3385 * 10^{10}$	26.91
9	$7.8044 * 10^{9}$	18.04

It is worthwhile mentioning that, as shown in the tables, the running time of the algorithm does not increase as the cluster number k increases. Actually, from a theoretical viewpoint, the complexity of the algorithm for solving (14) is independent of k. This indicates our algorithm is scalable to large data set, while how to solve (14) efficiently still remains a challenge. In contrast, the complexity of the approximation algorithms in [22] increases with respect to k.

4 Relations to Other Clustering Methods

In the previous sections, we proposed and analyzed the 0–1 SDP model for MSSC. In this section, we consider the more general 0–1 SDP model for clustering

$$\max\ \mathrm{Tr}(WZ) \tag{15}$$
$$Ze = e,\ \mathrm{Tr}(Z) = k\ ,$$
$$Z \geq 0.\ Z^2 = Z, Z = Z^T,$$

where W is the so-called affinity matrix whose entries represent the similarities or closeness among the entities in the data set. In the MSSC model, we use the geometric distance between two points to characterize the similarity between them. In this case, we have $W_{ij} = s_i^T s_j$. However, we can also use a general function $\phi(s_i, s_j)$ to describe the similarity relationship between s_i and s_j. For example, let us choose

$$W_{ij} = \phi(s_i, s_j) = \exp^{-\frac{\|s_i - s_j\|^2}{\sigma}},\quad \sigma > 0\ . \tag{16}$$

In order to apply the classical K-means algorithm to (15), we first use the singular eigenvalue decomposition method to decompose the matrix W into the product of two matrices, i.e., $W = U^T U$. In this case, each column of U can be cast as a point in a suitable space. Then, we can apply the classical K-means method for MSSC model to solving problem (15). This is exactly the procedure what the recently proposed spectral clustering follows. However, now we can interpret the spectral clustering as a variant of MSSC in a different kernel space. It is worthwhile mentioning that certain variants of K-means can be adapted to solve (15) directly without using the SVD decomposition of the affinity matrix.

We note that recently, the k-ways normalized cut and spectral clustering received much attention in the machine learning community, and many interesting results about these two approaches have been reported [7,20,23,25,28–30]. In particular, Zha et al. [30] discussed the links between spectral relaxation and K-means. Similar ideas was also used in [23]. An SDP relaxation for normalized k-cut was discussed [29]. The relaxed SDP in [29] takes a form quite close to (10). As we pointed out in Sect. 3, the main difference between the relaxed model in [29] and (10) lies in the constraint.

In fact, with a closer look at the model for normalized k-cut in [29], one can find that it is a slight variant of the model (15). To see this, let us recall the exact model for normalized k-cut [29]. Let W be the affinity matrix defined by (16) and X be the assignment matrix in the set $\mathcal{F}_k$ defined by

$$\mathcal{F}_k = \{X : Xe^k = e^n, x_{ij} \in \{0,1\}\}\ .$$

Let $d = We^n$ and $D = \mathrm{diag}\,(d)$. The exact model for normalized k-cut in [29] can be rewritten as

$$\max_{X \in \mathcal{F}_k}\ \mathrm{Tr}\big((X^T DX)^{-1}X^T WX\big) \tag{17}$$

If we define

$$Z = D^{\frac{1}{2}}X(X^T DX)^{-1}X^T D^{\frac{1}{2}}\ ,$$

then we have

$$Z^2 = Z, \ Z^T = Z, \ Z \geq 0, \ Zd^{\frac{1}{2}} = d^{\frac{1}{2}} \ .$$

Following a similar process as in the proof of Theorem 2.1, we can show that the model (17) equals to the following 0–1 SDP:

$$\max \ \mathrm{Tr}\left(\mathrm{D}^{-\frac{1}{2}}\mathrm{W}\mathrm{D}^{-\frac{1}{2}}Z\right) \tag{18}$$
$$Zd^{\frac{1}{2}} = d^{\frac{1}{2}}, \mathrm{Tr}(Z) = k \ ,$$
$$Z \geq 0, Z^2 = Z, Z = Z^T.$$

The only difference between (15) and (18) is the introduction of the scaling matrix D. However, our new unified model (15) provides more insight for clustering problem and opens new avenues for designing new efficient clustering methods. It is also interesting to note that when we use SDP relaxation to solve (15), the constraint $Z \preceq I$ can be waived without any influence on the solution, while such a constraint should be kept in the SDP relaxation for (18). This will definitely impact the numerical efficiency of the approach. It will be helpful to compare these two models in real application to see what is the role of the scaling matrix D.

5 Conclusions

In this paper, we reformulated the classical MSSC as a 0–1 SDP. Our new model not only provides a unified framework for several existing clustering approaches, but also opens new avenues for clustering. Several LP/SDP relaxations are suggested to attack the underlying 0–1 SDP. Preliminary numerical tests indicate that these approaches are feasible, and have a lot of potential for further improvement.

Several important issues regarding the new framework remain open. The first is how to estimate the approximate rate of the approximation solution obtained from the relaxed LP/SDP problems. Secondly, the issue of how to design a rounding procedure without using the classical K-means heuristics to extract a feasible solution deserves further study. Thirdly, for specific clustering problem, how to choose a suitable affinity matrix, or in other words, how to find a suitable kernel space needs to be investigated. The last, but also the most important issue, is to develop efficient optimization algorithms for solving the relaxed problems so that these techniques can be applied to large size data set. We hope future study can help us to address these questions.

Acknowledgement

The authors thank the two anonymous referees for their useful comments.

References

1. Agarwal, P.K. and Procopiuc. (2002). Exact and approximation algorithms for clustering. *Algorithmica*, 33, 201–226.
2. Bradley, P.S., Fayyad, U.M., and Mangasarian, O.L. (1999). Mathematical Programming for data mining: formulations and challenges. *Informs J. Comput.*, 11. 217–238.
3. Du Merle, O., Hansen, P., Jaumard, B. and Mladenović, N. (2000). An interior-point algorithm for minimum sum of squares clustering. *SIAM J. Sci. Comput.*, 21. 1485–1505.
4. Ghosh J. (2003). Scalable Clustering. In N. Ye, Editor, The Handbook of Data Mining, Lawrence Erlbaum Associate, Inc, pp. 247–277.
5. Goemans, M.X. (1997). Semidefinite programming in combinatorial optimization. *Mathematical Programming*, 79, 143–161.
6. Gordon, A.D. and Henderson, J.T. (1977). Al algorithm for Euclidean sum of squares classification. *Biometrics*, 33, 355–362.
7. Gu, M., Zha, H., Ding, C., He, X. and Simon, H. (2001). Spectral relaxation models and structure analysis for k-way graph Clustering and bi-clustering. Penn State Univ Tech Report.
8. Hansen, P. and Jaumard B. (1997). Cluster analysis and mathematical programming. *Math. Programming*, 79(B), 191–215.
9. Hansen, P., Jaumard. B. and Mladenović, N. (1998). Minimum sum of squares clustering in a low dimensional space. *J. Classification*, 15, 37–55.
10. Hasegawa, S., Imai, H., Inaba, M., Katoh, N. and Nakano, J. (1993). Efficient algorithms for variance-based k-clustering. In *Proc. First Pacific Conf. Comput. Graphics Appl., Seoul. Korea*. 1, 75–89. World Scientific. Singapore.
11. Howard, H. (1966). Classifying a population into homogeneous groups. In Lawrence, J.R. Eds. *Operational Research in Social Science*, Tavistock Publ., London.
12. Jain, A.K., and Dubes, R.C. (1988). *Algorithms for clustering data*. Englewood Cliffs, NJ: Prentice Hall.
13. Jain, A.K., Murty, M.N. and Flynn, P.J. (1999). Data clustering: A review. *ACM Computing Surveys*, 31, 264–323.
14. Jancey, R.C. (1966). Multidimensional group analysis. *Australian J. Botany*, 14, 127–130.
15. Kaufman, L. and Peter Rousseeuw, P. (1990). Finding Groups in Data, an Introduction to Cluster Analysis, John Wiley.
16. Karisch. S.E. and Rendl, F. (1998). Semidefinite programming and graph equipartition. *Fields Institute Communications*. 18, 77–95.
17. Leisser, A. and Rendl, F. (2003). Graph partitioning using linear and semidefinite programming. *Mathematical Programming (B)*, 95,91–101.
18. McQueen, J. (1967). Some methods for classification and analysis of multivariate observations. *Computer and Chemistry*, 4, 257–272.
19. Mangasarian, O.L. (1997). Mathematical programming in data mining. *Data Min. Knowl. Discov.*. 1, 183–201.
20. Meila, M. and Shi, J. (2001). A random walks view of spectral segmentation. Int'l Workshop on AI & Stat.
21. Michalski, R.S. and Chilausky, R.L. (1980a). Learning by being told and learning from examples: An experimental comparison of the two methods of knowledge

acquisition in the context of developing an expert system for soybean disease diagnosis. *International Journal of Policy Analysis and Information Systems*, 4(2), 125–161.

22. Matousek, J. (2000). On approximate geometric k-clustering. *Discrete Comput. Geom.*, 24, 61–84.

23. Ng, A.Y., Jordan, M.I. and Weiss, Y. (2001). On spectral clustering: Analysis and an algorithm. *Proc. Neural Info. Processing Systems, NIPS*, 14.

24. Ruspini, E.H. (1970). Numerical methods for fuzzy clustering. *Inform. Sci.*, 2, 319–350.

25. Shi, J. and Malik, J. (2000). Normalized cuts and image segmentation. *IEEE. Trans. on Pattern Analysis and Machine Intelligence*, 22, 888–905.

26. Späth, H. (1980). *Algorithms for Data Reduction and Classification of Objects*, John Wiley & Sons, Ellis Horwood Ltd.

27. Ward, J.H. (1963). Hierarchical grouping to optimize an objective function. *J. Amer. Statist. Assoc.*, 58, 236–244.

28. Weiss, Y. (1999). Segmentation using eigenvectors: a unifying view. *Proceedings IEEE International Conference on Computer Vision*, 975–982.

29. Xing, E.P. and Jordan, M.I. (2003). On semidefinite relaxation for normalized k-cut and connections to spectral clustering. Tech Report CSD-03–1265, UC Berkeley.

30. Zha, H., Ding, C., Gu, M., He, X. and Simon, H. (2002). Spectral Relaxation for K-means Clustering. In Dietterich, T., Becker, S. and Ghahramani, Z. Eds., *Advances in Neural Information Processing Systems 14*, pp. 1057–1064. MIT Press.

Part II

Recent Advances in Data Mining

Clustering Via Decision Tree Construction

B. Liu[1], Y. Xia[2], and P.S. Yu[3]

[1] Department of Computer Science University of Illinois at Chicago 851 S. Morgan
Street Chicago, IL 60607-7053
liub@cs.uic.edu
[2] School of Computing National University of Singapore 3 Science Drive 2,
Singapore 117543
xiayy@comp.nus.edu.sg
[3] IBM T. J. Watson Research Center Yorktown Heights, NY 10598
psyu@us.ibm.com

Clustering is an exploratory data analysis task. It aims to find the intrinsic
structure of data by organizing data objects into similarity groups or clusters. It is often called unsupervised learning because no class labels denoting
an a priori partition of the objects are given. This is in contrast with supervised learning (e.g., classification) for which the data objects are already
labeled with known classes. Past research in clustering has produced many
algorithms. However, these algorithms have some shortcomings. In this paper, we propose a novel clustering technique, which is based on a supervised
learning technique called decision tree construction. The new technique is able
to overcome many of these shortcomings. The key idea is to use a decision
tree to partition the data space into cluster (or dense) regions and empty (or
sparse) regions (which produce outliers and anomalies). We achieve this by
introducing virtual data points into the space and then applying a modified
decision tree algorithm for the purpose. The technique is able to find "natural"
clusters in large high dimensional spaces efficiently. It is suitable for clustering
in the full dimensional space as well as in subspaces. It also provides easily
comprehensible descriptions of the resulting clusters. Experiments on both
synthetic data and real-life data show that the technique is effective and also
scales well for large high dimensional datasets.

1 Introduction

Clustering aims to find the intrinsic structure of data by organizing objects
(data records) into similarity groups or clusters. Clustering is often called unsupervised learning because no classes denoting an a priori partition of the
objects are known. This is in contrast with supervised learning, for which the

data records are already labeled with known classes. The objective of supervised learning is to find a set of characteristic descriptions of these classes.

In this paper, we study clustering in a numerical space, where each dimension (or attribute) has a bounded and totally ordered domain. Each data record is basically a point in the space. Clusters in such a space are commonly defined as connected regions in the space containing a relatively high *density* of points, separated from other such regions by a region containing a relatively low density of points [12].

Clustering has been studied extensively in statistics [5], pattern recognition [16], machine learning [15], and database and data mining (e.g., [1–3, 7, 8, 10, 11, 14, 20–23, 25, 29–32]). Existing algorithms in the literature can be broadly classified into two categories [24]: *partitional clustering* and *hierarchical clustering*. Partitional clustering determines a partitioning of data records into k groups or clusters such that the data records in a cluster are more similar or nearer to one another than the data records in different clusters. Hierarchical clustering is a nested sequence of partitions. It keeps merging the closest (or splitting the farthest) groups of data records to form clusters.

In this paper, we propose a novel clustering technique, which is based on a supervised learning method called decision tree construction [26]. The new technique, called CLTree (*CL*ustering based on decision *Trees*), is quite different from existing methods, and it has many distinctive advantages. To distinguish from decision trees for classification, we call the trees produced by CLTree the *cluster trees*.

Decision tree building is a popular technique for classifying data of various classes (at least two classes). Its algorithm uses a *purity function* to partition the data space into different class regions. The technique is not directly applicable to clustering because datasets for clustering have no pre-assigned class labels. We present a method to solve this problem.

The basic idea is that we regard each data record (or point) in the dataset to have a class Y. We then assume that the data space is uniformly distributed with another type of points, called *non-existing points*. We give them the class, N. With the N points added to the original data space, our problem of partitioning the data space into *data (dense) regions* and *empty (sparse) regions* becomes a classification problem. A decision tree algorithm can be applied to solve the problem. However, for the technique to work many important issues have to be addressed (see Sect. 2). The key issue is that the purity function used in decision tree building is not sufficient for clustering.

We use an example to show the intuition behind the proposed technique. Figure 1(A) gives a 2-dimensional space, which has 24 data (Y) points represented by filled rectangles. Two clusters exist in the space. We then add some uniformly distributed N points (represented by "o") to the data space (Fig. 1(B)). With the augmented dataset, we can run a decision tree algorithm to obtain a partitioning of the space (Fig. 1(B)). The two clusters are identified.

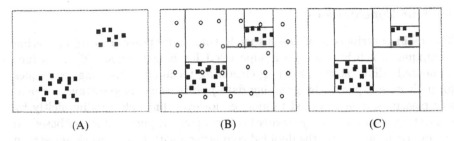

(A)   (B)   (C)

Fig. 1. Clustering using decision trees: an intuitive example

The reason that this technique works is that if there are clusters in the data, the data points cannot be uniformly distributed in the entire space. By adding some uniformly distributed N points, we can isolate the clusters because within each cluster region there are more Y points than N points. The decision tree technique is well known for this task.

We now answer two immediate questions: (1) how many N points should we add, and (2) can the same task be performed without physically adding the N points to the data? The answer to the first question is that it depends. The number changes as the tree grows. It is insufficient to add a fixed number of N points to the original dataset at the beginning (see Sect. 2.2). The answer to the second question is yes. Physically adding N points increases the size of the dataset and also the running time. A subtle but important issue is that it is unlikely that we can have points truly uniformly distributed in a very high dimensional space because we would need an exponential number of points [23]. We propose a technique to solve the problem, which guarantees the uniform distribution of the N points. This is done by not adding any N point to the space but computing them when needed. Hence, CLTree is able to produce the partition in Fig. 1(C) with no N point added to the original data.

The proposed CLTree technique consists of two steps:

1. Cluster tree construction: This step uses a modified decision tree algorithm with a new purity function to construct a cluster tree to capture the natural distribution of the data without making any prior assumptions.
2. Cluster tree pruning: After the tree is built, an interactive pruning step is performed to simplify the tree to find meaningful/useful clusters. The final clusters are expressed as a list of hyper-rectangular regions.

The rest of the paper develops the idea further. Experiment results on both synthetic data and real-life application data show that the proposed technique is very effective and scales well for large high dimensional datasets.

1.1 Our Contributions

The main contribution of this paper is that it proposes a novel clustering technique, which is based on a supervised learning method [26]. It is fundamentally different from existing clustering techniques. Existing techniques form clusters explicitly by grouping data points using some distance or density measures. The proposed technique, however, finds clusters implicitly by separating data and empty (sparse) regions using a purity function based on the information theory (the detailed comparison with related work appears in Sect. 5). The new method has many distinctive advantages over the existing methods (although some existing methods also have some of the advantages, there is no system that has all the advantages):

- CLTree is able to find "natural" or "true" clusters because its tree building process classifies the space into data (dense) and empty (sparse) regions without making any prior assumptions or using any input parameters. Most existing methods require the user to specify the number of clusters to be found and/or density thresholds (e.g., [1–3, 7, 10, 14, 21, 23, 25, 32]). Such values are normally difficult to provide, and can be quite arbitrary. As a result, the clusters found may not reflect the "true" grouping structure of the data.
- CLTree is able to find clusters in the full dimension space as well as in any subspaces. It is noted in [3] that many algorithms that work in the full space do not work well in subspaces of a high dimensional space. The opposite is also true, i.e., existing subspace clustering algorithms only find clusters in low dimension subspaces [1, 2, 3]. Our technique is suitable for both types of clustering because it aims to find simple descriptions of the data (using as fewer dimensions as possible), which may use all the dimensions or any subset.
- It provides descriptions of the resulting clusters in terms of hyper-rectangle regions. Most existing clustering methods only group data points together and give a centroid for each cluster with no detailed description. Since data mining applications typically require descriptions that can be easily assimilated by the user as insight and explanations, interpretability of clustering results is of critical importance.
- It comes with an important by-product, the empty (sparse) regions. Although clusters are important, empty regions can also be useful. For example, in a marketing application, clusters may represent different segments of existing customers of a company, while the empty regions are the profiles of non-customers. Knowing the profiles of non-customers allows the company to probe into the possibilities of modifying the services or products and/or of doing targeted marketing in order to attract these potential customers. Sparse regions also reveal outliers and anomalies, which are important for many applications.
- It deals with outliers effectively. Outliers are data points in a relatively empty region. CLTree is able to separate outliers from real clusters because

it naturally identifies sparse and dense regions. When outliers are concentrated in certain areas, it is possible that they will be identified as small clusters. If such outlier clusters are undesirable, we can use a simple threshold on the size of clusters to remove them. However, sometimes such small clusters can be very useful as they may represent exceptions (or unexpected cases) in the data. The interpretation of these small clusters is dependent on applications.

2 Building Cluster Trees

This section presents our cluster tree algorithm. Since a cluster tree is basically a decision tree for clustering, we first review the decision tree algorithm in [26]. We then modify the algorithm and its purity function for clustering.

2.1 Decision Tree Construction

Decision tree construction is a well-known technique for classification [26]. A database for decision tree classification consists of a set of data records, which are pre-classified into $q(\geq 2)$ known classes. The objective of decision tree construction is to partition the data to separate the q classes. A decision tree has two types of nodes, *decision nodes* and *leaf nodes*. A decision node specifies some test on a single attribute. A leaf node indicates the class.

From a geometric point of view, a decision tree represents a partitioning of the data space. A serial of tests (or cuts) from the root node to a leaf node represents a hyper-rectangle. For example, the four hyper-rectangular regions in Fig. 2(A) are produced by the tree in Fig. 2(B). A region represented by a leaf can also be expressed as a rule, e.g., the upper right region in Fig. 2(A) can be represented by $X > 3.5, Y > 3.5 \rightarrow O$, which is also the right most leaf in Fig. 2(B). Note that for a numeric attribute, the decision tree algorithm in [26] performs binary split, i.e., each cut splits the current space into two parts (see Fig. 2(B)).

The algorithm for building a decision tree typically uses the divide and conquer strategy to recursively partition the data to produce the tree. Each

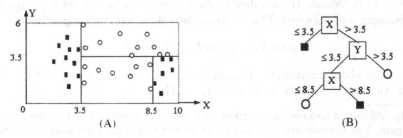

Fig. 2. An example partition of the data space and its corresponding decision tree

successive step greedily chooses the best cut to partition the space into two parts in order to obtain purer regions. A commonly used criterion (or *purity function*) for choosing the best cut is the *information gain* [26][4].

The *information gain* criterion is derived from *information theory*. The essential idea of information theory is that *the information conveyed by a message depends on its probability and can be measured in bits as minus the logarithm to base 2 of that probability.*

Suppose we have a dataset D with q classes, $C_1, \ldots, C_q$. Suppose further that we have a possible test x with m outcomes that partitions D into m subsets $D_1, \ldots, D_m$. For a numeric attribute, $m = 2$, since we only perform binary split. The probability that we select one record from the set D of data records and announce that it belongs to some class C_j is given by:

$$\frac{freq(C_j, D)}{|D|}$$

where $freq(C_j, D)$ represents the number of data records (points) of the class C_j in D, while $|D|$ is the total number of data records in D. So the information that it conveys is:

$$- \log_2 \left(\frac{freq(C_j, D)}{|D|} \right) bits$$

To find the expected information needed to identify the class of a data record in D before partitioning occurs, we sum over the classes in proportion to their frequencies in D, giving:

$$info(D) = - \sum_{j=1}^{q} \frac{freq(Cj, D)}{|D|} \times \log 2 \left(\frac{freq(Cj, D)}{|D|} \right)$$

Now, suppose that the dataset D has been partitioned in accordance with the m outcomes of the test x. The expected amount of information needed to identify the class of a data record in D after the partitioning had occurred can be found as the weighted sum over the subsets, as:

$$infoX(D) = - \sum_{i=1}^{m} \frac{|Di|}{|D|} \times info(Di)$$

where $|D_i|$ represents the number of data records in the subset D_i after the partitioning had occurred. The information gained due to the partition is:

$$gain(X) = info(D) - info_x(D)$$

Clearly, we should maximize the gain. The gain criterion is to select the test or cut that maximizes the gain to partition the current data (or space).

---

[4]In [26], it can also use the information gain ratio criterion, which is the normalized gain. The normalization is used to avoid favoring a categorical attribute that has many values. Since we have no categorical attribute, this is not a problem.

```
1  for each attribute A_i ∈ {A₁, A₂, , A_d} do
   /*A₁, A₂, and A_d are the attributes of D*/
2       for each value x of A_i in D do
        /*each value is considered as a possible cut*/
3            compute the information gain at x
4       end
5  end
6  Select the test or cut that gives the best information gain to partition the
   space
```

Fig. 3. The information gain evaluation

The procedure for information gain evaluation is given in Fig. 3. It evaluates every possible value (or cut point) on all dimensions to find the cut point that gives the best gain.

Scale-up Decision Tree Algorithms: Traditionally, a decision tree algorithm requires the whole data to reside in memory. When the dataset is too large, techniques from the database community can be used to scale up the algorithm so that the entire dataset is not required in memory. Reference [4] introduces an interval classifier that uses data indices to efficiently retrieve portions of data. SPRINT [27] and RainForest [18] propose two scalable techniques for decision tree building. For example, RainForest only keeps an AVC-set (attribute-value, classLabel and count) for each attribute in memory. This is sufficient for tree building and gain evaluation. It eliminates the need to have the entire dataset in memory. BOAT [19] uses statistical techniques to construct the tree based on a small subset of the data, and correct inconsistency due to sampling via a scan over the database.

2.2 Building Cluster Trees: Introducing N Points

We now present the modifications made to the decision tree algorithm in [26] for our clustering purpose. This sub-section focuses on introducing N points. The next sub-section discusses two changes that need to be made to the decision tree algorithm. The final sub-section describes the new cut selection criterion or purity function.

As mentioned before. we give each data point in the original dataset the class Y, and introduce some uniformly distributed "non-existing" N points. We do not physically add these N points to the original data, but only assume their existence.

We now determine how many N points to add. We add a different number of N points at each node in tree building. The number of N points for the current node E is determined by the following rule (note that at the root node, the number of inherited N points is 0):

Figure 4 gives an example. The (parent) node P has two children nodes L and R. Assume P has 1000 Y points and thus 1000 N points, stored in $P.Y$ and $P.N$ respectively. Assume after splitting, L has 20 Y points and 500 N

1 **If** the number of N points inherited from the parent node of E is less than the number of Y points in E **then**
2 the number of N points for E is increased to the number of Y points in E
3 **else** the number of inherited N points from its parent is used for E

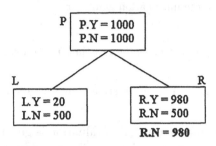

Fig. 4. Distributing N points

points, and R has 980 Y points and 500 N points. According to the above rule, for subsequent partitioning, we increase the number of N points at R to 980. The number of N points at L is unchanged.

The basic idea is that we use an equal number of N points to the number of Y (data) points (in fact, 1:1 ratio is not necessary, see Sect. 4.2.2). This is natural because it allows us to isolate those regions that are densely populated with data points. The reason that we increase the number of N points of a node (line 2) if it has more inherited Y points than N points is to avoid the situation where there may be too few N points left after some cuts or splits. If we fix the number of N points in the entire space to be the number of Y points in the original data, the number of N points at a later node can easily drop to a very small number for a high dimensional space. If there are too few N points, further splits become difficult, but such splits may still be necessary. Figure 5 gives an example.

In Fig. 5, the original space contains 32 data (Y) points. According to the above rule, it also has 32 N points. After two cuts, we are left with a smaller region (region 1). All the Y points are in this region. If we do not increase the number of N points for the region, we are left with only $32/2^2 = 8N$ points

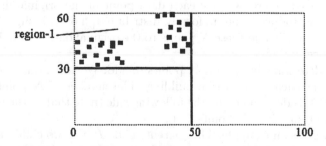

Fig. 5. The effect of using a fixed number of N points

in region 1. This is not so bad because the space has only two dimensions. If we have a very high dimensional space, the number of N points will drop drastically (close to 0) after some splits (as the number of N points drops exponentially).

The number of N points is not reduced if the current node is an N node (*an N node has more N points than Y points*) (line 3). A reduction may cause outlier Y points to form Y nodes or regions (*a Y node has an equal number of Y points as N points or more*). Then cluster regions and non-cluster regions may not be separated.

2.3 Building Cluster Trees: Two Modifications to the Decision Tree Algorithm

Since the N points are not physically added to the data, we need to make two modifications to the decision tree algorithm in [26] in order to build cluster trees:

1. *Compute the number of N points on the fly:* From the formulas in Sect. 2.1, we see that the gain evaluation needs the frequency or the number of points of each class on each side of a possible cut (or split). Since we do not have the N points in the data, we need to compute them. This is simple because we assume that the N points are uniformly distributed in the space. Figure 6 shows an example. The space has 25 data (Y) points and 25 N points. Assume the system is evaluating a possible cut P. The number of N points on the left-hand-side of P is $25 * 4/10 = 10$. The number of Y points is 3. Likewise. the number of N points on the right-hand-side of P is 15 (25–10), and the number of Y points is 22. With these numbers, the information gain at P can be computed. Note that by computing the number of N points. we essentially guarantee their uniform distribution.
2. *Evaluate on both sides of data points:* In the standard decision tree building, cuts only occur on one side of data points [26]. However, for our purpose, this is not adequate as the example in Fig. 7 shows. Figure 7 gives 3 possible cuts. cut_1 and cut_3 are on the right-hand-side of some data points, while cut_2 is on the left-hand-side. If we only allow cuts on the right-hand-side

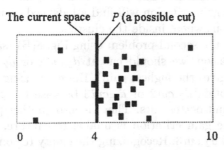

Fig. 6. Computing the number of N points

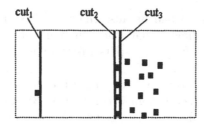

Fig. 7. Cutting on either side of data points

of data points, we will not be able to obtain a good cluster description. If we use cut_1, our cluster will contain a large empty region. If we use cut_3, we lose many data points. In this case, cut_2 is the best. It cuts on the left-hand-side of the data points.

2.4 Building Cluster Trees: The New Criterion for Selecting the Best Cut

Decision tree building for classification uses the gain criterion to select the best cut. For clustering, this is insufficient. The cut that gives the best gain may not be the best cut for clustering. There are two main problems with the gain criterion:

1. The cut with the best information gain tends to cut into clusters.
2. The gain criterion does not look ahead in deciding the best cut.

Let us see an example. Figure 8(A) shows a space with two clusters, which illustrates the first problem. Through gain computation, we find the best cuts for dimension 1 (d_1_cut), and for dimension 2 (d_2_cut) respectively. Clearly, both cuts are undesirable because they cut into clusters. Assume d_1_cut gives a better information gain than d_2_cut. We will use d_1_cut to partition the space. The cluster points on the right of d_1_cut from both clusters are lost.

This problem occurs because at cluster boundaries there is normally a higher proportion of N points than that of cluster centers for clusters whose data points follow a normal-like distribution (cluster centers are much denser than boundaries) as we assume that the N points are uniformly distributed in the entire area. The gain criterion will find a balanced point for partitioning, which tends to be somewhere inside the clusters.

Next, we look at the second problem using Fig. 8(B) (same as Fig. 8(A)). Ideally, in this situation, we should cut at d_2_cut2 or d_2_cut3, rather than d_1_cut (although it gives the highest gain). However, using the gain criterion, we are unable to obtain d_2_cut2 or d_2_cut3 because the gain criterion does not look ahead to find better cuts. There is also another piece of important information that the gain criterion is unable to capture, the empty region between d_2_cut2 and d_2_cut3. Recognizing the empty region is very important for clustering.

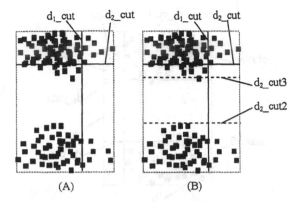

Fig. 8. Problems with the gain criterion

The two problems result in severe fragmentation of clusters (each cluster is cut into many pieces) and loss of data points. To overcome these problems, we designed a new criterion, which still uses information gain as the basis, but adds to it the ability to look ahead. We call the new criterion the lookahead gain criterion. For the example in Fig. 8(B), we aim to find a cut that is very close to d_2_cut2 or d_2_cut3.

The basic idea is as follows: For each dimension i, based on the first cut found using the gain criterion, we look ahead (at most 2 steps) along each dimension to find a better cut c_i that cuts less into cluster regions, and to find an associated region r_i that is relatively empty (measured by *relative density*, see below). c_i of the dimension i whose r_i has the lowest relative density is selected as the best cut. The intuition behind this modified criterion is clear. It tries to find the emptiest region along each dimension to separate clusters.

Definition (relative density): The *relative density* of a region r is computed with $r.Y/r.N$, where $r.Y$ and $r.N$ are the number of Y points and the number of N points in r respectively. We use the example in Fig. 9 (a reproduction of Fig. 8(A)) to introduce the lookahead gain criterion. The algorithm is given in Fig. 10. The new criterion consists of 3 steps:

1. **Find the initial cuts** (line 2. Fig. 10): For each dimension i, we use the gain criterion to find the first best cut point d_i_cut1. For example, in Fig. 9, for dimension 1 and 2, we find d_1_cut1, and d_2_cut1 respectively. If we cannot find d_i_cut1 with any gain for a dimension, we ignore this dimension subsequently.

2. **Look ahead to find better cuts** (lines 3 and 6, Fig. 10): Based on the first cut. we find a better cut on each dimension by further gain evaluation. Our objectives are to find:
   (a) a cut that cuts less into clusters (to reduce the number of lost points), and
   (b) an associated region with a low relative density (*relatively empty*).

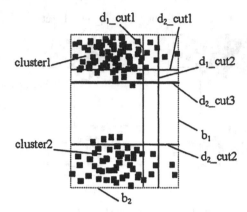

Fig. 9. Determining the best cut

Let us denote the two regions separated by d_i_cut1 along dimension i as L_i and H_i, where L_i has a lower relative density than H_i. d_i_cut1 forms one boundary of L_i along dimension i. We use b_i to denote the other boundary of L_i. To achieve both 2(a) and 2(b), we only find more cuts (at most two) in L_i. We do not go to H_i because it is unlikely that we can find better cuts there to achieve our objectives (since H_i is denser). This step goes as follows:

Along each dimension, we find another cut (d_i_cut2) in L_i that gives the best gain. After d_i_cut2 is found, if the relative density of the region between d_i_cut1 and d_i_cut2 is higher than that between d_i_cut2 and b_i, we stop because both objectives are achieved. If not, we find the third cut (d_i_cut3) by applying the same strategy. We seek the additional cut in this case because if the region between d_i_cut2 and b_i is denser, it means that there may be clusters in that region. Then, d_i_cut2 is likely to cut into these clusters.

For example, in Fig. 9, we first obtain d_1_cut2 and d_2_cut2. Since the relative density of the region between d_1_cut1 and d_1_cut2 is higher than that between d_1_cut2 and the boundary on the right (b_1), we stop for dimension 1. We have found a better cut d_1_cut2 and also a low density region between d_1_cut2 and the right space boundary (b_1).

However, for dimension 2, we now obtain a situation (in the region between d_2_cut1 and the bottom space boundary, b_2) like that for dimension 1 before d_1_cut2 is found. d_2_cut2 cuts into another cluster. We then apply the same method to the data points and the region between d_2_cut1 and d_2_cut2 since the relative density is lower between them, another local best cut d_2_cut3 is found, which is a much better cut, i.e., cutting almost at the cluster boundary. We now have two good cuts d_1_cut2 and d_2_cut3 for dimension 1 and 2 respectively. We also found two low density regions associated with the cuts, i.e., the region between d_1_cut2 and the right space boundary (b_1)

Algorithm evaluateCut(D)

1 **for** each attribute $A_i \in \{A1, A2, , Ad\}$ **do**

2   d_i_cut1 = the value (cut) of A_i that gives the best gain on dimension i;

3   d_i_cut2 = the value (cut) of A_i that gives the best gain in the L_i region produced by d_i_cut1;

4   **if** the relative density between d_i_cut1 and d_i_cut2 is higher than that between d_i_cut2 and b_i **then**

5          $r\_density_i = y_i/n_i$, where y_i and n_i are the numbers of Y points and N points between d_i_cut2 and b_i

6   **else** d_i_cut3 = the value (cut) that gives the best gain in the L_i region produced by d_i_cut2;
           /* L_i here is the region between d_i_cut1 and d_i_cut2 */

7          $r\_density_i = y_i/n_i$, where y_i and n_i are the numbers of Y points and N points in the region between d_i_cut1 and d_i_cut3 or d_i_cut2 and d_i_cut3 that has a lower proportion of Y points (or a lower relative density).

8   **end**

9 **end**

10 bestCut = d_i_cut3 (or d_i_cut2 if there is no d_i_cut3) of dimension i whose $r\_density_i$ is the minimal among the d dimensions.

Fig. 10. Determining the best cut in CLTree

for dimension 1, and the region between d_2_cut2 and d_2_cut3 for dimension 2.

3. **Select the overall best cut** (line 5, 7 and 10): We compare the relative densities ($r\_density_i$) of the low density regions identified in step 2 of all dimensions. The best cut in the dimension that gives the lowest $r\_density_i$ value is chosen as the best cut overall. In our example, the relative density between d_2_cut2 and d_2_cut3 is clearly lower than that between d_1_cut2 and the right space boundary, thus d_2_cut3 is the overall best cut.

The reason that we use relative density to select the overall best cut is because it is desirable to split at the cut point that may result in a big empty (N) region (e.g., between d_2_cut2 and d_2_cut3), which is more likely to separate clusters.

This algorithm can also be scaled up using the existing decision tree scale-up techniques in [18,27] since the essential computation here is the same as that in decision tree building, i.e., the gain evaluation. Our new criterion simply performs the gain evaluation more than once.

3 User-Oriented Pruning of Cluster Trees

The recursive partitioning method of building cluster trees will divide the data space until each partition contains only points of a single class, or until no test (or cut) offers any improvement[5]. The result is often a very complex tree that partitions the space more than necessary. This problem is the same as that in classification [26]. There are basically two ways to produce simpler trees:

1. Stopping: deciding not to divide a region any further, or
2. Pruning: removing some of the tree parts after the tree has been built.

The first approach has the attraction that time is not wasted in building the complex tree. However, it is known in classification research that stopping is often unreliable because it is very hard to get the stopping criterion right [26]. This has also been our experience. Thus, we adopted the pruning strategy. Pruning is more reliable because after tree building we have seen the complete structure of data. It is much easier to decide which parts are unnecessary.

The pruning method used for classification, however, cannot be applied here because clustering, to certain extent, is a subjective task. Whether a clustering is good or bad depends on the application and the user's subjective judgment of its usefulness [9, 24]. Thus, we use a subjective measure for pruning. We use the example in Fig. 11 to explain.

The original space is partitioned into 14 regions by the cluster tree. By simply looking at Fig. 11, it is not clear whether we should report one cluster (the whole space) or two clusters. If we are to report two clusters, should we report the region $C1$ and $C2$, or $S1$ and $S2$? The answers to these questions depend on the specific application.

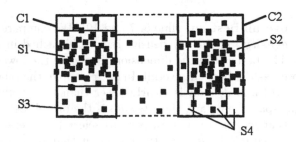

Fig. 11. How many clusters are there, and what are the cluster regions?

We propose two interactive approaches to allow the user to explore the cluster tree to find meaningful/useful clusters.

Browsing: The user simply explores the tree him/herself to find meaningful clusters (prune the rest). A user interface has been built to facilitate this exploration. This is not a difficult task because the major clusters are identified at the top levels of the tree.

---

[5]We use the same criterion as that in [26] to decide whether any improvement can be made.

User-Oriented Pruning: The tree is pruned using two user-specify parameters (see below). After pruning, we summarize the clusters by extracting
only those Y leaves from the tree and express them with hyper-rectangular
regions (each Y leaf naturally forms a region or a rule (see Sect. 2.1)). The
user can then view and study them. This is more convenient then viewing a
tree, as a tree with many internal nodes can be confusing.

The two parameters used in pruning are as follows:

min_y: It specifies the minimal number of Y points that a region must contain
(to be considered interesting). min_y is expressed as a percentage of $|D|$. That
is, a node with fewer than $min\_y * |D|$ number of Y points is not interesting.
For example, in Fig. 11, if $min\_y * |D| = 6$, the number of Y points (which is
4) in $S4$ (before it is cut into three smaller regions) is too few. Thus, further
cuts will not be considered, i.e., the two cuts in $S4$ are pruned. However, $S4$
may join S2 (see below) to form a bigger cluster. min_rd: It specifies whether
an N region (node) E should join an adjacent Y region F to form a bigger
cluster region. If the relative density of E, $E.Y/E.N$, is greater than min_rd,
where $E.Y$ (or $E.N$) gives the number of Y (or N) points contained in E,
then E and F should be combined to form a bigger cluster. For example, the
min_rd value will decide whether $S3$ should join S1 to form a bigger cluster.
If so, more data points are included in the cluster.

The pruning algorithm is given in Fig. 12. The basic idea is as follows: It
recursively descends down the tree in a depth first manner and then backs up
level-by-level to determine whether a cluster should be formed by a Y node
alone or by joining it with the neighboring node. If the two subtrees below a
node can be pruned, the algorithm assigns TRUE to the Stop field (*node.Stop*)
of the node data structure. Otherwise, it assigns FALSE to the field.

Once the algorithm is completed, we simply descend down the tree again
along each branch to find the first Y node whose Stop field is TRUE (not
shown in Fig. 12). These nodes are the clusters, which are presented to the
user as a list of cluster regions.

The evaluatePrune algorithm is linear to the number of nodes in the tree
as it traverses the tree only once.

4 Performance Experiments

In this section, we empirically evaluate CLTree using synthetic as well as
real-life datasets. Our experiments aim to establish the following:

- Efficiency: Determine how the execution time scales with, dimensionality of
  clusters, size of datasets, dimensionality of datasets, and number of clusters
  in the data space.
- Accuracy: Test if CLTree finds known clusters in subspaces as well as in the
  full space of a high dimensional space. Since CLTree provides descriptions
  of clusters, we test how accurate the descriptions are with different pruning

Algorithm evaluatePrune($Node$, min_y, min_rd)
/* $Node$ is the node being analyzed */

```
 1  if Node is a leaf then Node.Stop = TRUE
 2  else LeftChild = Node.left; RightChild = Node.right;
 3       if LeftChild.Y < min_y   D| then LeftChild.Stop = TRUE
 4       else evaluatePrune(LeftChild, min_y, min_rd);
 5       end
 6       if RightChild.Y < min_y   D| then RightChild.Stop = TRUE
 7       else evaluatePrune(RightChild, min_y, min_rd);
 8       end
 9       if LeftChild.Stop = TRUE then
            /* We assume that the relative density of LeftChild is */
10           if RightChild.Stop = TRUE then
               /* always higher than that of RightChild */
11             if RightChild.Y/RightChild.N > min_rd then
12               Node.Stop = TRUE
                  /* We can prune from Node either because we can join
                     or because both children are N nodes. */
13             elseif LeftChild is an N node then Node.Stop = TRUE
14             else Node.Stop = FALSE
15             end
16           else Node.Stop = FALSE
17           end
18       else Node.Stop = FALSE
19       end
20  end
```

Fig. 12. The cluster tree pruning algorithm

parameters. We also test how the ratio of the numbers of N and Y points affect the accuracy.

All the experiments were run on SUN E450 with one 250MHZ cpu and 512MB memory.

4.1 Synthetic Data Generation

We implemented two data generators for our experiments. One generates datasets following a normal distribution, and the other generates datasets following a uniform distribution. For both generators, all data points have coordinates in the range [0, 100] on each dimension. The percentage of noise or outliers (*noise level*) is a parameter. Outliers are distributed uniformly at random throughout the entire space.

Normal Distribution: The first data generator generates data points in each cluster following a normal distribution. Each cluster is described by the subset of dimensions, the number of data points, and the value range along each cluster dimension. The data points for a given cluster are generated as follows: The coordinates of the data points on the non-cluster dimensions are generated uniformly at random. For a cluster dimension, the coordinates of the data points projected onto the dimension follow a normal distribution. The data generator is able to generate clusters of elliptical shape and also has the flexibility to generate clusters of different sizes.

Uniform Distribution: The second data generator generates data points in each cluster following a uniform distribution. The clusters are hyper-rectangles in a subset (including the full set) of the dimensions. The surfaces of such a cluster are parallel to axes. The data points for a given cluster are generated as follows: The coordinates of the data points on non-cluster dimensions are generated uniformly at random over the entire value ranges of the dimensions. For a cluster dimension in the subspace in which the cluster is embedded, the value is drawn at random from a uniform distribution within the specified value range.

4.2 Synthetic Data Results

Scalability Results

For our experiments reported below, the noise level is set at 10%. The execution times (in sec.) do not include the time for pruning, but only tree building. Pruning is very efficient because we only need to traverse the tree once. The datasets reside in memory.

Dimensionality of Hidden Clusters: CLTree can be used for finding clusters in the full dimensional space as well as in any subspaces. Figure 13 shows the scalability as the dimensionality of the clusters is increased from 2 to 20 in a 20-dimensional space. In each case, 5 clusters are embedded in different subspaces of the 20-dimensional space. In the last case, the clusters are in the full space. Each dataset has 100,000 records. From the figure, we see that when the clusters are hyper-rectangles (in which the data points are uniformly distributed), CLTree takes less time to build the tree. This is because CLTree can naturally find hyper-rectangular clusters, and thus tree building stopped earlier. For both normal and uniform distribution data, we obtain better than linear scale-up.

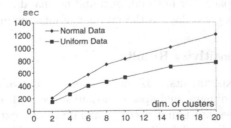

Fig. 13. Scalability with the dimensionality of hidden clusters

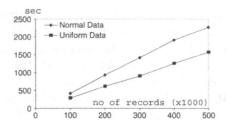

Fig. 14. Scalability with the dataset size

Dataset Size: Figure 14 shows the scalability as the size of the dataset is increased from 100,000 to 500,000 records. The data space has 20 dimensions, and 5 hidden clusters, each in a different 5-dimensional subspace. The execution time scales up linearly (although the time complexity of the algorithm is $O(nlogn)$).

Dimensionality of the Data Space: Figure 15 shows the scalability as the dimensionality of the data space is increased from 10 to 100. The dataset has 100,000 records and 5 clusters, each in a different 5-dimensional subspace. In both cases (normal distribution data and uniform distribution data), the algorithm scales up linearly.

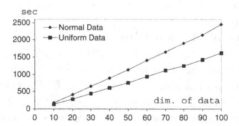

Fig. 15. Scalability with the dimensionality of the data space

Number of Clusters in the Space: Figure 16 shows the scalability as the number of clusters in the space is increased from 2 to 20. The data space has 20 dimensions and 100,000 records. Each cluster is embedded in a different 5-dimensional subspace. For both uniform and normal datasets, the execution times do not vary a great deal as the number of clusters increases.

Accuracy and Sensitivity Results

In all the above experiments, CLTree recovers all the original clusters embedded in the full dimensional space and subspaces. All cluster dimensions and their boundaries are found without including any extra dimension. For pruning, we use CLTree's default settings of $min\_y = 1\%$ and $min\_rd = 10\%$ (see below).

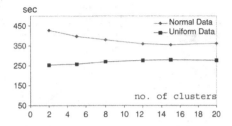

Fig. 16. Scalability with the number of clusters

Since CLTree provides precise cluster descriptions, which are represented
by hyper-rectangles and the number of data (Y) points contained in each of
them, below we show the percentage of data points recovered in the clusters using various min_y and min_rd values, and noise levels. Two sets of
experiments are conducted. In both sets, each data space has 20 dimensions,
and 100,000 data points. In the first set, the number of clusters is 5, and
in the other set, the number of clusters is 10. Each cluster is in a different
5-dimensional subspace. We will also show how the ratio of N and Y points
affects the accuracy. min_y: We vary the value of min_y from 0.05% to 5%,
and set $min\_rd = 10\%$ and noise level = 10%. Figure 17 gives the results. For
uniform distribution, even min_y is very low, all the data points in the clusters
are found for both 5 and 10 cluster cases. All the clusters and their dimensions are also recovered. For normal distribution, the percentage of data points
found in the clusters is relatively low when min_y is very small (0.05%, 0.1%
or 0.3%). It increases dramatically and stabilizes after min_y passes 0.5%.
From $min_y = 0.5$, the percentages of data points (outliers are not counted)
recovered are very high. around 95%.

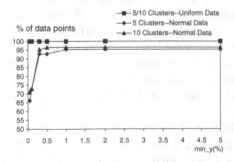

Fig. 17. Percentage of cluster data points found with min_y

min_rd: Figure 18 shows the percentages of data points found in the clusters with different values of min_rd. The noise level is set at 10% and min_y
at 1%. The min_rd values in the range of 2–30% do not affect the number
of cluster data points found in the uniform distribution datasets. The reason

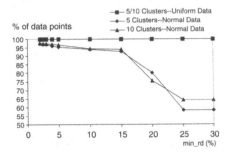

Fig. 18. Percentage of cluster data points found with min_rd

is that clusters in these datasets do not have the problem of low-density re-gions around the cluster boundaries as in the case of normal distribution. For the normal distribution datasets, when min_rd is small, more data points are found, 95% or more. When min_rd is too small, i.e., below 2% (not shown in Fig. 18), some clusters are merged, which is not desirable. When min_rd is more than 15%, the number of data points recovered in the clusters starts to drop and reaches 58–64% when min_rd is 25% and 30%. That is, when the min_rd value is very high, we only find the core regions of the clusters. In all these experiments (except those with min_rd below 2%), pruning finds the correct number of clusters, and also the cluster dimensions. Only the number of data points contained in each cluster region changes with different min_rd values. The results shown in Fig. 18 can be explained using Fig. 11 (Sect. 3). If min_rd is set low, we will find C1 and C2. If min_rd is set too low we will find the whole space as one cluster. If it is set very high, we only find S1 and S2.

Noise Level: Figure 19 shows how the noise level affects the percentage of data points recovered in the cluster regions (the noise data or outliers are not included in counting). We use $min\_y = 1\%$ and $min\_rd = 10\%$, which are the default settings for pruning as they perform very well. For the uniform distribution datasets, the noise level in the range of 5–30% does not affect the result. All data points are recovered. For the normal distribution datasets, when the noise level goes up, the percentages of data points found do not change a great deal for both 5 and 10 clusters. In all cases, the clusters are found.

$N:Y$ Ratio: In Sect. 2.2, we assume an equal number of N points as Y points. In fact, 1:1 ratio is not necessary. We experimented with different $N : Y$ ratios (using default min_y and min_rd values, and 10% noise). The results are shown in Fig. 20. In these experiments, all correct clusters are found. The percentage of cluster data points found hardly change.

min_y and min_rd in Applications: From the above experiment re-sults, we see that the clustering results are not very sensitive to the min_y and min_rd values. Although when min_rd is too high, we may lose many data points, we can still find the core regions of the clusters. Thus, in a real-life

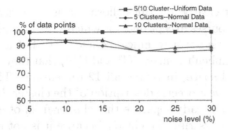

Fig. 19. Percentage of cluster data points found with the noise level

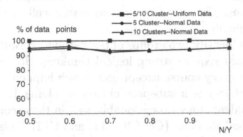

Fig. 20. Percentage of cluster data points found with N:Y ratio

application, it is rather safe to give both parameters high values, e.g., $min\_y = 1\text{--}5\%$ and $min\_rd = 10\text{--}30\%$. After we have found the core of each cluster, we can lower down the values to find bigger cluster regions. Alternatively, we can explore the cluster tree ourselves from those nodes representing the cluster cores.

4.3 Real-Life Data Results

We applied CLTree to three real life datasets. Due to space limitations, we can only describe one here. $min\_y = 1\%$ and $min\_rd = 10\%$ are used for pruning.

This dataset is obtained from an educational institution. It stores the examination grades (in A, B, C, D, and F) of their students in 12 courses. CLTree finds 4 main clusters (with some smaller ones). The descriptions of the 4 clusters and the dimensions used are shown in the table below. The last column gives the percentage of students that falls into each cluster.

Clusters	Courses												% of Students
	C1	C2	C3	C4	C5	C6	C7	C8	C9	C10	C11	C12	
1	B-A	C-A	C-A	C-A	C-A	C-A	C-A	C-A	D-A	B-A	C-A	B-A	14.2%
2		C-A								F-C			47.7%
3	F-C	D-D	F-C	F-D	D-D	F-D	D-D	F-C		D-D	F-D	F-D	6.5%
4	F-D	F-D	F-D	F-D	F-D	F-C	C-A			F-C			8.1%

The first cluster clearly identifies those consistently good students. They perform well in all courses except one, for which they may not do well. It turns out that this course is not a technical course (C9), but a course on writing. There are also two difficult courses (C1 and C10) that they do quite well. This cluster is in the full space, involving all 12 dimensions. The value ranges in the 12 dimensions give a precise description of the cluster. Note that although using C10 alone can already separate this cluster from others, we cannot use only B-A for C10 to describe the cluster because it is not precise enough.

The second cluster represents the average student population. Their examination scores can go up and down. This cluster is a subspace cluster, involving only two dimensions. They tend to do reasonably well in C2 and poorly in C10, which is a difficult course.

Cluster 3 and 4 identify two groups of poor students. Both groups perform badly in courses that require strong logical thinking. However, one group is bad consistently in every course except one, which happens to be the writing course, C9. This cluster is a subspace cluster involving 11 dimensions. The other cluster of students may do reasonably well in those courses that do not require strong logical thinking (C8, C9, C11, and C12). This cluster is also a subspace cluster involving 8 dimensions.

5 Related Work

This section consists of two sub-sections. In the first sub-section, we compare CLTree with existing clustering techniques. In the second sub-section, we compare the parameters used in CLTree with those used in existing methods.

5.1 Clustering Techniques

Traditional clustering techniques can be broadly categorized into partitional clustering and hierarchical clustering [12,24]. Given n data records, partitional clustering determines a partitioning of the records into k clusters such that the data records in a cluster are nearer to one another than the records in different clusters [12, 24]. The main problem with partitional techniques is that they are often very sensitive to the initial seeds, outliers and the order in which the data points are presented [7,14,24]. CLTree does not have these problems.

Hierarchical clustering is a nested sequence of partitions. A clustering can be created by building a tree (also called a dendrogram) either from leaves to the root (agglomerative approach) or from the root down to the leaves (divisive approach) by merging or dividing clusters at each step. Building a tree using either an agglomerative or a divisive approach can be prohibitively expensive for large datasets because the complexity of the algorithms is usually at least $O(n^2)$, where n is the number of data points or records.

Both partitional and hierarchical methods are based on distance comparison. It is shown in [23] that distance-based approaches are not effective for clustering in a very high dimensional space because it is unlikely that data points are nearer to each other than the average distance between data points [6, 23]. As a consequence, the difference between the distance to the nearest and the farthest neighbor of a data point goes to zero [6].

CLTree is different from partitional clustering because it does not explicitly group data points using distance comparison. It is different from hierarchical clustering because it does not merge the closest (or split the farthest) groups of records to form clusters. Instead, CLTree performs clustering by classifying data regions (Y regions) and empty regions (N regions) in the space using a purity function based on the information theory.

Most traditional clustering methods have very high computational complexities. They are not suitable for clustering of large high dimensional datasets. In the past few years, a number of studies were made to scale up these algorithms by using randomized search (CLARANS [25]), by condensing the data (BIRCH [32] and the system in [7]), by using grids (e.g., DENCLUE [22]), and by sampling (CURE [21]). These works are different from ours because our objective is not to scale up an existing algorithm, but to propose a new clustering technique that can overcome many problems with the existing methods.

Recently, some clustering algorithms based on local density comparisons and/or grids were reported, e.g., DBSCAN [10], DBCLASD [31], STING [30], WaveCluster [28] and DENCLUE [22]. Density-based approaches, however, are not effective in a high dimensional space because the space is too sparsely filled (see [23] for more discussions). It is noted in [3] that DBSCAN only runs with datasets having fewer than 10 dimensions. Furthermore, these methods cannot be used to find subspace clusters. OptiGrid [23] finds clusters in high dimension spaces by projecting the data onto each axis and then partitioning the data using cutting planes at low-density points. The approach will not work effectively in situations where some well-separated clusters in the full space may overlap when they are projected onto each axis. OptiGrid also cannot find subspace clusters.

CLIQUE [3] is a subspace clustering algorithm. It finds dense regions in each subspace of a high dimensional space. The algorithm uses equal-size cells and cell densities to determine clustering. The user needs to provide the cell size and the density threshold. This approach does not work effectively for clusters that involve many dimensions. According to the results reported in [3], the highest dimensionality of subspace clusters is only 10. Furthermore, CLIQUE does not produce disjoint clusters as normal clustering algorithms do. Dense regions at different subspaces typically overlap. This is due to the fact that for a given dense region all its projections on lower dimensionality subspaces are also dense, and get reported. [8] presents a system that uses the same approach as CLIQUE, but has a different measurement of good clustering. CLTree is different from this grid and density based approach because its cluster tree building does not depend on any input parameter. It is also able to find disjoint clusters of any dimensionality.

[1, 2] studies projected clustering, which is related to subspace clustering in [3], but find disjoint clusters as traditional clustering algorithms do. Unlike traditional algorithms, it is able to find clusters that use only a subset of the dimensions. The algorithm ORCLUS [2] is based on hierarchical merging clustering. In each clustering iteration, it reduces the number of clusters (by merging) and also reduces the number of dimensions. The algorithm assumes that the number of clusters and the number of projected dimensions are given beforehand. CLTree does not need such parameters. CLTree also does not require all projected clusters to have the same number of dimensions, i.e., different clusters may involve different numbers of dimensions (see the clustering results in Sect. 4.3).

Another body of existing work is on clustering of categorical data [15, 20, 29]. Since this paper focuses on clustering in a numerical space, we will not discuss these works further.

5.2 Input Parameters

Most existing cluster methods critically depend on input parameters, e.g., the number of clusters (e.g., [1, 2, 7, 25, 32]), the size and density of grid cells (e.g., [3, 8, 22]), and density thresholds (e.g., [10, 23]). Different parameter settings often result in completely different clustering. CLTree does not need any input parameter in its main clustering process, i.e., cluster tree building. It uses two parameters only in pruning. However, these parameters are quite different in nature from the parameters used in the existing algorithms. They only facilitate the user to explore the space of clusters in the cluster tree to find useful clusters.

In traditional hierarchical clustering, one can also save which clusters are merged and how far apart they are at each clustering step in a tree form. This information can be used by the user in deciding which level of clustering to make use of. However, as we discussed above, distance in a high dimensional space can be misleading. Furthermore, traditional hierarchical clustering methods do not give a precise description of each cluster. It is thus hard for the user to interpret the saved information. CLTree, on the other hand, gives precise cluster regions in terms of hyper-rectangles, which are easy to understand.

6 Conclusion

In this paper, we proposed a novel clustering technique, called CLTree, which is based on decision trees in classification research. CLTree performs clustering by partitioning the data space into data and empty regions at various levels of details. To make the decision tree algorithm work for clustering, we proposed a technique to introduce *non-existing* points to the data space, and also designed a new purity function that looks ahead in determining the best partition. Additionally, we devised a user-oriented pruning method in order to find subjectively interesting/useful clusters. The CLTree technique has many advantages over existing clustering methods. It is able to find "natural" clusters. It is suitable for subspace as well as full space clustering. It also provides descriptions of the resulting clusters. Finally, it comes with an important by-product, the empty (sparse) regions, which can be used to find outliers and anomalies. Extensive experiments have been conducted with the proposed technique. The results demonstrated its effectiveness.

References

1. C. Aggarwal, C. Propiuc, J. L. Wolf, P. S. Yu, and J. S. Park (1999) A framework for finding projected clusters in high dimensional spaces, SIGMOD-99
2. C. Aggarwal, and P. S. Yu (2000) Finding generalized projected clusters in high dimensional spaces, SIGMOD-00
3. R. Agrawal, J. Gehrke, D. Gunopulos, and P. Raghavan (1998) Automatic subspace clustering for high dimensional data for data mining applications, SIGMOD-98
4. R. Agrawal, S. Ghosh, T. Imielinski, B. Lyer, and A. Swami (1992) In interval classifier for database mining applications, VLDB-92
5. P. Arabie and L. J. Hubert (1996) An overview of combinatorial data analysis, In P. Arabie, L. Hubert, and G. D. Soets, editors, Clustering and Classification, pp. 5–63
6. K. Beyer, J. Goldstein, R. Ramakrishnan, and U. Shaft (1999) When is nearest neighbor meaningful?" Proc. 7th Int. Conf. on Database Theory (ICDT)
7. P. Bradley, U. Fayyad, and C. Reina (1998) Scaling clustering algorithms to large databases, KDD-98
8. C. H. Cheng, A. W. Fu, and Y. Zhang. "Entropy-based subspace clustering for mining numerical data." KDD-99
9. R. Dubes and A. K. Jain (1976) Clustering techniques: the user's dilemma, Pattern Recognition. 8:247–260
10. M. Ester, H.-P. Kriegal, J. Sander, and X. Xu (1996) A density-based algorithm for discovering clusters in large spatial databases with noise, KDD-96
11. M. Ester, H.-P. Kriegel, and X. Xu (1995). A database interface for clustering in large spatial data bases, KDD-95
12. B. S. Everitt (1974) Cluster analysis, Heinemann, London
13. C. Faloutsos and K. D. Lin (1995) FastMap: A fast algorithm for indexing, data-mining and visualisation of traditional and multimedia datasets, SIGMOD-95
14. U. Fayyad, C. Reina. and P. S. Bradley (1998), Initialization of iterative refinement clustering algorithms, KDD-98
15. D. Fisher (1987) Knowledge acquisition via incremental conceptual clustering, Machine Learning, 2:139–172
16. K. Fukunaga (1990) Introduction to statistical pattern recognition, Academic Press

17. V. Ganti, J. Gehrke, and R. Ramakrishnan (1999) CACTUS-Clustering categorical data using summaries, KDD-99
18. J. Gehrke, R. Ramakrishnan, and V. Ganti (1998) RainForest – A framework for fast decision tree construction of large datasets, VLDB-98
19. J. Gehrke, V. Ganti, R. Ramakrishnan, and W.-Y. Loh (1999) BOAT – Optimistic decision tree construction, SIGMOD-99
20. S. Guha, R. Rastogi, and K. Shim (1999) ROCK: a robust clustering algorithm for categorical attributes, ICDE-99
21. S. Guha, R. Rastogi, and K. Shim (1998) CURE: an efficient clustering algorithm for large databases, SIGMOD-98
22. A. Hinneburg and D. A. Keim (1998) An efficient approach to clustering in large multimedia databases with noise, KDD-98
23. A. Hinneburg and D. A. Keim (1999) An optimal grid-clustering: towards breaking the curse of dimensionality in high-dimensional clustering, VLDB-99
24. A. K. Jain and R. C. Dubes (1988) Algorithms for clustering data, Prentice Hall
25. R. Ng and J. Han (1994) Efficient and effective clustering methods for spatial data mining, VLDB-94
26. J. R. Quinlan (1992) C4.5: program for machine learning, Morgan Kaufmann
27. J. C. Shafer, R. Agrawal, and M. Mehta (1996) SPRINT: A scalable parallel classifier for data mining, VLDB-96
28. G. Sheikholeslami, S. Chatterjee and A. Zhang (1998) WaveCluster: a multi-resolution clustering Approach for Very Large Spatial Databases, VLDB-98
29. K. Wang, C. Xu, and B. Liu (1999) Clustering transactions using large items, CIKM-99
30. W. Wang, J. Yang, and R. Muntz (1997) STING: A statistical information grid approach to spatial data mining, VLDB-97
31. X. Xu, M. Ester, H.-P. Kriegel, and J. Sander (1998) A non-parametric clustering algorithm for knowledge discovery in large spatial databases, ICDE-98
32. T. Zhang, R. Ramakrishnan, and M. Linvy (1996) BIRCH: an efficient data clustering method for very large databases, SIGMOD-96, 103–114

Incremental Mining on Association Rules

W.-G. Teng and M.-S. Chen

Department of Electrical Engineering, National Taiwan University,
Taipei. Taiwan, ROC

Summary. The discovery of association rules has been known to be useful in selective marketing, decision analysis, and business management. An important application area of mining association rules is the market basket analysis, which studies the buying behaviors of customers by searching for sets of items that are frequently purchased together. With the increasing use of the record-based databases whose data is being continuously added, recent important applications have called for the need of incremental mining. In dynamic transaction databases, new transactions are appended and obsolete transactions are discarded as time advances. Several research works have developed feasible algorithms for deriving precise association rules efficiently and effectively in such dynamic databases. On the other hand, approaches to generate approximations from data streams have received a significant amount of research attention recently. In each scheme, previously proposed algorithms are explored with examples to illustrate their concepts and techniques in this chapter.

1 Introduction

Due to the increasing use of computing for various applications, the importance of data mining is growing at rapid pace recently. It is noted that analysis of past transaction data can provide very valuable information on customer buying behavior, and thus improve the quality of business decisions. In essence, it is necessary to collect and analyze a sufficient amount of sales data before any meaningful conclusion can be drawn therefrom. Since the amount of these processed data tends to be huge, it is important to devise efficient algorithms to conduct mining on these data. Various data mining capabilities have been explored in the literature [2,4,5,10–12,20,38,50,51]. Among them, the one receiving a significant amount of research attention is on mining association rules over basket data [2,3,16,21,24,32,35,40,43,49]. For example, given a database of sales transactions, it is desirable to discover all associations among items such that the presence of some items in a transaction will imply the presence of other items in the same transaction, e.g., 90% of customers that purchase milk and bread also purchase eggs at the same time.

Recent important applications have called for the need of incremental mining. This is due to the increasing use of the record-based databases whose data is being continuously added. Examples of such applications include Web log records, stock market data, grocery sales data, transactions in electronic commerce, and daily weather/traffic records, to name a few. In many applications, we would like to mine the transaction database for a fixed amount of most recent data (say, data in the last 12 months). That is, in the incremental mining, one has to not only include new data (i.e., data in the new month) into, but also remove the old data (i.e., data in the most obsolete month) from the mining process.

A naive approach to solve the incremental mining problem is to re-run the mining algorithm on the updated database. However, it obviously lacks of efficiency since previous results are not utilized on supporting the discovering of new results while the updated portion is usually small compared to the whole dataset. Consequently, the efficiency and the effectiveness of algorithms for incremental mining are both crucial issues which will be discussed in details in this chapter.

The rest of the chapter is organized as follows. Preliminaries including the mining of association rules and the need for performing incremental mining are given in Sect. 1. Algorithms for mining association rules incrementally from transactional databases that generate precise results are described in Sect. 2. On the other hand, algorithms dealing with data streams (i.e., online transaction flows) that generate approximate results are explored in Sect. 3. In each scheme, the algorithms can be further categorized according to their major principles. Namely, the Apriori-based algorithms, the partition-based algorithms and the pattern growth algorithms. In Sect. 4, remarks on the relationship between two schemes are made, giving the summary of the state of the art for incremental mining on association rules.

1.1 Mining Association Rules

Mining association rules was first introduced in [2], where the goal is to discover interesting relationships among items in a given transactional dataset.

A mathematical model was proposed in [2] to address the problem of mining association rules. Let $\mathcal{I} = \{i_1, i_2, \ldots, i_m\}$ be a set of literals, called items. Let D be a set of transactions, where each transaction T is a set of items such that $T \subseteq \mathcal{I}$. Note that the quantities of items bought in a transaction are not considered, meaning that each item is a binary variable representing if an item was bought. Each transaction is associated with an identifier, called TID. Let X be an itemset, i.e., a set of items. A transaction T is said to contain X if and only if $X \subseteq T$. An association rule is an implication of the form $X \Longrightarrow Y$, where $X \subset \mathcal{I}$, $Y \subset \mathcal{I}$ and $X \cap Y = \phi$. The rule $X \Longrightarrow Y$ has *support* s in the transaction set D if $s\%$ of transactions in D contain $X \cup Y$. The rule $X \Longrightarrow Y$ holds in the transaction set D with *confidence* c if $c\%$ of transactions in D that contain X also contain Y. By utilizing the notation of probability theory, the concepts of support and confidence for an association rule can be formulated as

$$support(X \implies Y) = P(X \cup Y) \text{ and,}$$
$$confidence(X \implies Y) = P(Y|X).$$

For a given pair of support and confidence thresholds, namely the minimum support S_{min} and the minimum confidence C_{min}, the problem of mining association rules is to find out all the association rules that have confidence and support greater than the corresponding thresholds. Moreover, the mining of association rules is a two-step process:

1. *Find all frequent itemsets*: Frequent itemsets are itemsets satisfying the support threshold, i.e., $\{X|X.support \geq S_{min}\}$. In some earlier literatures, frequent itemsets are also termed as *large* itemsets and the set of frequent k-itemsets (which are composed of k items) is thus commonly denoted by L_k. Consequently, the goal in this step is to discover the set $\{L_1, \ldots L_k\}$.
2. *Generate association rules from the frequent itemsets*: For any pair of frequent itemsets W and X satisfying $X \subset W$, if $\frac{X.support}{W.support} \geq C_{min}$, then $X \implies Y (= W - X)$ is identified as a valid rule.

The overall performance of mining association rules is in fact determined by the first step. After the frequent itemsets are identified, the corresponding association rules can be derived in a straightforward manner [2] as shown in the second step above. A numerous prior works including the Apriori [2], the DHP [40], and partition-based ones [31, 43] are proposed to solve the first subproblem efficiently. In addition, several novel mining techniques, including TreeProjection [1], FP-tree [23,24,42], and constraint-based ones [22,26,41,49] also received a significant amount of research attention. To clarify the idea of the mining of association rules before formally introducing the incremental mining, some representative algorithms along with illustrative examples are provided in following sections.

Apriori-Based Algorithms

Algorithm Apriori [2] is an influential algorithm for mining association rules. It uses *prior knowledge* of frequent itemset properties to help on narrowing the search space of required frequent itemsets. Specifically, k-itemsets are used to explore $(k + 1)$-itemsets during the *levelwise* process of frequent itemset generation. The set of frequent 1-itemsets (L_1) is firstly found by scanning the whole dataset once. L_1 is then used by performing join and prune actions to form the set of candidate 2-itemsets (C_2). After another data scan, the set of frequent 2-itemsets (L_2) are identified and extracted from C_2. The whole process continues iteratively until there is no more candidate itemsets which can be formed from previous L_k.

	T_1	A B C
	T_2	A F
	T_3	A B C E
	T_4	A B D F
D	T_5	C F
	T_6	A B C
	T_7	A B C E
	T_8	C D E
	T_9	B D E

Fig. 1. An illustrative transaction database

Example 1. Consider an example transaction database given in Fig. 1. In each iteration (or each pass), algorithm Apriori constructs a candidate set of large itemsets, counts the number of occurrences of each candidate itemset, and then determine large itemsets based on a predetermined minimum support S_{min}. In the first iteration, Apriori simply scans all the transactions to count the number of occurrences for each item. The set of candidate 1-itemsets, C_1, obtained is shown in Fig. 2. Assuming that $S_{min} = 40\%$, the set of large 1-itemsets (L_1) composed of candidate 1-itemsets with the minimum support required, can then be determined. To discover the set of large 2-itemsets, in view of the fact that any subset of a large itemset must also have minimum support, Apriori uses $L_1 * L_1$ to generate a candidate set of itemsets C_2 where * is an operation for concatenation in this case. C_2 consists of

$$\binom{|L_1|}{2}$$

2-itemsets. Next, the nine transactions in D are scanned and the support of each candidate itemset in C_2 is counted. The middle table of the second row in Fig. 2 represents the result from such counting in C_2. The set of large 2-itemsets, L_2, is therefore determined based on the support of each candidate 2-itemset in C_2.

The set of candidate 3-itemsets (C_3) is generated from L_2 as follows. From L_2, two large 2-itemsets with the same first item, such as {AB} and {AC}, are identified first. Then, Apriori tests whether the 2-itemset {BC}, which consists of their second items, contributes a frequent 2-itemset or not. Since {BC} is a frequent itemset by itself, we know that all the subsets of {ABC} are frequent and then {ABC} becomes a candidate 3-itemset. There is no other candidate 3-itemset from L_2. Apriori then scans all the transactions and discovers the large 3-itemsets L_3 in Fig. 2. Since there is no candidate 4-itemset to be constituted from L_3, Apriori ends the process of discovering frequent itemsets.

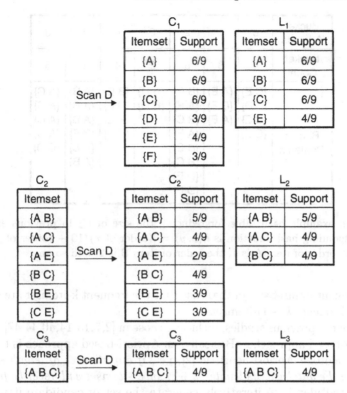

Fig. 2. Generation of candidate itemsets and frequent itemsets using algorithm Apriori

Similar to Apriori, another well known algorithm, DHP [39], also generates candidate k-itemsets from L_{k-1}. However, DHP employs a hash table, which is built in the previous pass, to test the eligibility of a k-itemset. Instead of including all k-itemset from $L_{k-1}*L_{k-1}$ into C_k, DHP adds a k-itemset into C_k only if that k-itemset is hashed into a hash entry whose value is larger than equal to the minimum transaction support required. As a result, the size of candidate set C_k can be reduced significantly. Such a filtering technique is particularly powerful in reducing the size of C_2.

Example 2. The effect of utilizing the hash table on helping to reduce the size of C_2 is provided in Fig. 3. It is noted that since the total counts for buckets 1 and 3 cannot satisfy the minimum support constraint, itemsets in these buckets, e.g., {A E}, should not be included in C_2. This improve the computing efficiency while the number of candidate itemsets to be checked is reduced.

DHP also reduces the database size progressively by not only trimming each individual transaction size but also pruning the number of transactions in the database. We note that both DHP and Apriori are iterative algorithms

Bucket Address	0	1	2	3	4	5	6
Bucket Count	4	3	8	2	4	6	5
Bucket Contents	{C E} {A D} {C E} {C E}	{A E} {C F} {A E}	{B C} {A F} {B C} {A F} {B C} {B C} {D E} {D E}	{B D} {B D}	{B E} {D F} {B E} {B E}	{A B} {A B} {A B} {B F} {A B} {A B}	{A C} {A C} {A C} {A C} {C D}

Fig. 3. An example hash table for reducing the size of C_2 in previous example. The corresponding hash function is $h(x, y) = [(\text{order of } x)*10 + (\text{order of } y)] \bmod 7$ where the order of item A is 1, the order of item B is 2, and so on

on the frequent itemset size in the sense that the frequent k-itemset are derived from the frequent $(k-1)$-itemsets.

Most of the previous studies, including those in [2,7,13,14,40,44,47], belong to Apriori-based approaches. Basically, an Apriori-based approach is based on an anti-monotone Apriori heuristic [2], i.e., *if any itemset of length k is not frequent in the database, its length (k+1) super-itemset will never be frequent.* The essential idea is to iteratively generate the set of candidate itemsets of length $(k+1)$ from the set of frequent itemsets of length k (for $k \geq 1$), and to check their corresponding occurrence frequencies in the database. As a result, if the largest *frequent* itemset is a j-itemset, then an Apriori-based algorithm may need to scan the database up to $(j+1)$ times.

In Apriori-based algorithms, C_3 is generated from $L_2 * L_2$. In fact, a C_2 can be used to generate the candidate 3-itemsets. This technique is referred to as scan reduction in [10]. Clearly, a C_3' generated from $C_2 * C_2$, instead of from $L_2 * L_2$, will have a size greater than $|C_3|$ where C_3 is generated from $L_2 * L_2$. However, if $|C_3'|$ is not much larger than $|C_3|$, and both C_2 and C_3 can be stored in main memory, we can find L_2 and L_3 together when the next scan of the database is performed, thereby saving one round of database scan. It can be seen that using this concept, one can determine all L_ks by as few as two scans of the database (i.e., one initial scan to determine L_1 and a final scan to determine all other frequent itemsets), assuming that C_k' for $k \geq 3$ is generated from C_{k-1}' and all C_k' for $k > 2$ can be kept in the memory. In [11], the technique of scan-reduction was utilized and shown to result in prominent performance improvement.

Partition-Based Algorithms

There are several techniques developed in prior works to improve the efficiency of algorithm Apriori, e.g., hashing itemset counts, transaction reduction, data sampling, data partitioning and so on. Among the various techniques, the data partitioning is the one with great importance since the goal in this chapter is on the incremental mining where bulks of transactions may be appended or discarded as time advances.

The works in [31,37,43] are essentially based on a partition-based heuristic, i.e., *if X is a frequent itemset in database D which is divided into n partitions $p_1, p_2, \ldots, p_n$, then X must be a frequent itemset in at least one of the n partitions.* The partition algorithm in [43] divides D into n partitions, and processes one partition in main memory at a time. The algorithm first scans partition p_i, for $i = 1$ to n, to find the set of all local frequent itemsets in p_i, denoted as L^{p_i}. Then, by taking the union of L^{p_i} for $i = 1$ to n, a set of candidate itemsets over D is constructed, denoted as C^G. Based on the above partition-based heuristic, C^G is a superset of the set of all frequent itemsets in D. Finally, the algorithm scans each partition for the second time to calculate the support of each itemset in C^G and to find out which candidate itemsets are really frequent itemsets in D.

Example 3. The flow of algorithm Partition is shown in Fig. 4. In this example, D is divided into three partitions, i.e., P_1, P_2 and P_3, and each partition contains three transactions. The sets of locally frequent itemsets L^{P_i} are discovered based on transactions in each partition. For example, $L^{P_2} = \{\{A\}, \{B\}, \{C\}, \{F\}, \{A\,B\}\}$. As is also shown in Fig. 4, the set of global candidate itemsets C^G is then generated by taking the union of L^{P_i}. Finally, these candidate itemsets are verified by scanning the whole dataset D once more.

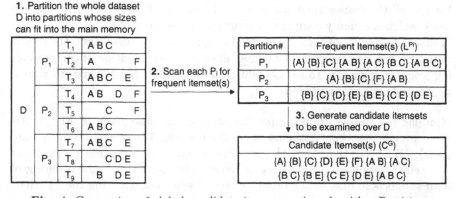

Fig. 4. Generation of global candidate itemsets using algorithm Partition

Instead of constructing C^G by taking the union of L^{p_i}, for $i = 1$ to n, at the end of the first scan, some variations of the above partition algorithm are proposed in [31, 37]. In [37], algorithm SPINC constructs C^G incrementally by adding L^{p_i} to C^G whenever L^{p_i} is available. SPINC starts the counting of occurrences for each candidate itemset $c \in C^G$ as soon as c is added to C^G. In [31], algorithm AS-CPA employs prior knowledge collected during the mining process to further reduce the number of candidate itemsets and to overcome the problem of data skew. However, these works were not devised to handle incremental updating of association rule.

Pattern Growth Algorithms

It is noted that the generation of frequent itemsets in both the Apriori-based algorithms and the partition-based algorithms is in the style of candidate generate-and-test. No matter how the search space for candidate itemsets is narrowed, in some cases, it may still need to generate a huge number of candidate itemsets. In addition, the number of database scans is limited to be at least twice, and usually some extra scans are needed to avoid unreasonable computing overheads. These two problems are nontrivial and are resulted from the utilization of the Apriori approach.

To overcome these difficulties, the tree structure which stores projected information of large datasets are utilized in some prior works [1, 24]. In [1], the algorithm TreeProjection constructs a lexicographical tree and has the whole database projected based on the frequent itemsets mined so far. The transaction projection can limit the support counting in a relatively small space and the lexicographical tree can facilitate the management of candidate itemsets. These features of algorithm TreeProjection provide a great improvement in computing efficiency when mining association rules.

An influential algorithm which further attempts to avoid the generation of candidate itemsets is proposed in [24]. Specifically, the proposed algorithm FP-growth (frequent pattern growth) adopts a *divide-and-conquer* strategy. Firstly, all the transactions are projected into an FP-tree (frequent pattern tree), which is a highly compressed structure, based on the frequent 1-itemsets in descending order of their supports. Then, for each frequent item, the conditional FP-tree can be extracted to be mined for the generation of corresponding frequent itemsets. To further illustrate how algorithm FP-growth works, an example is provided below.

Example 4. Consider the example transaction database given in Fig. 1. After scanning the whole database once, the frequent 1-itemsets discovered are {A:6}, {B:6}, {C:4} and {E:4} where the numbers are occurrence counts for the items. Note that the ordering of frequent items, which is the ordering of item supports in descending order, is crucial when constructing the FP-tree.

The construction of the FP-tree is shown in Fig. 5. The first transaction T_1:{A B C} is mapped to the single branch of the FP-tree in Fig. 5(a). In

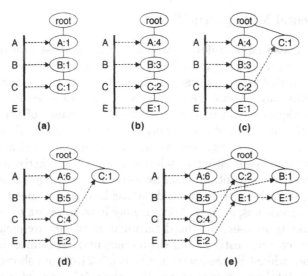

Fig. 5. Building the FP-tree based on the dataset in Fig. 1: (**a**) T_1 is added; (**b**) T_2, T_3 and T_4 are added; (**c**) T_5 is added; (**d**) T_6 and T_7 are added; (**e**) T_8 and T_9 are added (final FP-tree)

Fig. 5(b), the next three transactions are added by simply extending the branch and increasing the corresponding counts of existing item nodes since these transactions can fit into that branch when infrequent items, i.e., items D and F, are discarded. However, the transaction T_5 contains only one frequent item C and a new branch is thus created. Note that the node-link (composing of arrows with dotted line) for item C is extended in Fig. 5(c) for tracking the information of frequent itemsets containing item C. Specifically, all the possible frequent itemsets can be obtained by following the node-links for the items. In Fig. 5(d), transaction T_6 and T_7 are added in a similar way as the addition of transactions T_2 to T_4. Finally, the complete FP-tree in Fig. 5(e) is constructed by adding the last two transactions T_8 and T_9. It can be noted from this example that since the heuristic that popular items tend to occur together works in most cases, the resulting FP-tree can achieve a very high data compression ratio, showing the compactness of algorithm FP-growth.

To discover all the frequent itemsets, each frequent item along with its conditional FP-tree are mined separately and iteratively, i.e., the *divide-and-conquer* strategy. For example, by firstly tracking the node-link of item A, the 3-itemset {A B C} is found to frequent if $S_{min} = 40\%$. Its subsets which contain the item A, i.e., {A B} and {A C}, can then be discovered with correct supports easily. In addition, by tracking the node-links of both items B and C, only one more frequent itemset {B C} is generated, showing the completeness of algorithm FP-growth.

1.2 Incremental Mining Primitives

The mining of association rules on transactional database is usually an offline process since it is costly to find the association rules in large databases. With usual market-basket applications, new transactions are generated and old transactions may be obsolete as time advances. As a result, incremental updating techniques should be developed for maintenance of the discovered association rules to avoid redoing mining on the whole updated database.

A database may allow frequent or occasional updates and such updates may not only invalidate existing association rules but also activate new rules. Thus it is nontrivial to maintain such discovered rules in large databases. Note that since the underlying transaction database has been changed as time advances, some algorithms, such as Apriori, may have to resort to the regeneration of candidate itemsets for the determination of new frequent itemsets, which is, however, very costly even if the incremental data subset is small. On the other hand, while FP-tree-based methods [23, 24, 42] are shown to be efficient for small databases, it is expected that their deficiency of memory overhead due to the need of keeping a portion of database in memory, as indicated in [25], could become more severe in the presence of a large database upon which an incremental mining process is usually performed. Consequently, ordinary approaches for mining association rules are closely related to solving the problem of incremental mining. However, these algorithms cannot be applied directly without taking the incremental characteristics into consideration.

The concept of incremental mining on transaction databases in further illustrated in Fig. 6. For a dynamic database, old transactions ($\triangle^-$) are deleted from the database D and new transactions ($\triangle^+$) are added as time advances. Naturally, $\triangle^- \subseteq D$. Denote the updated database by D', $D' = (D - \triangle^-) \cup \triangle^+$. We also denote the set of unchanged transactions by $D^- = D - \triangle^-$.

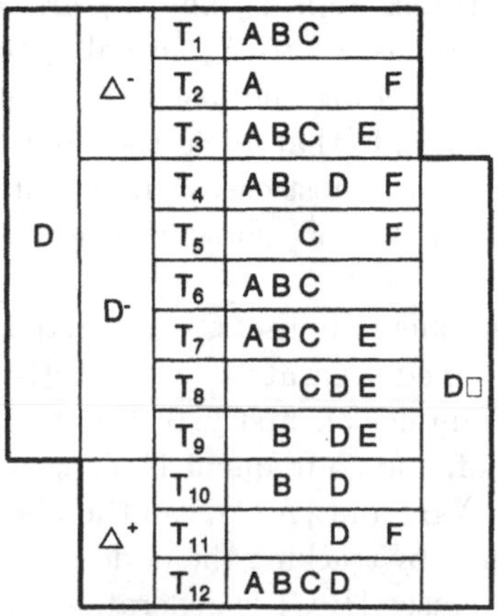

Group	Trans.	Items
$\triangle^-$	T_1	A B C
$\triangle^-$	T_2	A F
$\triangle^-$	T_3	A B C E
D^-	T_4	A B D F
D^-	T_5	C F
D^-	T_6	A B C
D^-	T_7	A B C E
D^-	T_8	C D E
D^-	T_9	B D E
$\triangle^+$	T_{10}	B D
$\triangle^+$	T_{11}	D F
$\triangle^+$	T_{12}	A B C D

(The groups D spans T_1–T_9; D' spans T_4–T_{12}.)

Fig. 6. An illustrative transactional database for incremental mining

Generally speaking, the goal is to solve the efficient update problem of association rules after a nontrivial number of new records have been added to or removed from a database. Assuming that the two thresholds, minimum support S_{min} and minimum confidence C_{min}, do not change, there are several important characteristics in the update problem.

1. The update problem can be reduced to finding the new set of frequent itemsets. After that, the new association rules can be computed from the new frequent itemsets.
2. An old frequent itemset has the potential to become infrequent in the updated database.
3. Similarly, an old infrequent itemset could become frequent in the new database.
4. In order to find the new frequent itemsets "exactly", all the records in the updated database, including those from the original database, have to be checked against every candidate set.

Note that the fourth characteristic is generally accepted when looking for exact frequent itemsets (and thus the association rules) from updated databases. On the contrary, in the data stream environment, the approximation of exact frequent itemsets is a key ingredient due to the high speed and huge volume of input transactions. Consequently, only increment portion of data rather than the whole unchanged portion has to be scanned, leading to an efficient way in performing updates of frequent itemsets. However, it is noted that the quality of approximation should be guaranteed within a probabilistic or deterministic error range, showing that the task of incremental mining on data streams is of more challenge.

To further understand the incremental mining techniques from either transaction databases or data streams, details are generally discussed in following sections. The algorithms of incremental mining which seek for exactly updated association rules from transactional databases are presented in Sect. 2. The importance and approaches targeting at mining and maintaining approximations incrementally from data streams are explored in Sect. 3.

2 Mining Association Rules Incrementally from Transactional Databases

Since database updates may introduce new association rules and invalidate some existing ones, it is important to study efficient algorithms for incremental update of association rules in large databases. In this scheme, a major portion of the whole dataset is remain unchanged while new transactions are appended and obsolete transactions may be discarded. By utilizing different core techniques, algorithms for incremental mining from transactional databases can be categorized into Apriori-based, partition-based or pattern growth algorithms. which will be fully explored in this section.

2.1 Apriori-Based Algorithms for Incremental Mining

As mentioned earlier, the Apriori heuristic is an anti-monotone principle. Specifically, *if any itemset is not frequent in the database, its super-itemset will never be frequent.* Consequently, algorithms belonging to this category adopt a levelwise approach, i.e., from shorter itemsets to longer itemsets, on generating frequent itemsets.

Algorithm FUP (Fast UPdate)

Algorithm FUP (Fast UPdate) [13] is the first algorithm proposed to solve the problem of incremental mining of association rules. It handles databases with transaction insertion only, but is not able to deal with transaction deletion. Specifically, given the original database D and its corresponding frequent itemsets $L = \{L_1, \ldots, L_k\}$. The goal is to reuse the information to efficiently obtain the new frequent itemsets $L' = \{L'_1, \ldots, L'_k\}$ on the new database $D' = D \cup \Delta^+$.

By utilizing the definition of support and the constraint of minimum support S_{min}. The following lemmas are generally used in algorithm FUP.

1. An original frequent itemset X, i.e., $X \in L$, becomes infrequent in D' if and only if $X.support_{D'} < S_{min}$.
2. An original infrequent itemset X, i.e., $X \notin L$, may become frequent in D' only if $X.support_{\Delta^+} \geq S_{min}$.
3. If a k-itemset X whose $(k-1)$-subset(s) becomes infrequent, i.e., the subset is in L_{k-1} but not in L'_{k-1}, X must be infrequent in D'.

Basically, similarly to that of Apriori, the framework of FUP, which can update the association rules in a database when new transactions are added to the database, contains a number of iterations [13,14]. The candidate sets at each iteration are generated based on the frequent itemsets found in the previous iteration. At the k-th iteration of FUP, Δ^+ is scanned exactly once. For the original frequent itemsets, i.e., $\{X|X \in L_k\}$, they only have to be checked against the small increment Δ^+. To discover the new frequent itemsets, the set of candidate itemsets C_k is firstly extracted from Δ^+, and then be pruned according to the support count of each candidate itemset in Δ^+. Moreover, the pool for candidate itemsets can be further reduced by discarding itemsets whose $(k-1)$-subsets are becoming infrequent.

The flows of FUP can be best understood by the following example. The dataset with the increment portion labeled as Δ^+ is shown in Fig. 7. Note that the first nine transactions are identical to those shown in earlier examples. In addition, the frequent itemsets of the unchanged portion D is also shown in Fig. 7, where the generation process is described earlier and is thus omitted here.

Example 5. The first iteration of algorithm FUP when performing incremental mining is represented in Fig. 8. The original frequent 1-itemsets are firstly

	T_1	A B C			
	T_2	A		F	
	T_3	A B C	E		
	T_4	A B	D	F	
D	T_5	C		F	
	T_6	A B C			
	T_7	A B C	E		
	T_8	C D E			
	T_9	B	D E		
	T_{10}	B	D		
Δ^+	T_{11}		D	F	
	T_{12}	A B C D			

(Original) frequent
itemsets $\{L_k\}$ of D

Itemset	Support
{A}	6/9
{B}	6/9
{C}	6/9
{E}	4/9
{A B}	5/9
{A C}	4/9
{B C}	4/9
{A B C}	4/9

Fig. 7. An illustrative database for performing algorithm FUP, and the original frequent itemsets generated using association rule mining algorithm(s)

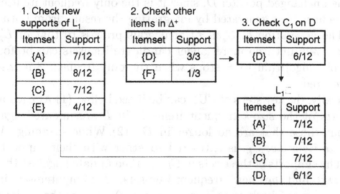

Fig. 8. The first iteration of algorithm FUP on the dataset in Fig. 6

verified on the increment portion Δ^+, and only itemsets with new supports no less than $S_{min}(=40\%)$ is retained as a part of new frequent 1-itemsets L_1', i.e., {A}, {B} and {C}. Then the supports of other possible items are also checked in Δ^+, leading to the construction of $C_1 : \{\{D\}\}$ to be further verified against the unchanged portion D. Finally, the new L_1' is generated by integrating the results from both possible sources.

The successive iterations work roughly the same as the first iteration. However, since the shorter frequent itemsets have already been discovered, the information can be utilized to further reduce the pool of candidate itemsets. Specifically, as shown in Fig. 9, since the 1-itemset {E} becomes infrequent as Δ^+ is considered, i.e., $\{E\} \in (L1 - L_1')$, all itemsets in L_2 containing a subset of {E} should be infrequent and are thus discarded with no doubt. Other itemsets in L_2 are then verified again Δ^+ to see if they are still frequent. The

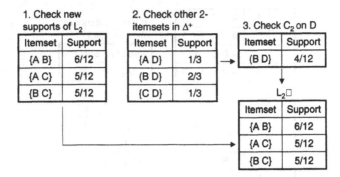

Fig. 9. The second iteration of algorithm FUP on the dataset in Fig. 6

set of candidate 2-itemsets C_2 is constructed by $(L_1' * L_1' - L_2)$ since itemsets in L_2 are already checked. In this example, the 2-itemsets {A D}, {B D} and {C D} are firstly checked against Δ^+. Afterward, only {B D} is being checked against the unchanged portion D, since it is the only frequent itemset in Δ^+. Finally, the new L_2' is generated by integrating the results from both possible sources. The generation of L_k' is of analogous process to that of L_2', and it works iteratively until no more longer candidate itemsets can be formed. In this example, algorithm FUP stops when the only frequent 3-itemset {A B C} is discovered.

Specifically, the key steps of FUP can be listed below. (1) At each iteration, the supports of the size-k frequent itemsets in L are updated against Δ^+ to filter out those that are no longer in D'. (2) While scanning Δ^+, a set of candidate itemsets, C_k, is extracted together with their supports in Δ^+ counted. The supports of these sets in C_k are then updated against the original database D to find the "new" frequent itemsets. (3) Many itemsets in C_k can be pruned away by checking their supports in Δ^+ before the update against the original database starts. (4) The size of C_k is further reduced at each iteration by pruning away a few original frequent itemsets in Δ^+.

The major idea is to reuse the information of the old frequent itemsets and to integrate the support information of the new frequent itemsets in order to substantially reduce the pool of candidate sets to be re-examined. Consequently, as compared to that of Apriori or DHP, the number of candidate itemsets to be checked against the whole database $D' = D \cup \Delta^+$ is much smaller, showing the major advantage of algorithm FUP.

Algorithms FUP$_2$ and FUP$_2\mathcal{H}$

The algorithm FUP [13] updates the association rules in a database when new transactions are *added* to the database. An extension to algorithm FUP was reported in [14] where the authors propose an algorithm FUP$_2$ for updating the existing association rules when transactions are *added* to and *deleted* from the database. In essence. FUP_2 is equivalent to FUP for the case of insertion, and is, however, a complementary algorithm of FUP for the case of deletion.

For a general case that transactions are added and deleted, algorithm FUP_2 can work smoothly with both the deleted portion Δ^- and the added portion Δ^+ of the whole dataset. A very feature is that the old frequent k-itemsets L_k from the previous mining result is used for dividing the candidate set C_k into two parts: $P_k = C_k \cap L_k$ and $Q_k = C_k - P_k$. In other words, P_k (Q_k) is the set of candidate itemsets that are previously frequent (infrequent) with respect to D. For the candidate itemsets in Q_k, their supports are unknown since they were infrequent in the original database D, posing some difficulties in generating new frequent itemsets. Fortunately, it is noted that if a candidate itemset in Q_k is frequent in Δ^-, it must be infrequent in D^-. This itemset is further identified to be infrequent in the updated database D' if it is also infrequent in Δ^+. This technique helps on effectively reducing the number of candidate itemsets to be further checked against the unchanged portion D^- which is usually much larger than either Δ^- or Δ^+.

Example 6. To further illustrate the flow of algorithm FUP_2, an example is provided in Fig. 10. In the first iteration, C_1 is exactly the set of all items. In subsequent iterations, C_k is generated from L'_{k-1}, the frequent itemsets found in the previous iteration. As shown in Fig. 10, the set of candidate itemsets C_i can be further divided into P_i and Q_i. For each itemset in P_i, the corresponding support for D is known since the itemsets is previously frequent. Therefore, by scanning only the deleted portion Δ^- and the added portion Δ^+, the new support can be obtained. For example, $Count(\{A\})_{D'} = Count(\{A\})_D - Count(\{A\})_{\Delta^-} + Count(\{A\})_{\Delta^+} = 6 - 3 + 1 = 4$ where $Count(X)_{db}$ represents the occurrence counts for itemset X in dataset db. On the other hand, to verify if an itemset in Q_i is frequent or not, the cost of scanning the unchanged (and usually large) portion D^- could be required since the corresponding support is previously unknown. Fortunately, in some cases, only the scans of Δ^- and Δ^+ are required. For example, it can be easily observed that $Count(\{F\})_{\Delta^+} - Count(\{F\})_{\Delta^-} = 0 \le (|\Delta^+| - |\Delta^-|) \times S_{\min} = 0$, showing that the support of itemset $\{F\}$ could not be improved by introducing the updated transactions (both deleted and added transactions are considered.) Consequently, fewer itemsets have to be further scanned against the unchanged portion D^-. An iteration is finished when all the itemsets in P_i and Q_i are all verified and therefore the new set of frequent itemsets L'_i is generated.

Another FUP-based algorithm, call $\text{FUP}_2\mathcal{H}$, was also devised in [14] to utilize the hash technique for performance improvement. In a similar way to that of the algorithm DHP, the counts of itemsets in D^- can be hashed, leading to an immediate improvement on efficiency.

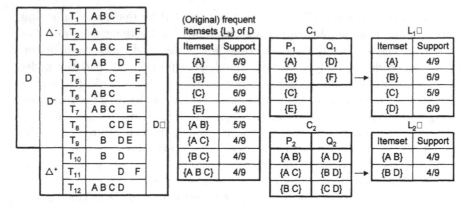

Fig. 10. Generation of new frequent itemsets from the dataset in Fig. 6 using algorithm FUP$_2$

Algorithm UWEP (Update with Early Pruning)

As pointed out earlier, the existing FUP-based algorithms in general suffer from two inherent problems, namely (1) the occurrence of a potentially huge set of candidate itemsets, which is particularly critical for incremental mining since the candidate sets for the original database and the incremental portion are generated separately, and (2) the need of multiple scans of database.

In [6], algorithm UWEP is proposed using the technique of update with early pruning. The major feature of algorithm UWEP over other FUP-based algorithms is that it prunes the supersets of an originally frequent itemset in D as soon as it becomes infrequent in the updated database D', rather than waiting until the k-th iteration. In addition, only itemsets which are frequent in both Δ^+ and $D'(= D \cup \Delta^+)$ are taken to generate candidate itemsets to be further checked against Δ^+. Specifically, if a k-itemset is frequent in Δ^+ but infrequent in D', it is not considered when generating C_{k+1}. This can significantly reduce the number of candidate itemsets in Δ^+ with the trade-off that an additional set of unchecked itemsets has to be maintained during the mining process. Consequently, these early pruning techniques can enhance the efficiency of FUP-based algorithms.

Algorithm Utilizing Negative Borders

Furthermore, the concept of *negative borders* [47] is utilized in [46] to improve the efficiency of FUP-based algorithms on incremental mining. Specifically, given a collection of frequent itemsets L, closed with respect to the set inclusion relation, the negative border $Bd^-(L)$ of L consists of the *minimal* itemsets $X \subseteq R$ not in L where R is the set of all items. In other words, the negative border consists of all itemsets that were candidates of the level-wise method which did not have enough support. That is, $Bd^-(L_k) = C_k - L_k$

where $Bd^-(L_k)$ is the set of k-itemsets in $Bd^-(L)$. The intuition behind the concept is that given a collection of frequent itemsets, the negative border contains the "closest" itemsets that could be frequent, too. For example, assume the collection of frequent itemsets is $L = \{\{A\}, \{B\}, \{C\}, \{F\}, \{A\ B\}, \{A\ C\}, \{A\ F\}, \{C\ F\}, \{A\ C\ F\}\}$, by definition the negative border of L is $Bd^-(L) = \{\{B\ C\}, \{B\ F\}, \{D\}, \{E\}\}$.

The algorithm proposed in [46] first generate the frequent itemsets of the increment portion Δ^+. A full scan of the whole dataset is required only if the negative border of the frequent itemsets expands, that is, if an itemset outside the negative border gets added to the frequent itemsets or its negative border. Even in such cases, it requires only one scan over the whole dataset. The possible drawback is that to compute the negative border closure may increase the size of the candidate set. However, a majority of those itemsets would have been present in the original negative border or frequent itemset. Only those itemsets which were not covered by the negative border need to be checked against the whole dataset. As a result, the size of the candidate set in the final scan could potentially be much smaller as compared to algorithm FUP.

Algorithm DELI (Difference Estimation for Large Itemsets)

To reduce the efforts of applying algorithm FUP_2 when database update occurs, the algorithm DELI which utilizes sampling techniques is proposed in [30]. Algorithm DELI can estimate the difference between the old and the new frequent itemsets. Only if the estimated difference is large enough, the update operation using algorithm FUP_2 has to be performed on the updated database to get the exact frequent itemsets.

Recall that in each iteration of algorithm FUP_2, all candidate itemsets in $Q_k (= C_k - P_k)$, i.e., previously infrequent itemsets, have to be further verified. For each candidate itemset $X \in Q_k$, if $X.support_{\Delta^+} \geq S_{min}$, the exact occurrence count of X in D^- has to be further identified. This data scan process is saved by algorithm DELI in the following way. By drawing m transactions from D^- with replacement to form the sample S, the support of itemset X in D^-, i.e., $\widehat{\sigma_X}$, can be estimated as

$$\widehat{\sigma_X} = \frac{T_x}{m} \cdot |D^-|,$$

where T_x is the occurrence count of X in S. With m sufficiently large, the normal distribution can be utilized to approximate the support of X in D^- with a $100(1 - \alpha)\%$ confidence interval $[a_X, b_X]$:

$$a_X = \widehat{\sigma_X} - z_{\frac{\alpha}{2}} \cdot \sqrt{\frac{\widehat{\sigma_X}(|D^-| - \widehat{\sigma_X})}{m}}, \quad \text{and}$$

$$b_X = \widehat{\sigma_X} + z_{\frac{\alpha}{2}} \cdot \sqrt{\frac{\widehat{\sigma_X}(|D^-| - \widehat{\sigma_X})}{m}}$$

where $z_{\frac{\alpha}{2}}$ is the critical value such that the area under the standard normal curve beyond $z_{\frac{\alpha}{2}}$ is exactly $\frac{\alpha}{2}$. In other words, there is $100(1 - \alpha)\%$ chance that the actual value of $X.support_{D^-}$ lies on the interval $[a_X, b_X]$. If the upper bound does not exceed the support threshold, i.e., $b_X < S_{min}$, itemset X is very likely to be infrequent and is thus dropped. Consequently, the resulting estimation for updated set of frequent itemsets $\widehat{L'_k}$ and the previous set of frequent itemsets L_k are compared to see if algorithm FUP_2 has to be resorted to obtain an accurate update.

Algorithms MAAP (Maintaining Association Rules with Apriori Property) and PELICAN

Several other algorithms, including the MAAP algorithm [53], and the PELI-CAN algorithm [48], are proposed to solve the problem of incremental mining. Algorithm MAAP firstly finds the frequent itemset(s) of the largest size based on previously discovered frequent itemsets. If a k-itemset is found to be frequent, then all of its subsets are concluded to be frequent and are thus be added to the new set of frequent itemsets L'. This eliminates the need to compute some frequent itemsets of shorter sizes. The other frequent itemsets are then identified by following the levelwise style of itemset generation, i.e., from 1-itemsets to $(k - 1)$-itemsets.

Both algorithms MAAP and PELICAN are similar to algorithm FUP_2, but they only focus on how to maintain maximum frequent itemsets when the database are updated. In other words, they do not consider non-maximum frequent itemsets, and therefore, the counts of non-maximum frequent itemsets cannot be calculated. The difference of these two algorithms is that MAAP calculates maximum frequent itemsets by Apriori-based framework while PELI-CAN calculates maximum frequent itemsets based on vertical database format and lattice decomposition. Since these two algorithms maintain maximum frequent itemsets only, the storage space and the processing time for performing each update can be thus reduced.

2.2 Partition-Based Algorithms for Incremental Mining

In contrast to the Apriori heuristic, the partition-based technique well utilizes the partitioning on the whole transactional dataset. Moreover, after the partitioning, it is understood that *if X is a frequent itemset in database D which is divided into n partitions p_1, p_2, ..., p_n, then X must be a frequent itemset in at least one of the n partitions.* Consequently, algorithms belonging to this category work on each partition of data iteratively and gather the information obtained from the processing of each partition to generate the final (integrated) results.

Consider a partitioned transaction database in Fig. 11. Note that $db^{i,j}$ is the part of the transaction database formed by a continuous region from

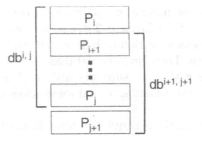

Fig. 11. Incremental mining for an ongoing time-variant transaction database

partition P_i to partition P_j. Suppose we have conducted the mining for the transaction database $db^{i,j}$. As time advances, we are given the new data of P_{j+1}, and are interested in conducting an incremental mining against the new data. With the model of sliding window which is usually adopted in temporal mining, our interest is limited to mining the data within a fixed period of time rather than taking all the past data into consideration. As a result, the mining of the transaction database $db^{i+1,j+1}$ is called for.

Algorithm SWF (Sliding–Window Filtering)

By partitioning a transaction database into several partitions, algorithm SWF (sliding-window filtering) [27] employs a filtering threshold in each partition to deal with the candidate itemset generation. Under SWF, the cumulative information of mining previous partitions is selectively carried over toward the generation of candidate itemsets for the subsequent partitions. After the processing of a partition, algorithm SWF outputs a *cumulative filter*, which consists of a progressive candidate set of itemsets, their occurrence counts and the corresponding partial support required. Specifically, the cumulative filter produced in processing each partition constitutes the key component to realize the incremental mining.

The key idea of algorithm SWF is to compute a set of candidate 2-itemsets as close to L_2 as possible. The concept of this algorithm is described as follows. Suppose the database is divided into n partitions $P_1, P_2, \ldots, P_n$, and processed one by one. *For each frequent itemset I, there must exist some partition P_k such that I is frequent from partition P_k to P_n.* A list of 2-itemsets CF is maintained by algorithm SWF to track the possible frequent 2-itemsets. For each partition P_i, algorithm SWF adds (locally) frequent 2-itemsets (together with its starting partition P_i and supports) that is not in CF and checks if the present 2-itemsets are continually frequent from its stating partition to the current partition. If a 2-itemset is no longer frequent, it is deleted from CF. However, if a deleted itemset is indeed frequent in the whole database, it must be frequent from some other partition P_j $(j > i)$, where we can add it to CF again.

It was shown that the number of the reserved candidate 2-itemsets will be close to the number of the frequent 2-itemsets. For a moderate number of candidate 2-itemsets, scan reduction technique [40] can be applied to generate all candidate k-itemsets. Therefore, one database scan is enough to calculate all candidate itemsets with their supports and to then determine frequent ones. In summary, the total number of database scans can be kept as small as two.

The flows of algorithm SWF is fully explored in the following example.

Example 7. Consider the transaction database in Fig. 6. A partitioned version is shown in Fig. 12.

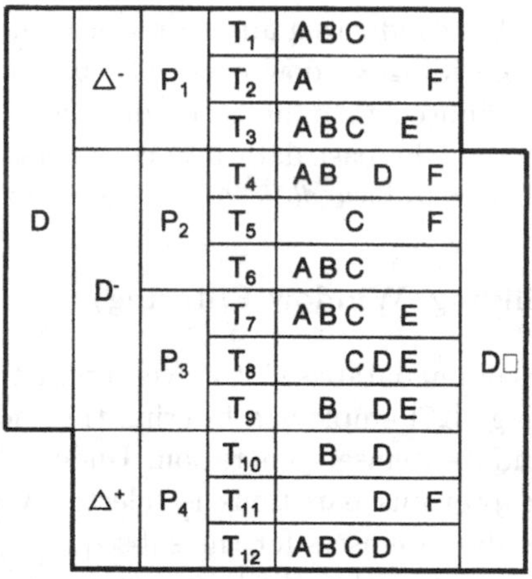

Fig. 12. A partitioned transaction database whose data records are identical to those in Fig. 6

Firstly, the goal is to generate all frequent itemsets in the original database D, that is defined as the *preprocessing* step in algorithm SWF. With $S_{min} = 40\%$, the generation of frequent 2-itemsets in each partition is shown in Fig. 13. After scanning the first three transactions, partition P_1, 2-itemsets {A B}, {A C} and {B C} are found to be frequent in P_1 and are thus potential candidates for D. It can be noted that each itemset shown in Fig. 13 has two attributes, i.e., "*Start*" contains the identity of the starting partition when the itemset was added, and "*Count*" contains the number of occurrences of this itemset since it was added. In addition the filtering threshold is $\lceil 3 * 0.4 \rceil = 2$ which is the minimal count required in P_1.

Note that the frequent itemsets discovered in each partition are carried to subsequent partitions. New itemsets could be added into the set of potential candidate itemsets, i.e., C_2, in a new partition while the counts of existing itemsets are increased and then be verified to see if these itemsets are still

	P_1			P_2			P_3	
Itemset	Start	Count	Itemset	Start	Count	Itemset	Start	Count
{A B}	1	2	{A B}	1	4	{A B}	1	5
{A C}	1	2	{A C}	1	3	{A C}	1	4
{B C}	1	2	{B C}	1	3	{B C}	1	4
						{B E}	3	2
						{C E}	3	2
						{D E}	3	2

Candidate itemsets in $db^{1,3}$:
{A} {B} {C} {D} {E} {F} {A B} {A C} {B C} {B E} {C E} {D E} {A B C} {B C E}
Frequent itemsets in $db^{1,3}$:
{A} {B} {C} {E} {A B} {A C} {B C} {A B C}

Fig. 13. Generation of frequent 2-itemsets in each partition using algorithm SWF

frequent from their starting partition to the current one. For example, there is no new 2-itemset added when processing P_2 since no extra frequent 2-itemsets in addition to the ones carried from P_1. Moreover, the counts for itemsets {A B}, {A C} and {B C} are all increased, making these itemsets frequent since their counts are no less than $\lceil 6 * 0.4 \rceil = 3$ when there are total six transactions, i.e., $P_1 \cup P_2$, are considered. Finally, after the analogous process for P_3, the resulting $C_2 = \{\{A\ B\},\ \{A\ C\},\ \{B\ C\},\ \{B\ E\},\ \{C\ E\},\ \{D\ E\}\}$ which are also shown in Fig. 13.

After generating C_2 from the first scan of database $db^{1,3}$, we employ the scan reduction technique and use C_2 to generate C_k $(k = 2, 3, \ldots, n)$, where C_n is the candidate *last*-itemsets. It can be verified that a C_2 generated by SWF can be used to generate the candidate 3-itemsets and its sequential C'_{k-1} can be utilized to generate C'_k. Clearly, a C'_3 generated from $C_2 * C_2$, instead of from $L_2 * L_2$, will have a size greater than $|C_3|$ where C_3 is generated from $L_2 * L_2$. However, since the $|C_2|$ generated by SWF is very close to the the-oretical minimum, i.e., $|L_2|$, the $|C'_3|$ is not much larger than $|C_3|$. Similarly, the $|C'_k|$ is close to $|C_k|$. All C'_k can be stored in main memory, and we can find L_k $(k = 1, 2, \ldots, n)$ together when the second scan of the database $db^{1,3}$ is performed. Thus, only two scans of the original database $db^{1,3}$ are required in the preprocessing step. In addition, instead of recording all L_ks in main memory, we only have to keep C_2 in main memory for the subsequent incre-mental mining of an ongoing time variant transaction database. The itemsets of C_2 and L_2 in $db^{1,3}$ are also shown in Fig. 13 for references.

The merit of SWF mainly lies in its incremental procedure. As depicted in Fig. 12, the mining database will be moved from $db^{1,3}$ to $db^{2,4}$. Thus, some transactions, i.e., t_1, t_2, and t_3, are *deleted* from the mining database and other transactions, i.e., t_{10}, t_{11}, and t_{12}, are *added*. For ease of exposition, this incremental step can also be divided into three sub-steps: (1) generating C_2 in $D^- = db^{1,3} - \Delta^-$, (2) generating C_2 in $db^{2,4} = D^- + \Delta^+$ and (3) scanning

$$db^{1,3} - \Delta^- = D^- \qquad\qquad D^- + \Delta^+ = D\square$$

Itemset	Start	Count
{A B}	2	3
{A C}	2	2
{B C}	2	2
{B E}	3	2
{C E}	3	2
{D E}	3	2

Itemset	Start	Count
{A B}	2	4
{B E}	3	2
{C E}	3	2
{D E}	3	2
(B D)	4	2

Candidate itemsets in $db^{2,4}$:
{A} {B} {C} {D} {E} {F} {A B} {B D}
Frequent itemsets in $db^{2,4}$:
{A} {B} {C} {D} {A B} {B D}

Fig. 14. Incremental mining using algorithm SWF where *shaded* itemsets are identified to be infrequent when Δ^- is deleted or Δ^+ is added

the database $db^{2,4}$ only once for the generation of all frequent itemsets L_k. The flows of these steps are presented in Fig. 14 where shaded itemsets are identified to be infrequent and are thus discarded during the mining process. Specifically, the counts of existing itemsets in Δ^- are firstly subtracted, and the processing of Δ^+ is then trivial and easy to achieve. After the new potential candidates for $D'(= db^{2,4})$ are generated, algorithm SWF can obtain new frequent itemsets with scanning $db^{2,4}$ only once.

The advantages of algorithm SWF not only can be fully exploited in the problem of incremental mining, but also are beneficial to the development of weighted mining [28].

Algorithms FI_SWF and CI_SWF

In [8], the algorithm SWF is extended by incorporating previous discovered information. Two enhancements are proposed, namely the algorithm FI_SWF (SWF with Frequent Itemset) and the algorithm CI_SWF (SWF with Candidate Itemset). These two algorithms reuse either the frequent itemsets or the candidate itemsets of previous mining task to reduce the number of new candidate itemsets. Therefore, the execution time for both algorithms can be improved and is thus better than that of algorithm SWF.

Consider the example provided earlier, it is easily observed that there are several (candidate or frequent) itemsets are identical for both the preprocessing procedure, and the incremental procedure. Therefore, if the previous mining result, i.e., the counts of the frequent itemsets, is incorporated, new counts can be obtained by only scanning the changed portions of the whole dataset in the incremental procedure.

2.3 Pattern Growth Algorithms for Incremental Mining

Both the Apriori-based algorithms and the partition-based algorithms aim at the goal of reducing the number of scans on the entire dataset when updates occur. Generally speaking, the updated portions, i.e., Δ^- and Δ^+, could be scanned several times during the levelwise generation of frequent itemsets in works belonging to these two categories. On the contrary, the algorithm FP-growth [24] (frequent pattern growth) along with the FP-tree structure adopts the *divide-and-conquer* strategy to mine association rules. The major difficulties that FP-tree cannot be directly applied to the problem of incremental mining are eased in some recent works [15, 18]. These works, in general, utilize alternative forms of FP-tree to store required data to achieve the goal of avoiding the overhead resulting from extra database scans.

Algorithms DB-tree and PotFp-tree (Potential Frequent Pattern)

In [18], two alternative forms of FP-tree are proposed to solve the problem of incremental mining. One is the algorithm DB-tree, which stores all the items in an FP-tree rather than only frequent 1-itemsets in the database. Besides, the construction of a DB-tree is exactly the same way as that of a FP-tree. Consequently, the DB-tree can be seen as an FP-tree with $S_{min} = 0$, and is thus a generalized form of FP-tree. When new transactions are added, corresponding branches of the DB-tree could be adjusted or new branches may be created. On the other hand, when old transactions are deleted, corresponding branches are also adjusted or removed. This retains the flexibility to accommodate the FP-tree to database changes when performing incremental mining. However, since the whole dataset being considered could be quite large, a much more space could be needed to maintain this DB-tree structure even a high compression is made by the nature of tree projection. This drawback may cause the problem of insufficient memory even more severe when the size of the DB-tree is far above the memory capacity.

The other algorithm proposed in [18] is the PotFp-tree, which stores only some potentially frequent items in addition to the frequent 1-itemsets at present. A tolerance parameter (or alternatively the watermark [24]) t is proposed to decide if an item is with the potential. Namely, for items with supports s where $t \leq s \leq S_{min}$ are defined to be potentially frequent items. Therefore, the need to scan the whole old database in order to update the FP-tree when updates occur is likely to be effectively reduced. Generally speaking, the PotFp-tree is seeking for the balance of required extra storage and possibility of re-scanning the dataset.

It is noted that the FP-tree is a subset of either the DB-tree or the PotFp-tree. To mine frequent itemsets, the FP-tree is firstly projected from either the DB-tree or the PotFp-tree. The frequent itemsets are then extracted from the FP-tree in the way described in [24]. The flows of utilizing algorithms DB-tree and PotFp-tree to mine association rules incrementally are best understood in the following example.

Example 8. Consider the transaction database in Fig. 6. By utilizing the algorithm DB-tree, the resulting DB-tree is shown in Fig. 15 where all the items, i.e., both the frequent ones (circles with solid lines) and the infrequent ones (circles with dotted lines), are included. It is noted that the corresponding FP-tree (which is shown in Fig. 5 earlier) forms a subset in the DB-tree and is located in the top portion of the DB-tree. For a clearer presentation, the node-links of individual items are ignored in Fig. 15.

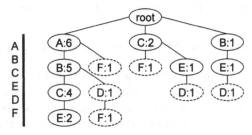

Fig. 15. Construction of the DB-tree based on the transaction database D in Fig. 6

On the other hand, by setting the tolerance $t = 30\%$ which is slightly lower than S_{min}, the PotFp-tree can be also built by including items with supports no less than t. The resulting PotFp-tree happens to be identical to the DB-tree in Fig. 15.

After constructing either the DB-tree or the PotFp-tree, the generation of frequent itemsets is very simple. Firstly, the corresponding FP-tree is extracted. Then, the frequent itemsets can be then discovered from the FP-tree in the way as described in earlier section.

When the updated database D' in Fig. 6 is being considered, some old transactions have to be removed from the DB-tree and some new transactions have to be included in the DB-tree. Note that the operations of removal and inclusion could cause the ordering of some items to be changed since some originally frequent items become infrequent while some originally infrequent items become frequent. For example, the ordering of items D and E in Fig. 16 are exchanged since item D becomes frequent while item E becomes infrequent. On the other hand, it is not so necessary to change the item order in the DB-tree when only the ordering of supports of frequent itemsets changes.

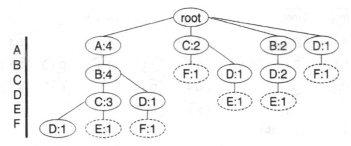

Fig. 16. Construction of the DB-tree based on the transaction database D' in Fig. 6

Algorithm FELINE (FrEquent/Large Patterns mINing with CATS trEe)

In [15], another alternative form of FP-tree is proposed to aim at the problem of incremental mining. Namely, the CATS tree (compressed and arranged transaction sequences tree) is with several common properties of FP-tree. Also, the CATS tree and the DB-tree are very alike since they both store all the items no matter they are frequent or not. This feature enables the CATS tree to be capable of avoiding re-scans of databases when updates occur. However, the construction of the CATS tree is different to that of an FP-tree and a DB-tree. The items along a path in the CATS-tree can be re-ordered to achieve locally optimized. Specifically, the FP-tree is built based on the ordering of global supports of all frequent items, while the CATS-tree is built based on the ordering of local supports of items in its path. Consequently, the CATS-tree is sensitive to the ordering of input transactions, making the CATS-tree not optimal since no preliminary analysis is done before the tree construction. This in turns can reduce the data scan required to only once, showing the advantage of this algorithm.

Example 9. Consider the transaction database in Fig. 6. The construction of CATS-tree for the nine transactions in D is shown in Fig. 17.

By removing Δ^- ($T_1 \tilde{} T_3$) and adding Δ^+ ($T_{10} \tilde{} T_{12}$), the incremental modification of the CATS-tree is shown in Fig. 18. Since no specific ordering is required when there is no conflict among items in a path, the CATS-tree can be built and modified easily.

3 Mining Frequent Patterns from Data Streams

In several emerging applications, data is in the form of continuous *data streams*, as opposed to finite stored databases. Examples include stock tickers, network traffic measurements, web logs, click streams, data captured from

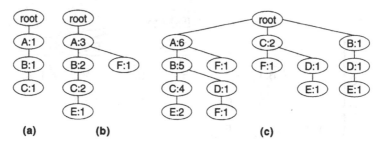

Fig. 17. Construction of the CATS-tree: (a) T_1 is added; (b) T_2 and T_3 are added; (c) T_4~T_9 are added

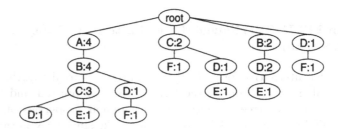

Fig. 18. The modified CATS-tree when database updates occur

sensor networks and call records. Consequently, mining association rules (or frequent patterns) incrementally in such a data stream environment could encounter more challenges than that in a static database with occasional updates. To explore related issues, several techniques proposed in recent works are summarized in this section.

3.1 Mining from Online Transaction Flows

For data stream applications, the volume of data is usually too huge to be stored on permanent devices or to be scanned thoroughly for more than once. It is hence recognized that both approximation and adaptability are key ingredients for executing queries and performing mining tasks over rapid data streams. With the computation model presented in Fig. 19 [19], a stream processor and the synopsis maintenance in memory are two major components for generating results in the data stream environment. Note that a buffer can be optionally set for temporary storage of recent data from data streams.

For time-variant databases, there is a strong demand for developing an efficient and effective method to mine frequent patterns [17]. However, most methods which were designed for a traditional database cannot be directly applied to a dynamic data stream due not only to the high complexity of mining temporal patterns but also to the pass-through nature of data streams. Without loss of generality, a typical market-basket application is used here for illustrative purposes. The transaction flow in such an application is shown

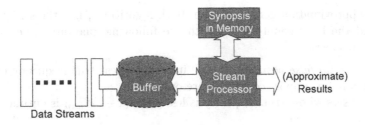

Fig. 19. Computation model for data streams

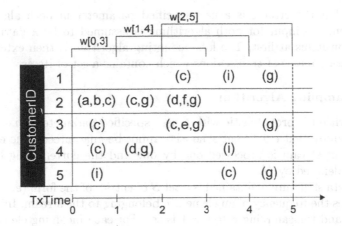

Fig. 20. An example of online transaction flows

in Fig. 20 where items a to g stand for items purchased by customers. For example, the third customer bought item c during time $t = [0, 1)$, items c, e and g during $t = [2, 3)$, and item g during $t = [4, 5)$. It can be seen that in such a data stream environment it is intrinsically very difficult to conduct the frequent pattern identification due to the limited time and space constraints.

3.2 Approximate Frequency Counts

As is mentioned in Sect. 1 in this chapter, the critical step to discover association rules lies in the frequency counting for all frequent itemsets. To obtain precise supports for each itemset, the fastest algorithms to date must employ two data scans. However, for the data stream environment, it is required to have only one data scan for online (and incremental) maintenance of association rules. Due to this limitation in processing data streams, two algorithms are proposed in [34] to generate approximate frequency counts for itemsets. One algorithm is based on the sampling technique to obtain a probabilistic error bound while the other is based on the data partitioning technique to obtain a deterministic error bound.

The approximation generated from both algorithms, i.e., the sticky sampling and the lossy counting, are with the following guarantees, i.e., the ϵ-deficient synopsis:

1. There are *no false positives*, i.e., all itemsets whose true frequency exceed the minimum support constraint are output.
2. No itemsets whose true frequency is less than $(1 - \epsilon)S_{min}$ is output where ϵ is a error parameter.
3. Estimated frequencies are less than the true frequencies by at most ϵ (in percentage).

Note that the error ϵ is a user-specified parameter in both algorithms. In addition, the input for both algorithms is assumed to be a data stream of singleton items at first. The lossy counting algorithm is then extended to handle data streams of transactions which contains a set of items.

Sticky Sampling Algorithm

This algorithm is probabilistic with a user-specified parameter δ, i.e., probability of failure that ϵ-deficient synopsis cannot be guaranteed. The elements of the input stream is processed one by one and the current length of the stream is denoted by N.

The data structure maintained is a set S of entries of the form (e, f), where f estimates the frequency of an element e belonging to the stream. Initially, S is empty, and the sampling rate $r = 1$ is set. For each incoming element e, if an entry for e already exists in S, the corresponding frequency f is increased by one. Otherwise, if the element is selected (with probability $\frac{1}{r}$) by sampling, an entry $(e, 1)$ is added to S.

The sampling rate r varies over the lifetime of a stream. Let $t = \frac{1}{\epsilon} \log(\frac{1}{S \times \delta})$. The first $2t$ elements are sampled at rate $r = 1$, the next $2t$ elements are sampled at rate $r = 2$, the next $4t$ elements are sampled at rate $r = 4$, and so on. Whenever the sampling rate changes, the entries in S is scanned and updated through a coin-tossing process, i.e., f is diminished by one for every unsuccessful outcome until the repeatedly toss is successful. If f becomes 0 during this process, the corresponding entry is deleted from S.

When a user requests the list of frequent items, the entries in S with $f \geq N(1 - \epsilon)S_{min}$ are output. It is proved in [34] that true supports of these frequent items are underestimated by at most ϵ with probability $1 - \delta$.

Intuitively, S sweeps over the stream like a magnet, attracting all elements which already have an entry in S. The sample rate r increases logarithmically proportional to the size of the stream. The most advantage of this algorithm is that the space complexity is independent of the current length of the stream.

Lossy Counting Algorithm

In contrast to the sticky sampling algorithm, the incoming data stream is conceptually divided into buckets of width $w = \lceil \frac{1}{\epsilon} \rceil$ transactions in the lossy counting algorithm. Each bucket is labeled with a bucket id, starting from (1). The data structure maintained is a set S of entries of the form $(e, f, \triangle)$, where f estimates the frequency of an element e belonging to the stream and $\triangle$ is the maximum possible error in f. Initially, S is empty. Whenever a new element e arrives, if the corresponding entry already exists, its frequency f is increased by one. Otherwise, a new entry of the form $(e, 1, b_{current} - 1)$ is created where $b_{current}$ denotes the current bucket id $(= \lceil \frac{N}{w} \rceil)$. The idea behind the lossy counting algorithm is to set a criterion that *a tracked item should at least occur once (in average) in each bucket*. Consequently, the maximum possible error $\triangle = b_{current} - 1$ for an entry added in bucket $b_{current}$. At bucket boundaries, S is pruned by deleting entries satisfying $f + \triangle \leq b_{current}$.

When a user requests the list of frequent items, the entries in S with $f \geq N(1 - \epsilon)S_{min}$ are output. It is proved in [34] that true supports of these frequent items are underestimated by at most ϵ.

To further discover frequent itemsets from data streams of transactions, the lossy counting algorithm is also extended in [34]. A significant difference is that the input stream is not processed transaction by transaction. Instead, the available main memory is filled with as many buckets of transactions as possible, and such a *batch* of transactions is processed together. If the number of buckets in main memory in the current batch being processed is β, then the new entry $(set, 1, b_{current} - \beta)$ is created when the itemset set is not present in S. Other methods are nearly identical to the original ones for streams of singleton items. Similarly, when a user requests the list of frequent itemsets, the entries in S with $f \geq N(1 - \epsilon)S_{min}$ are output.

3.3 Finding Recent Frequent Itemsets Adaptively

In order to differentiate the information of recently generated transactions from the obsolete information of old transactions, a weighting scheme and the *estDec* method for discovering recent frequent itemsets from data streams are developed in [9]. A delay rate d $(0 < d < 1)$ is introduced in this weighting scheme that the weight value of an old transaction T_k with id k (starting from 1) is $d^{(current-k)}$ where *current* is the current transaction id being processed. Consequently, when the first transaction is looked up, the total number of transactions is obviously 1 since there is no previous transaction whose weight should be decayed. As time advances, when k transactions are processed, the total number of transactions can be expressed by

$$|D|_k = d^{k-1} + d^{k-2} + \cdots + d + 1 = \frac{1 - d^k}{1 - d}.$$

Also, it is obvious that

$$|D|_k = \begin{cases} 1, & \text{if } k = 1, \\ |D|_{k-1} \times d + 1, & \text{if } k \geq 2. \end{cases}$$

Moreover, the *estDec* method proposed in [9] maintains a monitoring lattice to track frequent itemsets. Each node in this lattice is an entry of the form $(cnt, err, MRtid)$ for a corresponding itemset X, where cnt is the occurrence count, err is the maximum error count and $MRtid$ is the transaction id of the most recent transaction containing X. The critical step in this approach is to update occurrence counts for tracked itemsets. If an itemset X is contained in the current transaction T_k and the previous entry is $(cnt_{pre}, err_{pre}, MRtid_{pre})$, its current entry $(cnt_k, err_k, MRtid_k)$ can be updated as

$$cnt_k = cnt_{pre} \times d^{(k - MRtid_{pre})} + 1 \, ,$$
$$err_k = err_{pre} \times d^{(k - MRtid_{pre})}, \text{ and}$$
$$MRtid_k = k \, .$$

Therefore, when the updated support of an itemset becomes less than the threshold, i.e., $\frac{cnt_k}{|D|_k} < S_{min}$, the entry corresponding to this itemset is dropped. The threshold for pruning here can be adjusted to be slightly less than the minimum support to reduce the overhead of reinserting a previously deleted itemset.

For inserting newly discovered itemsets, another feature of the *estDec* method is that only when all of the $(n-1)$-subsets of an n-itemset are frequent, this n-itemset is able to become a candidate. In other words, a frequent itemset may not be identified immediately since all its subsets have to become frequent first, leading to a possible delayed insertion.

3.4 A Scheme for Mining Temporal Patterns

In a temporal database, frequent patterns are usually targets of mining tasks. In addition to the mining of association rules, similar techniques can be extended to facilitate the mining of other types of temporal patterns.

Mining of Temporal Patterns

Prior works have developed several models of temporal patterns, including the inter-transaction association rule [33], the causality rule [29], the episode [36] and the sequential pattern [4]. Note that the very difference among the above temporal patterns lies the ordering of occurrences. Mining of sequences corresponds to the one with strict order of events, while mining inter-transaction associations corresponds to the one without limitation on order of events. Between these two extremes, mining of causalities and episodes mainly emphasizes the ordering of triggering events and consequential events. Although the mining procedures may vary when being applied to discover different types of temporal patterns, a typical Apriori framework is commonly adopted. By utilizing the downward closure property in this framework [52], a fundamental issue of mining frequent temporal patterns is the frequency counting of patterns.

In many applications, a time-constraint is usually imposed during the mining process to meet the respective constraint. Specifically, the sliding window model is introduced here, i.e., data expires after exactly N time units after its arrival where N is the user-specified window size. Consequently, a temporal pattern is frequent if its support, i.e., occurrence frequency, in the current window is no less than the threshold.

Support Framework for Temporal Patterns

To evaluate the importance of a temporal pattern, the support, i.e., occurrence frequency, is a metric commonly used. However, the definition of support for a pattern may vary from one application to another. Consider again the market-basket database as an example. In mining sequential patterns [4], all the transactions of a customer can be viewed as a sequence together and the support for a sequential pattern is the fraction of customers whose purchasing sequences contain that pattern. Analogously, we have the model of frequency counting in mining causality rules [29]. On the other hand, in mining inter-transaction association rules [33], the repetitive occurrences of a pattern from an identical customer are counted cumulatively. Moreover, when the sliding window constraint is introduced in mining episodes [36], the support is defined to be the fraction of windows in which an episode occurs.

To deal with data streams, problems arise due to different support definitions. Specifically, since it is not possible to store all the historical data in the memory, to identify repetitive occurrences of a pattern is difficult. As a result, it is very important to properly formulate the support of temporal patterns. With the sliding window model, the support or the occurrence frequency of a temporal pattern X at a specific time t is denoted by *the ratio of the number of customers having pattern X in the current time window to the total number of customers*.

Example 10. Given the window size $N = 3$, three sliding windows, i.e., $w[0,3]$, $w[1,4]$ and $w[2,5]$, are shown in Fig. 20 for the transaction flows. For example, according to the support definition, supports of the inter-transaction itemset $\{c, g\}$ from TxTime $t = 1$ to $t = 5$ are obtained as in Table 1. Accordingly, the support variations can be presented as a time series as shown in Fig. 21. For simplicity, the total number of customers is a constant in this example, and could be a variable as time advances in real applications.

Algorithm FTP-DS (Frequent Temporal Patterns of Data Streams)

Note that although the approaches proposed in [34] work successfully for counting supports of singleton items, as the number of items increases, the

Table 1. The support values of the inter-transaction itemset $\{c, g\}$

TxTime		Occurrence(s) of $\{c, g\}$	Support
$t = 1$	w[0,1]	none	0
$t = 2$	w[0,2]	CustomerID $= \{2, 4\}$	$2/5 = 0.4$
$t = 3$	w[0,3]	CustomerID $= \{2, 3, 4\}$	$3/5 = 0.6$
$t = 4$	w[1,4]	CustomerID $= \{2, 3\}$	$2/5 = 0.4$
$t = 5$	w[2,5]	CustomerID $= \{1, 3, 5\}$	$3/5 = 0.6$

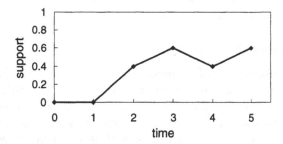

Fig. 21. Support variations of the inter-transaction itemset $\{c, g\}$

rapidly increasing number of temporal patterns can cause problems of prohibitive storage and computing overheads. Explicitly, if the lossy counting scheme proposed in [34] is adopted, patterns with supports no less than ϵ are maintained during the mining process to guarantee the error range to be within ϵ. However, since the threshold ϵ, whose value could be one-tenth of $S_{\min}$, is usually too small to filter out uninteresting patterns, the storage space could be quite large.

To address this point, the algorithm FTP-DS is developed in [45]. With the sliding window model employed, only the occurrences of singleton items are being counted in the first time window. After the counting iteration, frequent items which have supports no less than the specified threshold are identified. These frequent items can be joined to generate candidate patterns of size two, which are then being counted in later iterations. After some patterns of size two are identified frequent, the candidate patterns of size three are generated and counted subsequently. As a result, longer candidate patterns are gradually generated, counted and verified to be frequent during the counting iterations. From the downward closure property [52], it follows that only patterns whose sub-patterns are all frequent are taken as candidates and to be counted subsequently.

Example 11. Given the support threshold $S_{min} = 40\%$, the window size $N = 3$ and the transaction flows in Fig. 20, suppose the frequent inter-transaction associations are being generated. Since the temporal order is not required for inter-transaction associations, we have the frequent temporal itemset generation shown in Table 2. The support calculation of each itemset is the same

Table 2. Generation of frequent temporal itemsets ($S_{min} = 40\%$)

$t = 1$	$t = 2$	$t = 3$
{c} 0.6/1	{c} 1.2/2	{c} 2/3
(a)	{g} 0.4/1	{d} 0.4/1
	(b)	{g} 1/2
		{c,g} 0.6/1
		(c)

$t = 4$	$t = 5$
{c} 2.8/4	
{d} 0.8/2	
{g} 1.6/3	{c} 3.4/5
{i} 0.4/1	{g} 2.4/4
{c,g} 1/2	{i} 0.8/2
{d,g} 0.4/1	{c,g} 1.6/3
(d)	(e)

as the process in Table 1. The averaged support value is represented by (accumulated supports over windows)/(number of recorded windows) in Table 2 where only itemsets with supports no less than $S_{min} = 40\%$ are listed. In addition, frequent itemsets generated in previous time window are used to generate longer candidates to be examined later. For example, according to Definition 1, the supports of itemset {d} during $t = 1$ to $t = 5$ are 0, 0.2, 0.4, 0.4 and 0.2, respectively. Not until $t = 3$ does the support value satisfy the threshold, meaning that itemset {d} is being tracked since $t = 3$ as shown in Table 2(c) and Table 2(d). However, the averaged support of itemset {d} is $(0.4 + 0.4 + 0.2)/3 = 1.0/3$ which is less than the $S_{min} = 40\%$, making this itemset discarded in Table 2(e). Moreover, the inclusion of itemset {d} at $t = 3$ results in the generation of related candidate itemsets, i.e., {c,d} and {d,g}, to be examined at $t = 4$. However, only itemset {d, g} satisfies the support threshold and is included in Table 2(d).

It can be seen that this approach can generate patterns of various lengths as time advances. However, as pointed out earlier, since a pattern is not taken as a candidate to accumulate its occurrence counts before all its subsets are found frequent, the phenomenon of *delayed pattern recognition* exists, i.e., some patterns are recognized with delays due to the candidate forming process in the data stream. For example, since items c and g are not both identified frequent until $t = 2$, the candidate itemset {c, g} is generated and counted at $t = 3$. However, it can be verified from Table 1 that {c, g} is actually frequent at $t = 2$. Therefore, a delay of one time unit is introduced for discovering this itemset {c, g}. It is worth mentioning that only long transient frequent patterns could be neglected in this pattern generation process. As time advances, patterns with supports near the threshold will be further examined and identified to be frequent if so qualified.

With a support threshold S_{min} to filter out uninteresting patterns, only new patterns whose frequencies in the current time unit meet this threshold are being recorded. Supports of existing patterns, i.e., patterns which were already being recorded in previous time unit, are updated according to their support values in the current time unit. Note that, as time advances, patterns whose averaged supports fall below the threshold are removed from the records. Therefore, only frequent patterns are monitored and recorded. In practice, since a frequent pattern is not always with a very steady frequency, the above mentioned removal can be delayed to allow a pattern whose statistics are already recorded to stay in the system longer with an expectation that this pattern will become frequent again soon. This will be an application-dependent design alternative.

4 Concluding Remarks

The mining of association rules among huge amounts of transactional data can provide very valuable in information on customer buying behavior, and thus improve the quality of business decisions. This *market basket analysis* is a crucial area of application when performing data mining techniques. Specifically, the mining of association rules is a two-step process where the first step is to find all *frequent* itemsets satisfying the minimum support constraint, and the second step is to generate association rules satisfying the minimum confidence constraint from the frequent itemsets. Since to identify the frequent itemsets is of great computational complexity, usually the problem of mining association rules can be reduced to the problem of discovering frequent itemsets.

According to the numerous works on mining association rules, the approaches utilized for efficiently and effectively discovering frequent itemsets can be categorized into three types. Namely, they are the *Apriori-based* algorithms, the *partition-based* algorithms and the *pattern growth* algorithms. The Apriori-based algorithms adopt a levelwise style of itemset generation while the database may need to be scanned for a few times. The partition-based algorithms firstly divide the whole database into several partitions and then perform the mining task on each partition separably and iteratively. The pattern growth algorithms usually project the whole database into a highly compressed tree structure and then utilize the divide-and-conquer strategy to mine required frequent itemsets.

Recent important applications have called for the need of incremental mining. This is due to the increasing use of the record-based databases whose data are being continuously added. As time advances, old transactions may become obsolete and are thus discarded from the database of interests to people. For the purpose of maintaining the discovered association rules, some previously valid rules may become invalid while some other new rules may show up. To efficiently reflect these changes, how to utilize discovered information well is undoubtedly an important issue. By extending the techniques used in mining ordinary association rules, several works reached a great achievement by developing either the Apriori-based algorithms, the partition-based algorithms and the pattern-growth algorithms.

Moreover, a challenging and interesting area of conducting the mining capabilities in a data stream environment is becoming popular in data mining society. To further extend the concept of mining and maintaining association rules from data streams, some recent works are also included in this chapter.

References

1. R. C. Agarwal, C. C. Aggarwal, and V. V. V. Prasad. A Tree Projection Algorithm for Generation of Frequent Itemsets. *Journal of Parallel and Distributed Computing (Special Issue on High Performance Data Mining)*, 61(3):350–371, 2001.

2. R. Agrawal, T. Imielinski, and A. Swami. Mining Association Rules between Sets of Items in Large Databases. *Proceedings of the 1993 ACM SIGMOD International Conference on Management of Data*, pp. 207–216, May 1993.

3. R. Agrawal and R. Srikant. Fast Algorithms for Mining Association Rules in Large Databases. *Proceedings of the 20th International Conference on Very Large Data Bases*, pp. 478–499, September 1994.

4. R. Agrawal and R. Srikant. Mining Sequential Patterns. *Proceedings of the 11th International Conference on Data Engineering*, pp. 3–14, March 1995.

5. J. M. Ale and G. Rossi. An Approach to Discovering Temporal Association Rules. *Proceedings of the 2000 ACM Symposium on Applied Computing*, pp. 294–300, March 2000.

6. N. F. Ayan, A. U. Tansel, and M. E. Arkun. An Efficient Algorithm to Update Large Itemsets with Early Pruning. *Proceedings of the 5th ACM SIGKDD International Conference on Knowledge Discovery and Data Mining*, pp. 287–291, August 1999.

7. S. Brin, R. Motwani, J. D. Ullman, and S. Tsur. Dynamic Itemset Counting and Implication Rules for Market Basket Data. *Proceedings of the 1997 ACM SIGMOD International Conference on Management of Data*, pp. 255–264, May 1997.

8. C.-H. Chang and S.-H. Yang. Enhancing SWF for Incremental Association Mining by Itemset Maintenance. *Proceedings of the 7th Pacific-Asia Conference on Knowledge Discovery and Data Mining*, April 2003.

9. J. H. Chang and W. S. Lee. Finding Recent Frequent Itemsets Adaptively over Online Data Streams. *Proceedings of the 9th ACM SIGKDD International Conference on Knowledge Discovery and Data Mining*, pp. 487–492, August 2003.

10. M.-S. Chen, J. Han, and P. S. Yu. Data Mining: An Overview from Database Perspective. *IEEE Transactions on Knowledge and Data Engineering*, 8(6):866–883, December 1996.

11. M.-S. Chen, J.-S. Park, and P. S. Yu. Efficient Data Mining for Path Traversal Patterns. *IEEE Transactions on Knowledge and Data Engineering*, 10(2):209–221, April 1998.

12. X. Chen and I. Petr. Discovering Temporal Association Rules: Algorithms, Language and System. *Proceedings of the 16th International Conference on Data Engineering*, pp. 306. February 2000.

13. D. Cheung, J. Han, V. Ng, and C. Y. Wong. Maintenance of Discovered Association Rules in Large Databases: An Incremental Updating Technique. *Proceedings of the 12th International Conference on Data Engineering*, pp. 106–114, February 1996.

14. D. Cheung, S. D. Lee, and B. Kao. A General Incremental Technique for Updating Discovered Association Rules. *Proceedings of the Fifth International Conference On Database Systems for Advanced Applications*, pp. 185–194, April 1997.

15. W. Cheung and O. R. Zaiane. Incremental Mining of Frequent Patterns without Candidate Generation or Support Constraint. *Proceedings of the 7th International Database Engineering and Application Symposium*, pp. 111–116, July 2003.

16. E. Cohen, M. Datar, S. Fujiwara, A. Gionis, P. Indyk, R. Motwani, J. D. Ullman, and C. Yang. Finding Interesting Associations without Support Pruning. *IEEE Transactions on Knowledge and Data Engineering*, 13(1):64–78, January/February 2001.

17. G. Das, K.-I. Lin, H. Mannila, G. Renganathan, and P. Smyth. Rule Discovery from Time Series. *Proceedings of the 4th ACM SIGKDD International Conference on Knowledge Discovery and Data Mining*, pp. 16–22, August 1998.

18. C. I. Ezeife and Y. Su. Mining Incremental Association Rules with Generalized FP-Tree. *Proceedings of the 15th Conference of the Canadian Society for Computational Studies of Intelligence on Advances in Artificial Intelligence*, pp. 147–160, May 2002.

19. M. N. Garofalakis, J. Gehrke, and R. Rastogi. Querying and Mining Data Streams: You Only Get One Look. *Proceedings of the 2002 ACM SIGMOD International Conference on Management of Data*, p. 635, June 2002.

20. J. Han, G. Dong, and Y. Yin. Efficient Mining of Partial Periodic Patterns in Time Series Database. *Proceeding of the 15th International Conference on Data Engineering*, pp. 106–115, March 1999.

21. J. Han and Y. Fu. Discovery of Multiple-Level Association Rules from Large Databases. *Proceedings of the 21th International Conference on Very Large Data Bases*, pp. 420–431, September 1995.

22. J. Han, L. V. S. Lakshmanan, and R. T. Ng. Constraint-Based, Multidimensional Data Mining. *COMPUTER (Special Issue on Data Mining)*, pp. 46–50, 1999.

23. J. Han, J. Pei, B. Mortazavi-Asl, Q. Chen, U. Dayal, and M. C. Hsu. FreeSpan: Frequent Pattern-Projected Sequential Pattern Mining. *Proceedings of the 6th ACM SIGKDD International Conference on Knowledge Discovery and Data Mining*, pp. 355–359, August 2000.

24. J. Han, J. Pei, and Y. Yin. Mining Frequent Patterns without Candidate Generation. *Proceedings of the 2000 ACM-SIGMOD International Conference on Management of Data*, pp. 1–12, May 2000.

25. J. Hipp, U. Guntzer, and G. Nakhaeizadeh. Algorithms for Association Rule Mining – A General Survey and Comparison. *SIGKDD Explorations*, 2(1):58–64, July 2000.

26. L. V. S. Lakshmanan, R. Ng, J. Han, and A. Pang. Optimization of Constrained Frequent Set Queries with 2-Variable Constraints. *Proceedings of the 1999 ACM SIGMOD International Conference on Management of Data*, pp. 157–168, June 1999.

27. C.-H. Lee, C.-R. Lin, and M.-S. Chen. Sliding-Window Filtering: An Efficient Algorithm for Incremental Mining. *Proceeding of the ACM 10th International*

Conference on Information and Knowledge Management, pp. 263–270, November 2001.

28. C.-H. Lee, J.-C. Ou, and M.-S. Chen. Progressive Weighted Miner: An Efficient Method for Time-Constraint Mining. *Proceedings of the 7th Pacific-Asia Conference on Knowledge Discovery and Data Mining*, pp. 449–460, April 2003.

29. C.-H. Lee, P. S. Yu. and M.-S. Chen. Causality Rules: Exploring the Relationship between Triggering and Consequential Events in a Database of Short Transactions. *Proceedings of the 2nd SIAM International Conference on Data Mining*, pp. 449–460, April 2002.

30. S. D. Lee, D. W. Cheung, and B. Kao. Is Sampling Useful in Data Mining? A Case Study in the Maintenance of Discovered Association Rules. *Data Mining and Knowledge Discovery*, 2(3):233–262, September 1998.

31. J. L. Lin and M. H. Dunham. Mining Association Rules: Anti-Skew Algorithms. *Proceedings of the 14th International Conference on Data Engineering*, pp. 486–493, February 1998.

32. B. Liu, W. Hsu, and Y. Ma. Mining Association Rules with Multiple Minimum Supports. *Proceedings of the 5th ACM SIGKDD International Conference on Knowledge Discovery and Data Mining*, pp. 337–341, August 1999.

33. H. Lu, J. Han, and L. Feng. Stock Movement Prediction and N-Dimensional Inter-Transaction Association Rules. *Proceedings of the 1998 ACM SIGMOD Workshop on Research Issues on Data Mining and Knowledge Discovery*, pp. 12:1–12:7, June 1998.

34. G. S. Manku and R. Motwani. Approximate Frequency Counts over Streaming Data. *Proceedings of the 28th International Conference on Very Large Data Bases*, pp. 346–357. August 2002.

35. H. Mannila, H. Toivonen, and A. I. Verkamo. Efficient Algorithms for Discovering Association Rules. *Proceedings of AAAI Workshop on Knowledge Discovery in Databases*, pp. 181–192, July 1994.

36. H. Mannila, H. Toivonen, and A. I. Verkamo. Discovery of Frequent Episodes in Event Sequences. *Data Mining and Knowledge Discovery*, 1(3):259–289, September 1997.

37. A. Mueller. Fast Sequential and Parallel Algorithms for Association Rule Mining: A Comparison. *Technical Report CS-TR-3515, Dept. of Computer Science, Univ. of Maryland, College Park. MD*, 1995.

38. R. T. Ng and J. Han. Efficient and Effective Clustering Methods for Spatial Data Mining. *Proceedings of the 20th International Conference on Very Large Data Bases*, pp. 144–155. September 1994.

39. J.-S. Park, M.-S. Chen, and P. S. Yu. An Effective Hash-Based Algorithm for Mining Association Rules. *Proceedings of the 1995 ACM-SIGMOD International Conference on Management of Data*, pp. 175–186, May 1995.

40. J.-S. Park, M.-S. Chen, and P. S. Yu. Using a Hash-Based Method with Transaction Trimming for Mining Association Rules. *IEEE Transactions on Knowledge and Data Engineering*. 9(5):813–825, October 1997.

41. J. Pei and J. Han. Can We Push More Constraints into Frequent Pattern Mining? *Proceedings of the 6th ACM SIGKDD International Conference on Knowledge Discovery and Data Mining*. pp. 350–354, August 2000.

42. J. Pei. J. Han, B. Mortazavi-Asl. H. Pinto, Q. Chen, U. Dayal, and M. C. Hsu. PrefixSpan: Mining Sequential Patterns Efficiently by Prefix-Projected Pattern Growth. *Proceedings of the 17th International Conference on Data Engineering*, pp. 215–226, April 2001.

43. A. Savasere, E. Omiecinski, and S. Navathe. An Efficient Algorithm for Mining Association Rules in Large Databases. *Proceedings of the 21th International Conference on Very Large Data Bases*, pp. 432–444, September 1995.
44. R. Srikant and R. Agrawal. Mining Generalized Association Rules. *Proceedings of the 21th International Conference on Very Large Data Bases*, pp. 407–419, September 1995.
45. W.-G. Teng, M.-S. Chen, and P. S. Yu. A Regression-Based Temporal Pattern Mining Scheme for Data Streams. *Proceedings of the 29th International Conference on Very Large Data Bases*, pp. 93–104, September 2003.
46. S. Thomas, S. Bodagala, K. Alsabti, and S. Ranka. An Efficient Algorithm for the Incremental Updation of Association Rules in Large Databases. *Proceedings of the 3rd ACM SIGKDD International Conference on Knowledge Discovery and Data Mining*, pp. 263–266, August 1997.
47. H. Toivonen. Sampling Large Databases for Association Rules. *Proceedings of the 22th International Conference on Very Large Data Bases*, pp. 134–145, September 1996.
48. A. Veloso, B. Possas, W. M. Jr., and M. B. de Carvalho. Knowledge Management in Association Rule Mining. *Workshop on Integrating Data Mining and Knowledge Management (in conjuction with ICDM2001)*, November 2001.
49. K. Wang, Y. He, and J. Han. Mining Frequent Itemsets Using Support Constraints. *Proceedings of the 26th International Conference on Very Large Data Bases*, pp. 43–52, September 2000.
50. K. Wang, S. Q. Zhou, and S. C. Liew. Building Hierarchical Classifiers Using Class Proximity. *Proceedings of the 25th International Conference on Very Large Data Bases*, pp. 363–374, September 1999.
51. C. Yang, U. Fayyad, and P. Bradley. Efficient Discovery of Error-Tolerant Frequent Itemsets in High Dimensions. *Proceedings of the 7th ACM SIGKDD International Conference on Knowledge Discovery and Data Mining*, pp. 194–203, August 2001.
52. M. J. Zaki, S. Parthasarathy, M. Ogihara, and W. Li. New Algorithms for Fast Discovery of Association Rules. *Proceedings of the 3rd ACM SIGKDD International Conference on Knowledge Discovery and Data Mining*, pp. 283–286, August 1997.
53. Z. Zhou and C. I. Ezeife. A Low-Scan Incremental Association Rule Maintenance Method Based on the Apriori Property. *Proceedings of the 14th Conference of the Canadian Society for Computational Studies of Intelligence on Advances in Artificial Intelligence*, pp. 26–35, June 2001.

Mining Association Rules from Tabular Data Guided by Maximal Frequent Itemsets

Q. Zou[1], Y. Chen[1], W.W. Chu[1], and X. Lu[2]

[1] Computer Science Department, University of California, Los Angeles, California, 90095
{zou,chenyu,wwc}@cs.ucla.edu
[2] Shandong University, Jinan, China
luxc@sdu.edu.cn

Summary. We propose the use of maximal frequent itemsets (MFIs) to derive association rules from tabular datasets. We first present an efficient method to derive MFIs directly from tabular data using the information from previous search, known as tail information. Then we utilize tabular format to derive MFI, which can reduce the search space and the time needed for support-counting. Tabular data allows us to use spreadsheet as a user interface. The spreadsheet functions enable users to conveniently search and sort rules. To effectively present large numbers of rules, we organize rules into hierarchical trees from general to specific on the spreadsheet Experimental results reveal that our proposed method of using tail information to generate MFI yields significant improvements over conventional methods. Using inverted indices to compute supports for itemsets is faster than the hash tree counting method. We have applied the proposed technique to a set of tabular data that was collected from surgery outcomes and that contains a large number of dependent attributes. The application of our technique was able to derive rules for physicians in assisting their clinical decisions.

1 Introduction

Many algorithms have been proposed on mining association rules in the past decade. Most of them are based on the transaction-type dataset. In the real world, a huge amount of data has been collected and stored in tabular datasets, which usually are dense datasets with a relatively small number of rows but a large number of columns. To mine a tabular dataset, previous approaches required transforming it into a transaction-type dataset in which column structures are removed. In contrast, we propose a new method that takes advantage of the tabular structure and that can mine association rules directly from tabular data.

Many previous approaches take four steps to generate association rules from a tabular dataset: (1) transforming a tabular dataset T into a transaction-type dataset D; (2) mining frequent itemsets (FI) or frequent closed itemsets (FCI) from D; (3) Generating rules from FI or FCI; (4) presenting rules to the user. This strategy achieves a certain degree of success but has three shortcomings:

First, transforming a tabular dataset into a transaction-type dataset increases the search space and the time required for support-counting, since column structures are removed. Figure 1a shows a table with five rows and five distinct columns, column A to E. The *Occ* column indicates the number of occurrences of a row. Thus, the *Occ* value can be used to count support of an itemset. For example, given a minimal support 3, "$A = 3$" is a frequent item since its support is 4, which can be obtained by adding the *Occ* values of rows 3 and 4. Likewise, both "$A = 1$" and "$A = 2$" have a support of 2, so they are infrequent. Figure 1b shows the corresponding transaction-type dataset where each column value is mapped to an item, e.g. $A = 1$ to a_1. There are 9 frequent items in Fig. 1b. Since any combination of the nine items can be a possible FI, FCI, or MFI, the search space is as large as $2^9 = 512$. In contrast, by keeping column structure, the search space in Fig. 1a can be significantly reduced to 162 (2*3*3*3*3) since the combinations of items on the same column can be excluded. Note that in Apriori-like level-wised approaches, keeping column structures reduces the number of candidates generated at all levels. For instance, in Fig. 1b, Apriori-like approaches would generate $b_1 d_1 d_2$ as a 3-item candidate since $b_1 d_1$ and $b_1 d_2$ are frequent. By using column constraints, $d_1 d_2$ are in the same column, so any candidate containing $d_1 d_2$ can be pruned. Furthermore, keeping column structure reduces the time needed for support-counting. For example, in vertical data representation, the supports of all items in one column can be counted in a single scan of the column.

Row No.	A	B	C	D	E	Occ
1	~~1~~	2	1	1	1	2
2	~~2~~	1	2	1	1	1
3	3	2	2	2	1	2
4	3	1	2	2	2	2
5	~~2~~	1	1	2	2	1
Frequent item#	1	2	2	2	2	

TID	Item set	Occ
1	$a_1 b_2 c_1 d_1 e_1$	2
2	$a_2 b_1 c_2 d_1 e_1$	1
3	$a_3 b_2 c_2 d_2 e_1$	2
4	$a_3 b_1 c_2 d_2 e_2$	2
5	$a_2 b_1 c_1 d_2 e_2$	1
	Totally 9 frequent items	

(a). A source tabulardata   (b). A transaction-type dataset

Fig. 1. Tabular data *vs.* transaction data

Second, in certain situations, mining frequent itemsets (FI) or frequent closed itemsets (FCI) becomes difficult since the number of FIs or FCIs can be very large. Researchers have realized this problem and recently proposed a number of algorithms for mining maximal frequent itemsets (MFI) [3,4,6,21], which achieve orders of magnitudes of improvement over mining FI or FCI.

When mining a dense tabular dataset, it is desirable to mine MFIs first and use them as a roadmap for rule mining.

Finally, a challenging problem is how to present a large number of rules effectively to domain experts. Some approaches prune and summarize rules into a small number of rules [10]; some cluster association rules into groups [14]. While reducing the number of rules or putting them into several clusters is desirable in many situations, these approaches are inconvenient when an expert needs to inspect more details or view the rules in different ways. Other approaches [11,17] developed their own tools or a new query language similar to SQL to select rules, which provides great flexibility in studying rules. But, in some situations, domain experts are reluctant to learn such a query language.

In this Chapter, we present a new approach to address these problems.

- We propose a method that can generate MFIs directly from tabular data, eliminating the need for conversion to transaction-type datasets. By taking advantage of column structures, our method can significantly reduce the search space and the support counting time.
- We introduce a framework that uses MFIs as a roadmap for rule mining. A user can select a subset of MFIs to including certain attributes known as targets (e.g., surgery outcomes) in rule generation.
- To derive rules from MFIs, we propose an efficient method for counting the supports for the subsets of user-selected MFIs. We first build inverted indices for the collection of itemsets and then use the indices for support counting. Experimental results show that our approach is notably faster than the conventional hash tree counting method.
- To handle the large number of rules generated, we hierarchically organize rules into trees and use spreadsheet to present the rule trees. In a rule tree, general rules can be extended into more specific rules. A user can first exam the general rules and then extend to specific rules in the same tree. Based on spreadsheet's rich functionality, domain experts can easily filter or extend branches in the rule trees to create the best view for their interest.

The organization of this chapter is as follows. Section 1 discusses related works. Section 2 presents the framework of our proposed algorithm called SmartRule and the method of mining MFIs directly from tabular data. A new support counting method is introduced in Sect. 3. Then rule trees are developed in Sect. 4 to represent a set of related rules and their representations on the spreadsheet user interface. Performance comparison with transaction datasets is given in Sect. 5. Finally, an application example of using this technique on a medical clinical dataset for surgery consultation is given.

1.1 Related Works

Let I be a set of items and D be a set of transactions, where each transaction is an itemset. The support of an itemset is the number of transactions containing the itemset. An itemset is frequent if its support is at least a user-specified minimum support. Let FI denote the set of all frequent itemsets. An itemset is closed if there is no superset with the same support. The set of all frequent closed itemsets is denoted by FCI. A frequent itemset is called maximal if it is not a subset of any other frequent itemset. Let MFI denote the set of all maximal frequent itemsets. Any maximal frequent itemset X is a frequent closed itemset since no nontrivial superset of X is frequent. Thus we have $MFI \subseteq FCI \subseteq FI$.

Many algorithms for association rule mining require discovering FI before forming rules. Most methods for generating FI can be classified into three groups. First is the candidate set generate-and-test approach [1,7,12], which finds FI in a bottom up fashion. Second, the sampling approach [16] reduces computation complexity but the results are incomplete. Third is the data transformation approach [8,20], which transforms a dataset to a new form for efficient mining. Some algorithms [13,19] use FCI to generate rules.

Recently many MFI mining algorithms have been proposed. MaxMiner [4] uses a breadth-first search and performs look-ahead pruning on tree branches. The developments in mining MFI, however, use a depth first search with dynamic reordering as in DepthProject [2], Mafia [3], GenMax [6], and Smart-Miner [21]. All of these methods for mining MFI have reported orders of magnitudes faster than methods mining FI/FCI. SmartMiner uses tail information to prune the search space without superset checking. Little research has been done on using MFIs for mining association rules since no rule can be generated from MFIs. However, MFIs can serve as roadmaps for rule mining. For example, given the set of MFIs, we can analyze many interesting properties of the dataset such as the longest pattern, the distribution and the overlap of the MFIs.

Extensive research has been done on association rule mining. For example, there has been research on mining multi-level association rules [9], on selecting the right interestingness measure [15], on synthesizing high frequency rules [18], etc.

Association rule mining techniques often produce a large number of rules that are hard to comprehend without further processing. Many techniques [10,11,14,17] are proposed to prune, cluster, or query rules. The rules are usually presented in free text format, where it is easy to read a single rule but difficult to compare multiple rules at the same time. In this chapter, we propose using spreadsheets to present hierarchically organized rules to remedy this problem.

2 Methodology

Figure 2 illustrates SmartRule using an Excel book (or other spreadsheet software) to store data and mining results. The Excel book also serves as an interface for interacting with users. There are three functions in the system:

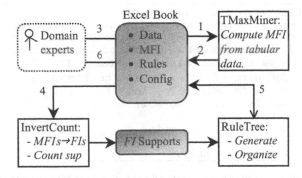

Fig. 2. System overview of SmartRule

TMaxMiner, InvertCount, and RuleTree. TMaxMiner directly mines MFIs for a given minimal support from tabular data. InvertCount builds the FI list contained by the user-selected MFIs and counts the supports for the FIs (Sect. 3). RuleTree is used to generate rules and to organize them hierarchically (Sect. 4).

In this chapter, we assume that our data only contains categorical values. In the situation that a column contains continuous values, we use clustering technique (e.g., [5, 22]) to partition the values into several groups.

2.1 Tail Information

Let $N = X : Y$ be a node where X is the head of N and Y is the tail of N. Let M be known frequent itemsets and $N = X : Y$ be a node. The **tail information** of M to N is denoted as $Inf(N|M)$, and is defined as the tail parts of the frequent itemsets in $\{X : Y\}$ that can be inferred from M, that is, $Inf(N|M) = \{Y \cap Z | \forall Z \in M, X \subseteq Z\}$.

For example, $Inf(e : bcd | \{abcd, abe, ace\}) = \{b, c\}$, which means that eb and ec are frequent given $\{abcd, abe, ace\}$ frequent. For simplicity, we call tail information "information".

Since the tail of a node may contain many infrequent items, pure depth-first search is inefficient. Hence, dynamic reordering is used to prune away infrequent items from the tail of a node before exploring its sub nodes.

2.2 Mining MFI from Tabular Data

We now present TmaxMiner, which computes MFIs directly from a tabular dataset. TMaxMiner uses tail information to eliminate superset checking. TMaxMiner uses tables as its data model and selects the column with the least entropy for the next search.

To process the tabular data more efficiently, we start with the column of the least entropy. The entropy of a column can be computed by the probability of each item in that column. For example, for column A in Fig. 4a,

```
TMaxMiner(table T, inf)
1   Count sup and remove infrequent items from T;
2   Find pep, remove pep's columns, update inf;;
3   while(select x on a column of least entropy){
4       Rows in T containing x →table T';
5       The part of inf relevant to T'→inf';
6       mfi'= TMaxMiner(T', inf');
7       Add (x + mfi') to mfi;
8       Remove x from T;
9       Update inf with mfi';
10 }
11 return (pep + mfi);
```

Fig. 3. TMaxMiner – a depth-first method that discovers MFI directly from a table guided by tail information

the probability of a_3 is $\{P(a_3) = 1\}$ where empty cells are ignored; thus the entropy of column A is $I(\{P(a_3) = 1\})$ is 0. The entropy of column B is $I(\{P(b_1) = 0.6, P(b_2) = 0.4\}) = -0.6 * \log_2 0.6 - 0.4 * \log_2 0.4 \approx 0.97$. So we start mining MFI from column A.

As shown in Fig. 3, TMaxMiner takes two parameters, a table T and information *inf*. It returns the discovered MFI in a depth-first fashion. The parameter T is the current table for processing and *inf* specifies the known frequent itemsets from the previous search. The TMaxMiner algorithm that uses tail information to derive MFIs is shown in Fig. 3. Line 1 computes the support for each item in T and then removes infrequent items. Parent equivalence pruning (PEP) [3] is defined as follows: Let x be a node's head and y be an element in its tail, if any transaction containing x also contains y, then move item y from the tail to the head. Line 2 finds the PEP items that appear in every row of T and remove the corresponding columns. Line 3 selects an item x in the column with the least entropy. Lines 4 to 10 find the MFIs containing x. Specifically, a new table T' is formed by selecting those rows of T containing x, and then removing the column of x. At Line 5, the part of *inf* mentioned x is assigned to *inf'*. Line 6 discovers MFI in the new table T', which is extended by the item x and added to *mfi*. Since we discovered the MFI containing x, we can remove x from T as in Line 8. Then *mfi'* is added into *inf* to inform the succeeding steps of the known frequent itemsets. Finally, the answer is returned at Line 11.

2.3 An Example of TMaxMiner

We shall show how TMaxMiner derives MFIs from the table as shown in Fig. 4a. Let minimum support threshold be equal to 3, we first select item a_3 as the first node since column A has the least entropy, and then yield the table $T(a_3)$, where items c_2 and d_2 appear in every row, i.e., $pep = c_2d_2$, and all other items are infrequent. Therefore we obtain MFI $a_3c_2d_2$ for $T(a_3)$.

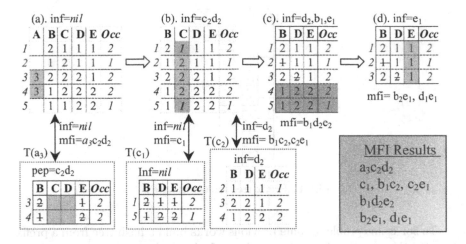

Fig. 4. An example of TMaxMiner: discover MFI directly from the tabular dataset

Removing column A from Fig. 4a, we have Fig. 4b with the tail information c_2d_2.

Next, column C in Fig. 4b is selected. We find a MFI c_1 from $T(c_1)$ and two MFIs b_1c_2 and c_2e_1 from $T(c_2)$. After removing column C, we have Fig. 4c with the tail information $\{d_2, b_1, e_1\}$. The rows 4 and 5 contain the same items and can be combined to form a MFI $b_1d_2e_2$. Then the two rows can be removed and the table shown in Fig. 4d is formed.

We find two more MFIs, b_2e_1 and d_1e_1, from Fig. 4d and the search process is now completed. The final results are the union of the MFIs in Fig. 4a–4d. They are $a_3c_2d_2$, c_1. b_1c_2. c_2e_1, $b_1d_2e_2$, b_2e_1, and d_1e_1.

3 Counting Support for Targeted Association Rules

Domain experts often wish to derive rules that contain a particular attribute. This can be accomplished by selecting a set of MFIs that contain such a target column. When the MFIs are very long, columns of less interest can be excluded.

The counting itemset C can be formed from the selected MFIs. For example, if the target column is column E in Fig. 1a, we can select the MFIs containing e_1 or e_2 and have $\{b_1d_2e_2, b_2e_1, c_2e_1, d_1e_1\}$. Then the counting itemsets C are all the subsets of the selected MFIs, $C = \{b_1, b_2, c_2, d_1, d_2, e_1, e_2. b_1d_2. b_1e_2, b_2e_1. c_2e_1, d_1e_1, d_2e_2, d_2b_1e_2\}$.

In order to generate rules, we need to count the support for the counting itemsets C from the source table T. There are two ways to determine which itemsets in C are contained in a row i in T: (1) for each subset s of i, check if s exists in C: (2) for each itemset c in C, check if c is contained in i. Clearly, both approaches are not efficient.

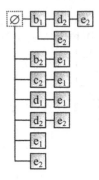

Fig. 5. Prefix tree for counting FIs

To solve this problem, some previous works build a hash tree from a prefix tree for the counting itemsets C. Figure 5 shows the prefix tree containing all the 14 itemsets in C. Then a hash table is added to each internal node to facilitate the search for a particular child. To find the itemset in C contained in a row i in T, we need to determine all the nodes in the prefix tree that match i. This approach is faster than the above two approaches, but the tree operations are relatively expensive.

We propose an algorithm InvertCount for counting supports that use only native CPU instructions such as "+1" and "mov".

Figure 6 illustrates that InvertCount employs inverted indices to count supports. Figure 6a shows the 14 itemsets for counting where the column len is the number of items in an itemset. The column sup stores the current counting results. The column cnt resets to 0 whenever the counting for one row is done. Figure 6b is the inverted indices $itm2ids$ that map an $item$ to a list of itemsets identifiers ids.

For a given row i, we can quickly determine the itemsets contained in i and increase their support by $Occ(i)$. For example, for the first row of Fig. 1a, $i = a_1b_2c_1d_1e_1$, we get the inverted indices $(b_2 : 1, 9)$, $(d_1 : 3, 11)$, and $(e_1 : 5, 9, 10, 11)$. There is no inverted index for a_1 and c_1. Then we use the column cnt to count the occurrence of each id in the above three indices, e.g., 9 occurring twice. Finally we determine that itemsets 1, 3, 5, 9, 11 are contained in i since their occurrences are equal to the values on column len (itemsets' lengths). Therefore we increase their support by $Occ(i)$, i.e. 2, as shown on the column sup.

The algorithm InvertCount (shown in Fig. 7) counts the support for itemsets in C from source tabular data T without any expensive function calls. Line 1 builds the data structure as shown in Fig. 6. For each row i in T, we first find the itemsets in C that are contained in i and then increase their support by $Occ(i)$. Specifically, Lines 3 and 4 count the occurrence of ids in the inverted lists of the items in i. Line 7 increases the support of an itemset

id	itemset	sup	len	cnt
0	b_1	4	1	
1	b_2	4	1	1
2	c_2	5	1	
3	d_1	3	1	1
4	d_2	5	1	
5	e_1	5	1	1
6	e_2	3	1	
7	b_1d_2	3	2	
8	b_1e_2	3	2	
9	b_2e_1	4	2	2
10	c_2e_1	3	2	1
11	d_1e_1	3	2	2
12	d_2e_2	3	2	
13	$b_1d_2e_2$	3	3	

a. FI-table

item	ids
b_1	0, 7, 8, 13
b_2	1, 9
c_2	2, 10
d_1	3, 11
d_2	4, 7, 12, 13
e_1	5, 9, 10, 11
e_2	6, 8, 12, 13

b. itm2ids

Example:
Count from row 1:
$a_1b_2c_1d_1e_1$: $occ=2$
Get inverted index
b_2: 1, 9
d_1: 3, 11
e_1: 5, 9, 10, 11

Fig. 6. Example of InvertCount for efficient computing FI supports

```
InvertCount(table T, C)
1   build FI-table F and itm2ids from C;
2   foreach(row i in T) do
3       foreach(item x in i) do
4           foreach(id in itm2ids[x].ids) do
F.cnt[id]++;
5       foreach(item x in i) do
6           foreach(id in itm2ids[x].ids) do
7               if(F.cnt[id]>=F.len[id])
F.sup[id]+=Occ(i);
8                   F.cnt[id]=0;
9   return F.sup;
```

Fig. 7. The InvertCount Algorithm – using inverted indices to count supports for the itemsets in C

by $Occ(i)$ if the itemset is a subset of i. Line 8 clears the count for item id in FI-table for later use. We return the counting results at Line 9.

Our preliminary experimental results reveal that InvertCount is notably faster than previous hash tree counting.

4 Generating Rule Trees

Using the final support counting results, we are able to generate association rules. For example, Fig. 6a shows the final counting results in the column sup. Since we are looking for rules with e_1 or e_2 as rule head, we can build a rule for every itemset containing e_1 or e_2. For instance, for itemset $b_1d_2e_2$ containing

id	rule	Sup	Conf
5	$nil \rightarrow e_1$	5	1.0
6	$nil \rightarrow e_2$	3	1.0
8	$b_1 \rightarrow e_2$	3	0.75
9	$b_2 \rightarrow e_1$	4	1.0
10	$c_2 \rightarrow e_1$	3	0.6
11	$d_1 \rightarrow e_1$	3	1.0
12	$d_2 \rightarrow e_2$	3	0.6
13	$b_1 d_2 \rightarrow e_2$	3	1.0

Fig. 8. A table of rules

e_2, a rule $b_1 d_2 \rightarrow e_2$ is created. Figure 8 shows the list of rules created from Fig. 6a.

The support and confidence of a rule can be easily derived from the final counting result. For example, for the itemset $d_1 e_1$ in Fig. 6a, we created a rule $d_1 \rightarrow e_1$ whose support is 3 and whose confidence is the support of $d_1 e_1$ divided by that of d_1, which is equal to 1.0.

Association rule mining usually generates too many rules for the user to comprehend, so we need to organize rules into a hierarchical structure so that users can study the rules at different levels with varying degrees of specificity. Trees are built from the list of rules, where each tree represents rules sharing the same rule head. The hierarchical relationship on a tree is determined by the containment relationship among rules' bodies. A rule r_1 is an ancestor of r_2 if and only if the head of r_1 is equal to the head of r_2 and where the body of r_1 is a subset of the body of r_2.

Figure 9 shows the algorithm TreeRule that takes a list of rules R as input, sorted by increasing length, and returns trees with hierarchically organized rules. Specifically, it builds a tree for each target column value, as shown in Lines 2 and 3 where rules have empty bodies. For a rule with a non-empty body, we add it into the tree corresponding with its head, as shown in Lines 4 to 5.

Figure 10 shows a recursive method AddRule that adds a rule r into a tree. If a more specific rule r has similar support and confidence with the current

```
vector TreeRule(List R)
1   int i=0;
2   for(; R[i].body.len==0; i++) //nil→h
3       new Tree(R[i]) →gtr[R[i].head];
4   for(; i<R.Count; i++)
5       tr[R[i].head].AddRule(R[i]);
6   return tr;
```

Fig. 9. TreeRule – building a forest of trees from a list of rules

```
AddRule(rule r)
1  if(r.sup, r.conf)≈(this.sup, this.conf);
2     this.optItms += r.body-this.body;
3     return;
4  r.body -= this.optItms;
5  foreach(subtree t & t.body is a subset of
r.body)
6     t.AddRule(r);
7  if(no such a subtree)
8     t'=new Tree(r);
9     Add t' to this.subTrees;
10 return;
```

Fig. 10. AddRule – recursively adding a rule into a rule tree

rule node referred by *this*, then we simply add the extra items as optional items (optItms) without creating a new node as in Lines 1 to 3.

If the rule r is not similar to the current rule node, then we add r into sub-trees of *this* whose bodies are subsets of $r.body$ as in Lines 5 and 6. If no such sub tree exists, we create a new tree and add it into *this*.subTree as in Lines 7 to 9. Line 10 returns the updated tree r.

For example, Fig. 11a shows the rule trees for the rules in Fig. 8. Two rule trees are built: one for e_1 with 3 branches, another for e_2 with two levels of children. The second level node $b_1 \rightarrow e_2$ is the parent of $b_1 d_2 \rightarrow e_2$ because they both have the same head e_2, and b_1 is a subset of $b_1 d_2$. In general, a more specific rule gives higher confidence but lower support. Figure 11b represents the tree in table format. in which each node is represented as a row with its support, confidence, node depth and number information. To output trees into tabular format, we number nodes in the trees by their preorders. The node's number plus its depth can be used to represent the hierarchical relationship of the specific rules in a rule tree.

a. Example of rule trees.

Rules	Sup	Conf	Depth	Num
$nil \rightarrow e_1$	5	1.0	0	0
$b_2 \rightarrow e_1$	4	1.0	1	1
$d_1 \rightarrow e_1$	3	1.0	1	2
$c_2 \rightarrow e_1$	3	0.6	1	3
$nil \rightarrow e_2$	3	1.0	0	4
$b_1 \rightarrow e_2$	3	0.75	1	5
$b_1 d_2 \rightarrow e_2$	3	1.0	2	6
$d_2 \rightarrow e_2$	3	0.6	1	7

b. Rule trees represented by node *Depth* and *Number*

Fig. 11. Example of rule trees and their tabular representation

5 Hybrid Clustering Technique
for Partitioning Continuous Attributes

The number of cells and the cell size will affect the clustering results and the data mining outcome, both in support and confidence.

When the sample size is very small and number of attributes is large, conventional statistical classification techniques such as CART [22] fail to classify the continuous value of an attribute into cells. Using unsupervised clustering techniques [5] has the problem of not knowing the optimal number of cells to represent the variables. Therefore, we developed a hybrid technique that combines both statistical and data mining techniques iteratively in determining the optimal number of cells as well as the cell sizes.

The basic idea is to use data mining technique to select a small set of key attributes, and then use a statistical classification technique such as CART to determine the cell sizes and number of cells from the training set. Then we use the partitioning result for data mining. The procedure works as follows:

Perform the mining on the training set, and select a set of attributes with high confidence and support for statistical classification.

Perform statistical classification based on the training set for the selected attributes set from step (1). Since the attribute set is greatly reduced, statistical classification techniques such as CART [22] can be used to determine the optimal number of cells and their corresponding cell sizes for attribute.

Based on the optimal cell sizes for each attribute from step (2), data mining algorithms can then be used to generate the rules for this set of attributes.

With this hybrid clustering approach, we are able to generate optimal partitioning for the set of continuous attributes. Our experimental results reveal that deriving rules based on such partitioning yield better mining results than conventional unsupervised clustering techniques [5].

6 Performance Comparisons

SmartRule was implemented using Microsoft .Net C# and the Office XP primary interop assemblies so that our program could directly read from and write to Excel workbooks. Experimental results have shown that SmartMiner is close to one order of magnitude faster than Mafia and Genmax in generating MFI from transaction dataset [21]. Note SmartMiner, Mafia and GenMax do not keep column constraints during generating MFIs. By taking advantage of column constraints in tabular data format, TmaxMiner achieves performance gains over SmartMiner as shown in Fig. 12. Further, we note that the gain increases as the support decreases. The dataset Mushroom used for performance comparison is in tabular format and was downloaded from the UCI machine learning repository [23].

To evaluate the performance of InvertCount, we used the Mushroom dataset and selected the MFIs containing the class attributes (*edible* or

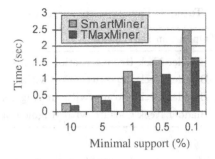

Fig. 12. Comparison of TMaxMiner with SmartMiner

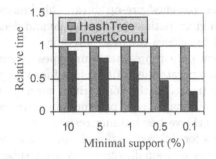

Fig. 13. Comparison of InvertCount with Hashtree

poisonous) to generate counting itemsets. Before testing, we build a hash tree and inverted indices for these itemsets. Then we compared the counting time of InvertCount and HashTree in the source Mushroom table.

Figure 13 shows the relative time of HashTree and InvertCount for the tests at varying minimal supports. When minimal support is 10%, they have similar performances because of the small number of counting itemsets. As the minimal support decreases, MFIs become longer, which results in an increase in the number of counting itemsets. In such a case, InvertCount is notably faster than HashTree.

7 An Application Example

We have applied SmartRule to mine a set of clinical data that was collected from urology surgeries during 1995 to 2002 at the UCLA Pediatric Urology Clinic. The dataset contains 130 rows (each row represents a patient) and 28 columns, which describe patient pre-operative conditions, type of surgery performed, post-op complications and final outcome of the surgeries. The pre-operative conditions include patient ambulatory status (A), catheterizing skills (CS), amount of creatinine in the blood (SerumCrPre), leak point pressure (LPP), and urodynamics, such as minimum volume infused into bladder

Table 1. Four types of surgeries

Operation Type	Operation Description
Op-1	Bladder Neck Reconstruction with Augmentation
Op-2	Bladder Neck Reconstruction without Augmentation
Op-3	Bladder Neck Closure without Augmentation
Op-4	Bladder Neck Closure with Augmentation

when pressure reached 20 cm of water (20%min). The data mining goal is to derive a set of rules from the clinical data set (training set) that summarize the outcome based on patients' pre-op data. This knowledge base can then be used to examine a given patient pre-op profile and decide which operation should be performed to achieve the best outcome.

This set of clinical data represents four types of surgery operations as shown in Table 1. We separate the patients into four groups based on the type of surgery they were treated with. For a given surgery type, we partitioned the continuous value attributes, e.g., patient urodynamics data, into discrete intervals or cells. To achieve best mining results, an attribute may partition into different cell sizes for different types of operations, as shown in Table 2. Since our sample size is very small, especially after subgrouping the dataset into different operations, we used the hybrid technique that combines both statistical and data mining techniques iteratively in determining the optimal number of cells as well as the cell sizes.

For the training set, we are able to generate optimal partitioning for the set of continuous attributes from each operation type as shown in Table 2. Sets of rules are generated based on the discretized variables for each type of operation, which can be viewed as the knowledge base for this type of operation. Our experimental results reveal that deriving rules based on such partitioning yield better mining results than conventional unsupervised clustering techniques [5].

For a given patient with a specific set of pre-op conditions, the generated rules from the training set can be used to predict success or failure rate for a specific operation.

To provide the user with a family of rules, SmartRule can organize matched rules into rule trees from general to specific rules. The rules closer to the root are more general, contain fewer constraints and yield higher support; the rules closer to the leaves are more specific, which contain more constraints and yield lower support. In case of multiple match rules, the quality of the rules in terms of confidence and support may be used in rule selection.

Given patient Matt's pre-op conditions as shown in Table 3(a), since the attributes in the rules are represented in discrete values, the continuous pre-op conditions are transformed into discrete values (Table 3(b)) based on the partitioning done on the attributes of operation types as shown in Table 2.

Table 2. Partition of continuous variables into optimal number of discrete intervals (cells) and cell sizes for four types of operations. Each table presents the partitioning for a specific operation type. The optimal number of cells for an attribute is represented by the number of rows in each table. The size for each cell is represented in the column. Since different attributes have different optimal number of cells, certain attributes may contain no values in certain rows, and we use n/a to designate such a undefined cell sizes

(a) Operation Type 1

Cell#	LPP	SerumCrPre
1	[0, 19]	[0, 0.75]
2	(19, 33.5]	[0.75, 2.2]
3	(33.5,40]	n/a
4	normal	n/a

(b) Operation Type 2

Cell#	20% min	20% mean	30% min	30% mean	LPP	SerumCrPre
1	[80, 118]	[50, 77]	[100, 170]	[51, 51]	[12, 20]	[0, 0.5]
2	[145, 178]	[88, 104]	[206, 241]	[94, 113]	[24, 36]	[0.7, 1.4]
3	[221, 264]	[135, 135]	n/a	[135, 135]	normal	n/a

(c) Operation Type 3

Cell#	20% min	20% mean	30% min	30% mean	LPP	SerumCrPre
1	[103,130]	[57, 75]	[129, 157]	[86, 93]	[6, 29]	[0.3, 0.7]
2	[156,225]	[92, 105]	[188, 223]	[100,121]	[30,40]	[1.0, 1.5]

(d) Operation Type 4

Cell#	LPP	20% mean
1	[0, 19]	[0, 33.37]
2	(19, 69]	(33.37, 37.5]
3	normal	(37.5, 52]
4	n/a	(52, 110]

Based on these attributes values of a given patient such as "Ambulatory Status=4" and "CathSkills=1", we can search for rules from the knowledge base that were generated from the training set to match Matt's pre-op profile as shown in Table 4.

Table 3.(a). Patient Matt's pre-operative conditions

Ambulatory Status (A)	Cath Skills (CS)	Serum CrPre	20% min	20% mean(M)	30% min	30% mean	LPP	UPP
4	1	0.5	31	20	50	33	27	unknown

Table 3.(b). Discretized pre-operative conditions of patient Matt's pre-op conditions. The attributes not used in rule generation are denoted as n/a

	Ambulatory Status (A)	Cath Skills (CS)	Serum CrPre	20% min	20% mean(M)	30% min	30% mean	LPP
Op-1	4	1	1	n/a	n/a	n/a	n/a	2
Op-2	4	1	1	<1	<1	<1	<1	2
Op-3	4	1	1	<1	<1	<1	<1	1
Op-4	4	1	n/a	n/a	1	n/a	n/a	2

Table 4. Rule trees selected from the knowledge base (derived form the training set) that match patient Matt's pre-op profile

Surgery	Conditions	Outcome	Support	Support(%)	Confidence
Op-1	CS = 1	Success	10	41.67	0.77
	CS = 1 and LPP = 2	Success	3	12.5	0.75
Op-2	CS = 1 and LPP = 2	Fail	2	16.67	0.67
	20%min = 1 and LPP = 2	Fail	2	16.67	0.67
Op-3	CS = 1 and SerumCrPre = 1	Success	5	50	0.83
	CS = 1, SerumCrPre = 1 and LPP = 1	Success	2	20	1
Op-4	A = 4	Success	14	32.55	0.78
	A = 4 and CS = 1	Success	11	25.58	0.79
	A = 4, CS = 1 and LPP = 2	Success	8	18.6	0.8
	A = 4, CS = 1 and M = 1	Success	6	13.95	1
	A = 4, CS = 1, M = 1 and LPP = 2	Success	6	13.95	1

Based on the rule tree, we note that Operations 3 and 4 both match patient Matt's pre-op conditions. However, Operation 4 matches more attributes in Matt's pre-op conditions than Operation 3. Thus, Operation 4 is more desirable for patient Matt. A screen shot of the corresponding spreadsheet user interface is shown in Fig. 14a and the corresponding rule tree representation is shown in Fig. 14b. We have received favorable user feedback in using the spreadsheet interface because of its ease in rule searching and sorting.

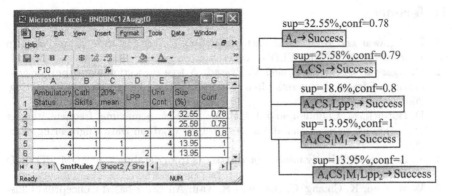

(a) Represent rule trees for Op-4 by spreadsheet   (b) Rule tree for Op-4

Fig. 14. Representing rules in a hierarchical structure for the example

8 Conclusion

In this chapter, we have proposed a method to derive association rules directly from tabular data with a large number of dependent variables. The SmartRule algorithm is able to use table structures to reduce the search space and the counting time for mining maximal frequent itemsets (MFI). Our experimental results reveal that using tabular data rather than transforming to transaction-type data can significantly improve the performance of mining MFIs. Using simple data structures and native CPU instructions, the proposed InvertCount is faster than hash tree for support counting. Finally, SmartRule organizes rules into hierarchical rule trees and uses spreadsheet as a user interface to sort, filter and select rules so that users can browse only a small number of interesting rules that they wish to study. We have successfully applied SmartRule to a set of medical clinical data and have derived useful rules for recommending the type of surgical operation for patients based on their pre-operative conditions and demography information.

Acknowledgements

The authors wish to thank Dr. B. Churchill and Dr. Andy Chen for providing the clinical data as well as collaborating in formulating the data mining model, and Professor James W. Sayre for his stimulating discussion in developing the statistical and data mining approach in determining the cell size. This research is supported by the NIH PPG Grant #4442511-33780 and the NSF IIS Grant #97438.

References

1. R. Agrawal and R. Srikant: Fast algorithms for mining association rules. In Proceedings of the 20th VLDB Conference, Santiago, Chile, 1994.
2. R. Agarwal, C. Aggarwal, and V. Prasad: A tree projection algorithm for generation of frequent itemsets. Journal of Parallel and Distributed Computing, 2001.
3. D. Burdick, M. Calimlim, and J. Gehrke: MAFIA: a maximal frequent itemset algorithm for transactional databases. In Intl. Conf. on Data Engineering, Apr. 2001.
4. R. Bayardo: Efficiently mining long patterns from databases. In ACM SIGMOD Conference, 1998.
5. W.W. Chu, K. Chiang, C. Hsu, and H. Yau: An Error-based Conceptual Clustering Method for Providing Approximate Query Answers Communications of ACM. 1996.
6. K. Gouda and M.J. Zaki: Efficiently Mining Maximal Frequent Itemsets. Proc. of the IEEE Int. Conference on Data Mining, San Jose, 2001.
7. H. Mannila, H. Toivonen, and A.I. Verkamo: Efficient algorithms for discovering association rules. In KDD-94: AAAI Workshop on Knowledge Discovery in Databases, pp. 181–192, Seattle, Washington, July 1994.
8. J. Han, J. Pei, and Y. Yin: Mining Frequent Patterns without Candidate Generation, Proc. 2000 ACM-SIGMOD Int. Conf. on Management of Data (SIGMOD'00), Dallas, TX, May 2000.
9. J. Han and Y. Fu: Discovery of Multiple-Level Association Rules from Large Databases. In Proc. of the 21th Int. Conf. on Very Large Databases, Zurich, Swizerland, 1995.
10. B. Liu, W. Hsu, and Y. Ma: Pruning and summarizing the discovered associations. In Proc. of the Fifth Int'l Conference on Knowledge Discovery and Data Mining, pp. 125–134, San Diego, CA, August 1999.
11. B. Liu, M. Hu, and W. Hsu: Multi-level organization and summarization of the discovered rules. Proc. ACM SIGKDD, 208–217, 2000.
12. J.S. Park, M. Chen, and P.S. Yu: An effective hash based algorithm for mining association rules. In Proc. ACM SIGMOD Intl. Conf. Management of Data, May 1995.
13. N. Pasquier, Y. Bastide, R. Taouil, and L. Lakhal: Discovering frequent closed itemsets for association rules. In 7th Intl. Conf. on Database Theory, January 1999.
14. B. Lent, A.N. Swami, and J. Widom: Clustering association rules. In Proceedings of International Conference on Data Engineering, 1997.
15. P. Tan, V. Kumar, and J. Srivastava: Selecting the Right Interestingness Measure for Association Patterns (2002). Proc of the Eighth ACM SIGKDD Int'l Conf. on Knowledge Discovery and Data Mining (KDD-2002).
16. Hannu Toivonen. Sampling large databases for association rules. In Proc. of the VLDB Conference, Bombay, India, September 1996.
17. A. Tuzhilin and B. Liu: Querying multiple sets of discovered rules. Proceedings of the ACM SIGKDD International Conference on Knowledge Discovery & Data Mining, Edmonton, Canada, July 23–26, 2002.
18. X. Wu and S. Zhang, Synthesizing High-Frequency Rules from Different Data Sources, IEEE Transactions on Knowledge and Data Engineering, Vol. 15, No. 2, March/April 2003, 353–367.

19. M.J. Zaki and C. Hsiao: Charm: An efficient algorithm for closed association rule mining. In Technical Report 99–10. Computer Science, Rensselaer Polytechnic Institute, 1999.
20. Q. Zou, W. Chu, D. Johnson, and H. Chiu: Pattern Decomposition Algorithm for Data Mining of Frequent Patterns. Journal of Knowledge and Information System, 2002.
21. Q. Zou, W. Chu, B. Lu: SmartMiner: A Depth First Algorithm Guided by Tail Information for Mining Maximal Frequent Itemsets. Proc. of the IEEE Int. Conference on Data Mining, Japan, 2002.
22. http://www.salford-systems.com/products-cart.html
23. http://www.ics.uci.edu/~mlearn/MLRepository.html

Sequential Pattern Mining by Pattern-Growth: Principles and Extensions*

J. Han[1], J. Pei[2], and X. Yan[1]

[1] University of Illinois at Urbana-Champaign
  {hanj, xyan}@cs.uiuc.edu
[2] State University of New York at Buffalo
  jianpei@cse.buffalo.edu

Summary. Sequential pattern mining is an important data mining problem with broad applications. However, it is also a challenging problem since the mining may have to generate or examine a combinatorially explosive number of intermediate subsequences. Recent studies have developed two major classes of sequential pattern mining methods: (1) a *candidate generation-and-test* approach, represented by (i) GSP [30], a horizontal format-based sequential pattern mining method, and (ii) SPADE [36], a vertical format-based method; and (2) a *sequential pattern growth* method, represented by PrefixSpan [26] and its further extensions, such as CloSpan for mining closed sequential patterns [35].

In this study, we perform a systematic introduction and presentation of the pattern-growth methodology and study its principles and extensions. We first introduce two interesting pattern growth algorithms, FreeSpan [11] and PrefixSpan [26], for efficient sequential pattern mining. Then we introduce CloSpan for mining closed sequential patterns. Their relative performance in large sequence databases is presented and analyzed. The various kinds of extension of these methods for (1) mining constraint-based sequential patterns, (2) mining multi-level, multi-dimensional sequential patterns, (3) mining top-k closed sequential patterns, and (4) their applications in bio-sequence pattern analysis and clustering sequences are also discussed in the paper.

Index terms. Data mining, sequential pattern mining algorithm, sequence database, scalability, performance analysis, application.

---

*The work was supported in part by the Natural Sciences and Engineering Research Council of Canada, the Networks of Centers of Excellence of Canada, the Hewlett-Packard Lab, the U.S. National Science Foundation NSF IIS-02-09199, NSF IIS-03-08001, and the University of Illinois. Any opinions, findings, and conclusions or recommendations expressed in this paper are those of the authors and do not necessarily reflect the views of the funding agencies.

1 Introduction

Sequential pattern mining, which discovers frequent subsequences as patterns in a sequence database, is an important data mining problem with broad applications, including the analysis of customer purchase patterns or Web access patterns, the analysis of sequencing or time-related processes such as scientific experiments, natural disasters, and disease treatments, the analysis of DNA sequences, and so on.

The sequential pattern mining problem was first introduced by Agrawal and Srikant in [2] based on their study of customer purchase sequences, as follows: *Given a set of sequences, where each sequence consists of a list of elements and each element consists of a set of items, and given a user-specified min_support threshold, sequential pattern mining is to find all frequent subsequences, i.e., the subsequences whose occurrence frequency in the set of sequences is no less than min_support.*

This problem is introduced from the examination of potential patterns in sequence databases, as follows.

Let $I = \{i_1, i_2, \ldots, i_n\}$ be a set of all **items**. An **itemset** is a subset of items. A **sequence** is an ordered list of itemsets. A sequence s is denoted by $\langle s_1 s_2 \cdots s_l \rangle$, where s_j is an itemset. s_j is also called an **element** of the sequence, and denoted as $(x_1 x_2 \cdots x_m)$, where x_k is an item. For brevity, the brackets are omitted if an element has only one item, i.e., element (x) is written as x. An item can occur at most once in an element of a sequence, but can occur multiple times in different elements of a sequence. The number of instances of items in a sequence is called the **length** of the sequence. A sequence with length l is called an l-**sequence**. A sequence $\alpha = \langle a_1 a_2 \cdots a_n \rangle$ is called a **subsequence** of another sequence $\beta = \langle b_1 b_2 \cdots b_m \rangle$ and β a **super-sequence** of α, denoted as $\alpha \sqsubseteq \beta$, if there exist integers $1 \leq j_1 < j_2 < \cdots < j_n \leq m$ such that $a_1 \subseteq b_{j_1}, a_2 \subseteq b_{j_2}, \ldots, a_n \subseteq b_{j_n}$.

A **sequence database** S is a set of tuples $\langle sid, s \rangle$, where sid is a **sequence_id** and s a sequence. A tuple $\langle sid, s \rangle$ is said to *contain* a sequence α, if α is a subsequence of s. The support of a sequence α in a sequence database S is the number of tuples in the database containing α, i.e., $support_S(\alpha) = | \{\langle sid, s \rangle | (\langle sid, s \rangle \in S) \wedge (\alpha \sqsubseteq s)\} |$. It can be denoted as $support(\alpha)$ if the sequence database is clear from the context. Given a positive integer $min\_support$ as the **support threshold**, a sequence α is called a **sequential pattern** in sequence database S if $support_S(\alpha) \geq min\_support$. A sequential pattern with length l is called an l-**pattern**.

Example 1. Let our running sequence database be S given in Table 1 and $min\_support = 2$. The set of *items* in the database is $\{a, b, c, d, e, f, g\}$.

A *sequence* $\langle a(abc)(ac)d(cf) \rangle$ has five *elements*: (a), (abc), (ac), (d) and (cf), where items a and c appear more than once respectively in different elements. It is a 9-*sequence* since there are 9 instances appearing in that sequence. Item a happens three times in this sequence, so it contributes 3 to

Table 1. A sequence database

Sequence_id	Sequence
1	$\langle a(abc)(ac)d(cf) \rangle$
2	$\langle (ad)c(bc)(ae) \rangle$
3	$\langle (ef)(ab)(df)cb \rangle$
4	$\langle eg(af)cbc \rangle$

the *length* of the sequence. However, the whole sequence $\langle a(abc)(ac)d(cf) \rangle$ contributes only one to the *support* of $\langle a \rangle$. Also, sequence $\langle a(bc)df \rangle$ is a *subsequence* of $\langle a(abc)(ac)d(cf) \rangle$. Since both sequences 10 and 30 *contain* subsequence $s = \langle (ab)c \rangle$, s is a *sequential pattern* of length 3 (i.e., *3-pattern*). □

From this example, one can see that **sequential pattern mining problem** can be stated as "*given a sequence database and the min_support threshold, **sequential pattern mining** is to find the complete set of sequential patterns in the database.*"

Notice that this model of sequential pattern mining is an abstraction from the customer shopping sequence analysis. However, this model may not cover a large set of requirements in sequential pattern mining. For example, for studying Web traversal sequences, gaps between traversals become important if one wants to predict what could be the next Web pages to be clicked. Many other applications may want to find gap-free or gap-sensitive sequential patterns as well, such as weather prediction, scientific, engineering and production processes, DNA sequence analysis, and so on. Moreover, one may like to find approximate sequential patterns instead of precise sequential patterns, such as in DNA sequence analysis where DNA sequences may contain nontrivial proportions of insertions, deletions, and mutations.

In our model of study, the gap between two consecutive elements in a sequence is unimportant. However, the gap-free or gap-sensitive frequent sequential patterns can be treated as special cases of our model since gaps are essentially constraints enforced on patterns. The efficient mining of gap-sensitive patterns will be discussed in our later section on constraint-based sequential pattern mining. Moreover, the mining of approximate sequential patterns is also treated as an extension of our basic mining methodology. Those and other related issues will be discussed in the later part of the paper.

Many previous studies contributed to the efficient mining of sequential patterns or other frequent patterns in time-related data [2, 4, 8, 20–22, 24, 29, 30, 32, 37]. Reference [30] generalized their definition of sequential patterns in [2] to include time constraints, sliding time window, and user-defined taxonomy and present an Apriori-based, improved algorithm GSP (i.e., *generalized sequential patterns*). Reference [21] presented a problem of mining frequent episodes in a sequence of events, where episodes are essentially acyclic graphs of events whose edges specify the temporal precedent-subsequent relationship without restriction on interval. Reference [4] considered a generalization of inter-transaction association rules. These are essentially rules whose left-hand and right-hand sides are episodes with time-interval restrictions. Reference [20] proposed inter-transaction association rules that are implication rules whose two sides are totally-ordered episodes with timing-interval restrictions. Reference [7] proposed the use of regular expressions as a flexible constraint specification tool that enables user-controlled focus to be incorporated into the sequential pattern mining process. Some other studies extended the scope from mining sequential patterns to mining partial periodic patterns. Reference [24] introduced cyclic association rules that are essentially partial periodic patterns with *perfect* periodicity in the sense that *each pattern reoccurs in every cycle*, with 100% confidence. Reference [8] developed a frequent pattern mining method for mining partial periodicity patterns that are frequent maximal patterns where each pattern appears in a fixed period with a fixed set of offsets, and with sufficient support. Reference [36] developed a vertical format-based sequential pattern mining method, called SPADE, which can be considered as an extension of vertical-format-based frequent itemset mining methods, such as [37, 39].

Almost all of the above proposed methods for mining sequential patterns and other time-related frequent patterns are Apriori-like, i.e., based on the Apriori principle, which states the fact that *any super-pattern of an infrequent pattern cannot be frequent*, and based on a candidate generation-and-test paradigm proposed in association mining [1].

In our recent studies, we have developed and systematically explored a pattern-growth approach for efficient mining of sequential patterns in large sequence database. The approach adopts a divide-and-conquer, pattern-growth principle as follows, *sequence databases are recursively projected into a set of smaller projected databases based on the current sequential pattern(s), and sequential patterns are grown in each projected database by exploring only locally frequent fragments.* Based on this philosophy, we first proposed a straight-forward pattern growth method, FreeSpan (for **Fre**quent pattern-projected **S**equential **patter**n mining) [11], which reduces the efforts of candidate subsequence generation. Then, we introduced another and more efficient method, called PrefixSpan (for **Prefix**-projected **S**equential **pattern** mining), which offers ordered growth and reduced projected databases. To further improve the performance, a *pseudo-projection* technique is developed in PrefixSpan. A comprehensive performance study shows that PrefixSpan in most cases outperforms the Apriori-based GSP algorithm, FreeSpan, and SPADE [36] (a sequential pattern mining algorithm that adopts vertical data format), and PrefixSpan integrated with pseudo-projection is the fastest among all the tested algorithms. Furthermore, our experiments show that PrefixSpan consumes a much smaller memory space in comparison with GSP and SPADE.

PrefixSpan is an efficient algorithm at mining the complete set of sequential patterns. However, a long sequential pattern may contain a combinatorial number of frequent subsequences. To avoid generating a large number by many of which are essentially redundant subsequences, our task becomes the mining of closed sequential pattern instead of the complete set of sequential patterns. An efficient algorithm called CloSpan [35] is developed based on the philosophy of (sequential) pattern-growth and by exploring sharing among generated or to be generated sequences. Our performance study shows that CloSpan may further reduce the cost at mining closed sequential patterns substantially in comparison with PrefixSpan.

This pattern-growth methodology has been further extended in various ways to cover the methods and applications of sequential and structured pattern mining. This includes (1) mining multi-level, multi-dimensional sequential patterns, (2) mining other structured patterns, such as graph patterns, (3) constraint-based sequential pattern mining, (4) mining closed sequential patterns, (5) mining top-k sequential patterns, (6) mining long sequences in the noise environment, (7) mining approximate consensus sequential patterns, and (8) clustering time-series gene expressions.

In this paper, we will systematically present the methods for pattern-growth-based sequential patterns, their principle and applications.

The remainder of the paper is organized as follows. In Sect. 2, we introduce the Apriori-based sequential pattern mining methods, GSP and SPADE, both relying on a candidate generation-and-test philosophy. In Sect. 3, our approach, projection-based sequential pattern growth, is introduced, by first summarizing FreeSpan, and then presenting PrefixSpan, associated with a pseudo-projection technique for performance improvement. In Sect. 4, we introduce CloSpan, an efficient method for mining closed sequential patterns. Some experimental results and performance analysis are summarized in Sect. 5. The extensions of the method in different directions are discussed in Sect. 6. We conclude our study in Sect. 7.

2 Previous Work:
The Candidate Generation-and-Test Approach

The candidate generation-and-test approach is an extension of the Apriori-based frequent pattern mining algorithm [1] to sequential pattern analysis. Similar to frequent patterns, sequential patterns has the anti-monotone (i.e., downward closure) property as follows: *every non-empty sub-sequence of a sequential pattern is a sequential pattern.*

Based on this property, there are two algorithms developed for efficient sequential pattern mining: (1) a horizontal data format based sequential pattern mining method: GSP [30], and (2) a vertical data format based sequential pattern mining method: SPADE [36]. We outline and analyze these two methods in this section.

2.1 GSP: A Horizontal Data Format
Based Sequential Pattern Mining Algorithm

From the sequential pattern mining point of view, a sequence database can be represented in two data formats: (1) a horizontal data format, and (2) a vertical data format. The former uses the natural representation of the data set as $\langle sequence\_id : a\_sequence\_of\_objects \rangle$, whereas the latter uses the vertical representation of the sequence database: $\langle object : (sequence\_id, time\_stamp) \rangle$, which can be obtained by transforming from a horizontal formatted sequence database.

GSP is a horizontal data format based sequential pattern mining developed by [30] by extension of their frequent itemset mining algorithm, Apriori [1]. Based on the downward closure property of a sequential pattern, GSP adopts a multiple-pass, candidate-generation-and-test approach in sequential pattern mining. The algorithm is outlined as follows. The first scan finds all of the frequent items which form the set of single item frequent sequences. Each subsequent pass starts with a *seed set* of sequential patterns, which is the set of sequential patterns found in the previous pass. This seed set is used to generate new potential patterns, called *candidate sequences*. Each candidate sequence contains one more item than a seed sequential pattern, where each element in the pattern may contain one or multiple items. The number of items in a sequence is called the *length* of the sequence. So, all the candidate sequences in a pass will have the same length. The scan of the database in one pass finds the support for each candidate sequence. All of the candidates whose support in the database is no less than *min_support* form the set of the newly found sequential patterns. This set then becomes the seed set for the next pass. The algorithm terminates when no new sequential pattern is found in a pass, or no candidate sequence can be generated.

The method is illustrated using the following example.

Example 2. (GSP) Given the database S and *min_support* in Example 1, GSP first scans S, collects the support for each item, and finds the set of frequent items, i.e., frequent length-1 subsequences (in the form of "*item : support*"): $\langle a \rangle : 4, \langle b \rangle : 4, \langle c \rangle : 3, \langle d \rangle : 3, \langle e \rangle : 3, \langle f \rangle : 3, \langle g \rangle : 1$.

By filtering the infrequent item g, we obtain the first seed set $L_1 = \{\langle a \rangle, \langle b \rangle, \langle c \rangle, \langle d \rangle, \langle e \rangle, \langle f \rangle\}$, each member in the set representing a 1-element sequential pattern. Each subsequent pass starts with the seed set found in the previous pass and uses it to generate new potential sequential patterns, called *candidate sequences*.

For L_1, a set of 6 length-1 sequential patterns generates a set of $6 \times 6 + \frac{6 \times 5}{2} = 51$ candidate sequences, $C_2 = \{\langle aa \rangle, \langle ab \rangle, \ldots, \langle af \rangle, \langle ba \rangle, \langle bb \rangle, \ldots, \langle ff \rangle, \langle (ab) \rangle, \langle (ac) \rangle, \ldots, \langle (ef) \rangle\}$.

The multi-scan mining process is shown in Fig. 1. The set of candidates is generated by a self-join of the sequential patterns found in the previous pass. In the k-th pass, a sequence is a candidate only if each of its length-$(k-1)$

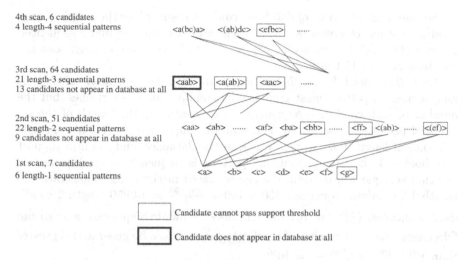

Fig. 1. Candidates. candidate generation, and sequential patterns in GSP

subsequences is a sequential pattern found at the $(k-1)$-th pass. A new scan of the database collects the support for each candidate sequence and finds the new set of sequential patterns. This set becomes the seed for the next pass. The algorithm terminates when no sequential pattern is found in a pass, or when there is no candidate sequence generated. Clearly, the number of scans is at least the maximum length of sequential patterns. It needs one more scan if the sequential patterns obtained in the last scan still generate new candidates.

GSP, though benefits from the Apriori pruning, still generates a large number of candidates. In this example. 6 length-1 sequential patterns generate 51 length-2 candidates. 22 length-2 sequential patterns generate 64 length-3 candidates. etc. Some candidates generated by GSP may not appear in the database at all. For example, 13 out of 64 length-3 candidates do not appear in the database. $\qquad\Box$

The example shows that an Apriori-like sequential pattern mining method, such as GSP. though reduces search space, bears three nontrivial, inherent costs which are independent of detailed implementation techniques.

First, *there are potentially huge sets of candidate sequences.* Since the set of candidate sequences includes all the possible permutations of the elements and repetition of items in a sequence, an Apriori-based method may generate a really large set of candidate sequences even for a moderate seed set. For example, if there are 1000 frequent sequences of length-1, such as $\langle a_1 \rangle, \langle a_2 \rangle, \ldots, \langle a_{1000} \rangle$, an Apriori-like algorithm will generate $1000 \times 1000 + \frac{1000 \times 999}{2} = 1,499,500$ candidate sequences, where the first term is derived from the set $\langle a_1 a_1 \rangle$, $\langle a_1 a_2 \rangle, \ldots, \langle a_1 a_{1000} \rangle, \langle a_2 a_1 \rangle, \langle a_2 a_2 \rangle, \ldots, \langle a_{1000} a_{1000} \rangle$, and the second term is derived from the set $\langle (a_1 a_2) \rangle, \langle (a_1 a_3) \rangle, \ldots, \langle (a_{999} a_{1000}) \rangle$.

Second, *multiple scans of databases could be costly.* Since the length of each candidate sequence grows by one at each database scan, to find a sequential pattern $\{(abc)(abc)\ (abc)(abc)(abc)\}$, an Apriori-based method must scan the database at least 15 times.

Last, *there are inherent difficulties at mining long sequential patterns.* A long sequential pattern must grow from a combination of short ones, but the number of such candidate sequences is exponential to the length of the sequential patterns to be mined. For example, suppose there is only a single sequence of length 100, $\langle a_1 a_2 \dots a_{100} \rangle$, in the database, and the min_support threshold is 1 (i.e., every occurring pattern is frequent), to (re-)derive this length-100 sequential pattern, the Apriori-based method has to generate 100 length-1 candidate sequences, $100 \times 100 + \frac{100 \times 99}{2} = 14,950$ length-2 candidate sequences, $\binom{100}{3} = 161,700$ length-3 candidate sequences, and so on. Obviously, the total number of candidate sequences to be generated is greater than $\Sigma_{i=1}^{100} \binom{100}{i} = 2^{100} - 1 \approx 10^{30}$.

In many applications, it is not rare that one may encounter a large number of sequential patterns and long sequences, such as stock sequence analysis. Therefore, it is important to re-examine the sequential pattern mining problem to explore more efficient and scalable methods. Based on our analysis, both the thrust and the bottleneck of an Apriori-based sequential pattern mining method come from its step-wise candidate sequence generation and test. Then the problem becomes, *"can we develop a method which may absorb the spirit of* Apriori *but avoid or substantially reduce the expensive candidate generation and test?"*

2.2 SPADE: An Apriori-Based Vertical Data Format Sequential Pattern Mining Algorithm

The Apriori-based sequential pattern mining can also be explored by mapping a sequence database into the vertical data format which takes each item as the center of observation and takes its associated sequence and event identifiers as data sets. To find sequence of length-2 items, one just needs to join two single items if they are frequent and they share the same sequence identifier and their event identifiers (which are essentially relative timestamps) follow the sequential ordering. Similarly, one can grow the length of itemsets from length two to length three, and so on. Such an Apriori-based vertical data format sequential pattern mining algorithm, called SPADE (Sequential PAttern Discovery using Equivalent classes) algorithm [36], is illustrated using the following example.

Example 3. (SPADE) Given our running sequence database S and *mins_support* in Example 1, SPADE first scans S, transforms the database into the vertical format by introducing EID (event ID) which is a (local) timestamp for each event. Each single item is associated with a set of SID (sequence_id) and

SID	EID	Items
1	1	a
1	2	abc
1	3	ac
1	4	d
1	5	cf
2	1	ad
2	2	c
2	3	bc
2	4	ae
3	1	ef
3	2	ab
3	3	df
3	4	c
3	5	b
4	1	e
4	2	g
4	3	af
4	4	c
4	5	b
4	6	c

a		b	···
SID	EID	SID	EID ···
1	1	1	2
1	2	2	3
1	3	3	2
2	1	3	5
2	4	4	5
3	2		
4	3		

ab			ba		···
SID	EID (a)	EID(b)	SID	EID (b)	EID(a) ···
1	1	2	1	2	3
2	1	3	2	3	4
3	2	5			
4	3	5			

aba				···
SID	EID (a)	EID(b)	EID(a) ···	
1	1	2	3	
2	1	3	4	

Fig. 2. Vertical format of the sequence database and fragments of the SPADE mining process

EID (event_id) pairs. For example, item "b" is associated with (SID, EID) pairs as follows: $\{(1,2),(2,3),(3,2),(3,5),(4,5)\}$, as shown in Fig. 2. This is because item b appears in sequence 1, event 2, and so on. Frequent single items "a" and "b" can be joined together to form a length-two subsequence by joining the same sequence_id with event_ids following the corresponding sequence order. For example, subsequence ab contains a set of triples $(SID, EID(a), EID(b))$, such as $(1,1,2)$, and so on. Furthermore, the frequent length-2 subsequences can be joined together based on the Apriori heuristic to form length-3 subsequences, and so on. The process continuous until no frequent sequences can be found or no such sequences can be formed by such joins.

Some fragments of the SPADE mining process are illustrated in Fig. 2. The detailed analysis of the method can be found in [36].   □

The SPADE algorithm may reduce the access of sequence databases since the information required to construct longer sequences are localized to the related items and/or subsequences represented by their associated sequence and event identifiers. However, the basic search methodology of SPADE is similar to GSP, exploring both breadth-first search and Apriori pruning. It has to generate a large set of candidates in breadth-first manner in order to grow longer subsequences. Thus most of the difficulties suffered in the GSP algorithm will reoccur in SPADE as well.

3 The Pattern-Growth Approach
for Sequential Pattern Mining

In this section, we introduce a pattern-growth methodology for mining sequential patterns. It is based on the methodology of pattern-growth mining of frequent patterns in transaction databases developed in the FP-growth algorithm [12]. We introduce first the FreeSpan algorithm and then a more efficient alternative, the PrefixSpan algorithm.

3.1 FreeSpan: Frequent Pattern-Projected
Sequential Pattern Mining

For a sequence $\alpha = \langle s_1 \cdots s_l \rangle$, the itemset $s_1 \cup \cdots \cup s_l$ is called α's *projected itemset*. FreeSpan is based on the following property: *if an itemset X is infrequent, any sequence whose projected itemset is a superset of X cannot be a sequential pattern*. FreeSpan mines sequential patterns by partitioning the search space and projecting the sequence sub-databases recursively based on the projected itemsets.

Let f_list $= \langle x_1, \ldots, x_n \rangle$ be a list of all frequent items in sequence database S. Then, the complete set of sequential patterns in S can be divided into n disjoint subsets: (1) the set of sequential patterns containing only item x_1, (2) those containing item x_2 but no item in $\{x_3, \ldots, x_n\}$, and so on. In general, the ith subset ($1 \le i \le n$) is the set of sequential patterns containing item x_i but no item in $\{x_{i+1}, \ldots, x_n\}$.

Then, the database projection can be performed as follows. At the time of deriving p's projected database from DB, the set of frequent items X of DB is already known. Only those items in X will need to be projected into p's projected database. This effectively discards irrelevant information and keeps the size of the projected database minimal. By recursively doing so, one can mine the projected databases and generate the complete set of sequential patterns in the given partition without duplication. The details are illustrated in the following example.

Example 4. (FreeSpan) Given the database S and *min_support* in Example 1, FreeSpan first scans S, collects the support for each item, and finds the set of frequent items. This step is similar to GSP. Frequent items are listed in support descending order (in the form of *"item : support"*), that is, f_list $= \langle a : 4, b : 4, c : 4, d : 3, e : 3, f : 3 \rangle$. They form six length-one sequential patterns: $\langle a \rangle$:4, $\langle b \rangle$:4, $\langle c \rangle$:4, $\langle d \rangle$:3, $\langle e \rangle$:3, $\langle f \rangle$:3.

According to the f_list, the complete set of sequential patterns in S can be divided into 6 disjoint subsets: (1) the ones containing only item a, (2) the ones containing item b but no item after b in f_list, (3) the ones containing item c but no item after c in f_list, and so on, and finally, (6) the ones containing item f.

The sequential patterns related to the six partitioned subsets can be mined by constructing six *projected databases* (obtained by one additional scan of the original database). Infrequent items, such as g in this example, are removed from the projected databases. The process for mining each projected database is detailed as follows.

- *Mining sequential patterns containing only item a.*
  The $\langle a \rangle$-*projected database* is $\{\langle aaa \rangle, \langle aa \rangle, \langle a \rangle, \langle a \rangle\}$. By mining this projected database, only one additional sequential pattern containing only item a, i.e., $\langle aa \rangle$:2, is found.
- *Mining sequential patterns containing item b but no item after b in the* f_list.
  By mining the $\langle b \rangle$-*projected database*: $\{\langle a(ab)a \rangle, \langle aba \rangle, \langle (ab)b \rangle, \langle ab \rangle\}$, four additional sequential patterns containing item b but no item after b in f_list are found. They are $\{\langle ab \rangle$:4, $\langle ba \rangle$:2, $\langle (ab) \rangle$:2, $\langle aba \rangle$:2$\}$.
- *Mining sequential patterns containing item c but no item after c in the* f_list.
  The mining of the $\langle c \rangle$-*projected database*: $\{\langle a(abc)(ac)c \rangle, \langle ac(bc)a \rangle, \langle (ab)cb \rangle, \langle acbc \rangle\}$, proceeds as follows. One scan of the projected database generates the set of length-2 frequent sequences, which are $\{\langle ac \rangle$:4, $\langle (bc) \rangle$:2, $\langle bc \rangle$:3, $\langle cc \rangle$:3, $\langle ca \rangle$:2, $\langle cb \rangle$:3$\}$. One additional scan of the $\langle c \rangle$-*projected database* generates all of its projected databases.
  The mining of the $\langle ac \rangle$-*projected database*: $\{\langle a(abc)(ac)c \rangle, \langle ac(bc)a \rangle, \langle (ab)cb \rangle, \langle acbc \rangle\}$ generates the set of length-3 patterns as follows: $\{\langle acb \rangle$:3, $\langle acc \rangle$:3, $\langle (ab)c \rangle$:2, $\langle aca \rangle$:2$\}$. Four projected database will be generated from them.
  The mining of the first one, the $\langle acb \rangle$-*projected database*: $\{\langle ac(bc)a \rangle, \langle (ab)cb \rangle, \langle acbc \rangle\}$ generates no length-4 pattern. The mining along this line terminates. Similarly, we can show that the mining of the other three projected databases terminates without generating any length-4 patterns for the $\langle ac \rangle$-*projected database*.
- *Mining other subsets of sequential patterns.*
  Other subsets of sequential patterns can be mined similarly on their corresponding projected databases. This mining process proceeds recursively, which derives the complete set of sequential patterns.   □

The detailed presentation of the FreeSpan algorithm, the proof of its completeness and correctness, and the performance study of the algorithm are in [11]. By the analysis of Example 4 and verified by our experimental study, we have the following observations on the strength and weakness of FreeSpan:

The strength of FreeSpan *is that it searches a smaller projected database than* GSP *in each subsequent database projection.* This is because that FreeSpan projects a large sequence database recursively into a set of small projected sequence databases based on the currently mined frequent item-patterns, and the subsequent mining is confined to each projected database relevant to a smaller set of candidates.

The major overhead of FreeSpan *is that it may have to generate many non-trivial projected databases.* If a pattern appears in each sequence of a database, its projected database does not shrink (except for the removal of some infrequent items). For example, the $\{f\}$-projected database in this example contains three same sequences as that in the original sequence database, except for the removal of the infrequent item g in sequence 4. Moreover, since a length-k subsequence may grow at any position, the search for length-$(k+1)$ candidate sequence will need to check every possible combination, which is costly.

3.2 PrefixSpan: **Prefix-Projected Sequential Patterns Mining**

Based on the analysis of the FreeSpan algorithm, one can see that one may still have to pay high cost at handling projected databases. To avoid checking every possible combination of a potential candidate sequence, one can first fix the order of items *within each element*. Since items within an element of a sequence can be listed in any order, without loss of generality, one can assume that they are always listed alphabetically. For example, the sequence in S with Sequence_id 1 in our running example is listed as $\langle a(abc)(ac)d(cf)\rangle$ instead of $\langle a(bac)(ca)d(fc)\rangle$. With such a convention, the expression of a sequence is unique.

Then, we examine whether one can fix the order of item projection in the generation of a projected database. Intuitively, if one follows the order of the prefix of a sequence and projects only the suffix of a sequence, one can examine in an orderly manner all the possible subsequences and their associated projected database. Thus we first introduce the concept of prefix and suffix.

Suppose all the items within an element are listed alphabetically. Given a sequence $\alpha = \langle e_1 e_2 \cdots e_n \rangle$ (where each e_i corresponds to a frequent element in S), a sequence $\beta = \langle e_1' e_2' \cdots e_m' \rangle$ $(m \leq n)$ is called a **prefix** of α if and only if (1) $e_i' = e_i$ for $(i \leq m - 1)$; (2) $e_m' \subseteq e_m$; and (3) all the frequent items in $(e_m - e_m')$ are alphabetically after those in e_m'. Sequence $\gamma = \langle e_m'' e_{m+1} \cdots e_n \rangle$ is called the **suffix** of α w.r.t. prefix β, denoted as $\gamma = \alpha/\beta$, where $e_m'' = (e_m - e_m')$.[1] We also denote $\alpha = \beta \cdot \gamma$. Note if β is not a subsequence of α, the suffix of α w.r.t. β is empty.

Example 5. For a sequence $s = \langle a(abc)(ac)d(cf)\rangle$, $\langle a \rangle$, $\langle aa \rangle$, $\langle a(ab)\rangle$ and $\langle a(abc)\rangle$ are *prefixes* of sequence $s = \langle a(abc)(ac)d(cf)\rangle$, but neither $\langle ab \rangle$ nor $\langle a(bc)\rangle$ is considered as a prefix if every item in the prefix $\langle a(abc)\rangle$ of sequence s is frequent in S. Also, $\langle (abc)(ac)d(cf)\rangle$ is the *suffix* w.r.t. the prefix $\langle a \rangle$, $\langle (\_bc)(ac)d(cf)\rangle$ is the *suffix* w.r.t. the prefix $\langle aa \rangle$, and $\langle (\_c)(ac)d(cf)\rangle$ is the *suffix* w.r.t. the prefix $\langle a(ab)\rangle$. □

Based on the concepts of prefix and suffix, the problem of mining sequential patterns can be decomposed into a set of subproblems as shown below.

---

[1] If e_m'' is not empty, the suffix is also denoted as $\langle (\_ \text{ items in } e_m'')e_{m+1} \cdots e_n \rangle$.

1. Let $\{\langle x_1 \rangle, \langle x_2 \rangle, \ldots, \langle x_n \rangle\}$ be the complete set of length-1 sequential patterns in a sequence database S. The complete set of sequential patterns in S can be divided into n disjoint subsets. The ith subset $(1 \leq i \leq n)$ is the set of sequential patterns with prefix $\langle x_i \rangle$.

2. Let α be a length-l sequential pattern and $\{\beta_1, \beta_2, \ldots, \beta_m\}$ be the set of all length-$(l+1)$ sequential patterns with prefix α. The complete set of sequential patterns with prefix α, except for α itself, can be divided into m disjoint subsets. The jth subset $(1 \leq j \leq m)$ is the set of sequential patterns prefixed with β_j.

Based on this observation, the problem can be partitioned recursively. That is, each subset of sequential patterns can be further divided when necessary. This forms a *divide-and-conquer* framework. To mine the subsets of sequential patterns, the corresponding projected databases can be constructed.

Let α be a sequential pattern in a sequence database S. The α-**projected database**, denoted as $S|_\alpha$, is the collection of suffixes of sequences in S w.r.t. prefix α. Let β be a sequence with prefix α. The **support count** of β in α-projected database $S|_\alpha$, denoted as $support_{S|_\alpha}(\beta)$, is the number of sequences γ in $S|_\alpha$ such that $\beta \sqsubseteq \alpha \cdot \gamma$.

We have the following lemma regarding to the projected databases.

Lemma 1. (Projected database) *Let α and β be two sequential patterns in a sequence database S such that α is a prefix of β.*

1. $S|_\beta = (S|_\alpha)|_\beta$;
2. for any sequence γ with prefix α, $support_S(\gamma) = support_{S|_\alpha}(\gamma)$; and
3. The size of α-projected database cannot exceed that of S.

Proof Sketch. The first part of the lemma follows the fact that, for a sequence γ, the suffix of γ w.r.t. β, γ/β, equals to the sequence resulted from first doing projection of γ w.r.t. α, i.e., γ/α, and then doing projection γ/α w.r.t. β. That is $\gamma/\beta = (\gamma/\alpha)/\beta$.

The second part of the lemma states that to collect support count of a sequence γ, only the sequences in the database sharing the same prefix should be considered. Furthermore, only those suffixes with the prefix being a super-sequence of γ should be counted. The claim follows the related definitions.

The third part of the lemma is on the size of a projected database. Obviously, the α-projected database can have the same number of sequences as S only if α appears in every sequence in S. Otherwise, only those sequences in S which are super-sequences of α appear in the α-projected database. So, the α-projected database cannot contain more sequences than S. For every sequence γ in S such that γ is a super-sequence of α, γ appears in the α-projected database in whole only if α is a prefix of γ. Otherwise, only a subsequence of γ appears in the α-projected database. Therefore, the size of α-projected database cannot exceed that of S. $\qquad\square$

Let us examine how to use the prefix-based projection approach for mining sequential patterns based our running example.

Example 6. (PrefixSpan) For the same sequence database S in Table 1 with $min\_sup = 2$, sequential patterns in S can be mined by a prefix-projection method in the following steps.

1. *Find length-1 sequential patterns.*

   Scan S once to find all the frequent items in sequences. Each of these frequent items is a length-1 sequential pattern. They are $\langle a \rangle$: 4, $\langle b \rangle$: 4, $\langle c \rangle$: 4, $\langle d \rangle$: 3, $\langle e \rangle$: 3, and $\langle f \rangle$: 3, where the notation "$\langle pattern \rangle$: $count$" represents the pattern and its associated support count.

2. *Divide search space.*

   The complete set of sequential patterns can be partitioned into the following six subsets according to the six prefixes: (1) the ones with prefix $\langle a \rangle$, (2) the ones with prefix $\langle b \rangle$, ..., and (6) the ones with prefix $\langle f \rangle$.

3. *Find subsets of sequential patterns.*

   The subsets of sequential patterns can be mined by constructing the corresponding set of *projected databases* and mining each recursively. The projected databases as well as sequential patterns found in them are listed in Table 2, while the mining process is explained as follows.

   (a) *Find sequential patterns with prefix $\langle a \rangle$.*

   Only the sequences containing $\langle a \rangle$ should be collected. Moreover, in a sequence containing $\langle a \rangle$, only the subsequence prefixed with the first occurrence of $\langle a \rangle$ should be considered. For example, in sequence $\langle (ef)(ab)(df)cb \rangle$, only the subsequence $\langle (\_b)(df)cb \rangle$ should be considered for mining sequential patterns prefixed with $\langle a \rangle$. Notice that $(\_b)$ means that the last element in the prefix, which is a, together with b, form one element.

   The sequences in S containing $\langle a \rangle$ are projected w.r.t. $\langle a \rangle$ to form the $\langle a \rangle$-*projected database*, which consists of four suffix sequences: $\langle (abc)(ac)d(cf) \rangle$, $\langle (\_d)c(bc)(ae) \rangle$, $\langle (\_b)(df)cb \rangle$ and $\langle (\_f)cbc \rangle$.

Table 2. Projected databases and sequential patterns

Prefix	Projected Database	Sequential Patterns
$\langle a \rangle$	$\langle (abc)(ac)d(cf) \rangle$, $\langle (\_d)c(bc)(ae) \rangle$, $\langle (\_b)(df)cb \rangle$, $\langle (\_f)cbc \rangle$	$\langle a \rangle$, $\langle aa \rangle$, $\langle ab \rangle$, $\langle a(bc) \rangle$, $\langle a(bc)a \rangle$, $\langle aba \rangle$, $\langle abc \rangle$, $\langle (ab) \rangle$, $\langle (ab)c \rangle$, $\langle (ab)d \rangle$, $\langle (ab)f \rangle$, $\langle (ab)dc \rangle$, $\langle ac \rangle$, $\langle aca \rangle$, $\langle acb \rangle$, $\langle acc \rangle$, $\langle ad \rangle$, $\langle adc \rangle$, $\langle af \rangle$
$\langle b \rangle$	$\langle (\_c)(ac)d(cf) \rangle$, $\langle (\_c)(ae) \rangle$, $\langle (df)cb \rangle$, $\langle c \rangle$	$\langle b \rangle$, $\langle ba \rangle$, $\langle bc \rangle$, $\langle (bc) \rangle$, $\langle (bc)a \rangle$, $\langle bd \rangle$, $\langle bdc \rangle$, $\langle bf \rangle$
$\langle c \rangle$	$\langle (ac)d(cf) \rangle$, $\langle (bc)(ae) \rangle$, $\langle b \rangle$, $\langle bc \rangle$	$\langle c \rangle$, $\langle ca \rangle$, $\langle cb \rangle$, $\langle cc \rangle$
$\langle d \rangle$	$\langle (cf) \rangle$, $\langle c(bc)(ae) \rangle$, $\langle (\_f)cb \rangle$	$\langle d \rangle$, $\langle db \rangle$, $\langle dc \rangle$, $\langle dcb \rangle$
$\langle e \rangle$	$\langle (\_f)(ab)(df)cb \rangle$, $\langle (af)cbc \rangle$	$\langle e \rangle$, $\langle ea \rangle$, $\langle eab \rangle$, $\langle eac \rangle$, $\langle eacb \rangle$, $\langle eb \rangle$, $\langle ebc \rangle$, $\langle ec \rangle$, $\langle ecb \rangle$, $\langle ef \rangle$, $\langle efb \rangle$, $\langle efc \rangle$, $\langle efcb \rangle$.
$\langle f \rangle$	$\langle (ab)(df)cb \rangle$, $\langle cbc \rangle$	$\langle f \rangle$, $\langle fb \rangle$, $\langle fbc \rangle$, $\langle fc \rangle$, $\langle fcb \rangle$

By scanning the $\langle a \rangle$-projected database once, its locally frequent items are $a : 2$, $b : 4$, $\_b : 2$, $c : 4$, $d : 2$, and $f : 2$. Thus all the length-2 sequential patterns prefixed with $\langle a \rangle$ are found, and they are: $\langle aa \rangle : 2$, $\langle ab \rangle : 4$, $\langle (ab) \rangle : 2$, $\langle ac \rangle : 4$, $\langle ad \rangle : 2$, and $\langle af \rangle : 2$.

Recursively, all sequential patterns with prefix $\langle a \rangle$ can be partitioned into 6 subsets: (1) those prefixed with $\langle aa \rangle$, (2) those with $\langle ab \rangle$, ..., and finally, (6) those with $\langle af \rangle$. These subsets can be mined by constructing respective projected databases and mining each recursively as follows.

i.   The $\langle aa \rangle$-projected database consists of two non-empty (suffix) subsequences prefixed with $\langle aa \rangle$: $\{\langle (\_bc)(ac)d(cf) \rangle\}$, $\{\langle (\_e) \rangle\}$. Since there is no hope to generate any frequent subsequence from this projected database, the processing of the $\langle aa \rangle$-projected database terminates.

ii. The $\langle ab \rangle$-projected database consists of three suffix sequences: $\langle (\_c)(ac)d(cf) \rangle$, $\langle (\_c)a \rangle$, and $\langle c \rangle$. Recursively mining the $\langle ab \rangle$-projected database returns four sequential patterns: $\langle (\_c) \rangle$, $\langle (\_c)a \rangle$, $\langle a \rangle$, and $\langle c \rangle$ (i.e., $\langle a(bc) \rangle$, $\langle a(bc)a \rangle$, $\langle aba \rangle$, and $\langle abc \rangle$.) They form the complete set of sequential patterns prefixed with $\langle ab \rangle$.

iii. The $\langle (ab) \rangle$-projected database contains only two sequences: $\langle (\_c)(ac) d(cf) \rangle$ and $\langle (df)cb \rangle$, which leads to the finding of the following sequential patterns prefixed with $\langle (ab) \rangle$: $\langle c \rangle$, $\langle d \rangle$, $\langle f \rangle$, and $\langle dc \rangle$.

iv. The $\langle ac \rangle$-, $\langle ad \rangle$- and $\langle af \rangle$- projected databases can be constructed and recursively mined similarly. The sequential patterns found are shown in Table 2.

(b) *Find sequential patterns with prefix $\langle b \rangle$, $\langle c \rangle$, $\langle d \rangle$, $\langle e \rangle$ and $\langle f \rangle$, respectively.*

This can be done by constructing the $\langle b \rangle$-, $\langle c \rangle$- $\langle d \rangle$-, $\langle e \rangle$- and $\langle f \rangle$-projected databases and mining them respectively. The projected databases as well as the sequential patterns found are shown in Table 2.

4. *The set of sequential patterns is the collection of patterns found in the above recursive mining process.*

One can verify that it returns exactly the same set of sequential patterns as what GSP and FreeSpan do.   □

Based on the above discussion, the algorithm of PrefixSpan is presented as follows.

3.3 Pseudo-Projection

The above analysis shows that the major cost of PrefixSpan is database projection, i.e., forming projected databases recursively. Usually, a large number of projected databases will be generated in sequential pattern mining. If the number and/or the size of projected databases can be reduced, the performance of sequential pattern mining can be further improved.

Algorithm 1.
(PrefixSpan) Prefix-projected sequential pattern mining.

Input: A sequence database S, and the minimum support threshold $min\_support$.
Output: The complete set of sequential patterns.
Method: Call PrefixSpan($\langle \rangle, 0, S$).

> **Subroutine** PrefixSpan($\alpha, l, S|_\alpha$)
> The parameters are (1) α is a sequential pattern; (2) l is the length of α; and
> (3) $S|_\alpha$ is the α-projected database if $\alpha \neq \langle \rangle$, otherwise, it is the sequence
> database S.
> **Method:**
> 1. Scan $S|_\alpha$ once, find each frequent item, b, such that
>   (a) b can be assembled to the last element of α to form a new sequential
>   pattern; or
>   (a) $\langle b \rangle$ can be appended to α to form a new sequential pattern.
> 2. For each new sequential pattern α', if α' is frequent, output α', construct
> α'-projected database $S|_{\alpha'}$, and call PrefixSpan($\alpha', l + 1, S|_{\alpha'}$).

Analysis. The correctness and completeness of the algorithm can be justified based
on Lemma 1. Here, we analyze the efficiency of the algorithm as follows.

- *No candidate sequence needs to be generated by* PrefixSpan.
  Unlike Apriori-like algorithms, PrefixSpan only grows longer sequential patterns
  from the shorter frequent ones. It neither generates nor tests any candidate sequence non-existent in a projected database. Comparing with GSP, which generates and tests a substantial number of candidate sequences, PrefixSpan searches a
  much smaller space.
- *Projected databases keep shrinking.*
  As indicated in Lemma 1, a projected database is smaller than the original one
  because only the suffix subsequences of a frequent prefix are projected into a projected database. In practice, the shrinking factors can be significant because (1)
  usually, only a small set of sequential patterns grow quite long in a sequence database, and thus the number of sequences in a projected database usually reduces
  substantially when prefix grows; and (2) projection only takes the suffix portion
  with respect to a prefix. Notice that FreeSpan also employs the idea of projected
  databases. However, the projection there often takes the whole string (not just
  suffix) and thus the shrinking factor is less than that of PrefixSpan.
- *The major cost of* PrefixSpan *is the construction of projected databases.*
  In the worst case, PrefixSpan constructs a projected database for every sequential
  pattern. If there exist a good number of sequential patterns, the cost is non-trivial.
  Techniques for reducing the number of projected databases will be discussed in
  the next subsection.   □

One technique which may reduce the number and size of projected databases is *pseudo-projection*. The idea is outlined as follows. Instead of performing physical projection, one can register the index (or identifier) of the corresponding sequence and the starting position of the projected suffix in the sequence. Then, a physical projection of a sequence is replaced by registering a sequence identifier and the projected position index point. *Pseudo-projection* reduces the cost of projection substantially when the projected database can fit in main memory.

This method is based on the following observation. For any sequence s, each projection can be represented by a corresponding projection position (an index point) instead of copying the whole suffix as a projected subsequence. Consider a sequence $\langle a(abc)(ac)d(cf)\rangle$. Physical projections may lead to repeated copying of different suffixes of the sequence. An index position pointer may save physical projection of the suffix and thus save both space and time of generating numerous physical projected databases.

Example 7. (Pseudo-projection) For the same sequence database S in Table 1 with $min\_sup = 2$, sequential patterns in S can be mined by pseudo-projection method as follows.

Suppose the sequence database S in Table 1 can be held in main memory. Instead of constructing the $\langle a \rangle$-projected database, one can represent the projected suffix sequences using pointer (sequence_id) and offset(s). For example, the projection of sequence $s_1 = \langle a(abc)d(ae)(cf)\rangle$ with regard to the $\langle a \rangle$-projection consists two pieces of information: (1) a *pointer* to s_1 which could be the string_id s_1, and (2) the *offset(s)*, which should be a single integer, such as 2, if there is a single projection point; and a set of integers, such as $\{2, 3, 6\}$, if there are multiple projection points. Each offset indicates at which position the projection starts in the sequence.

The projected databases for prefixes $\langle a \rangle$-, $\langle b \rangle$-, $\langle c \rangle$-, $\langle d \rangle$-, $\langle f \rangle$-, and $\langle aa \rangle$- are shown in Table 3, where $ indicates the prefix has an occurrence in the current sequence but its projected suffix is empty, whereas $\emptyset$ indicates that there is no occurrence of the prefix in the corresponding sequence. From Table 3, one can see that the pseudo-projected database usually takes much less space than its corresponding physically projected one.   □

Pseudo-projection avoids physically copying suffixes. Thus, it is efficient in terms of both running time and space. However, it may not be efficient

Table 3. A sequence database and some of its pseudo-projected databases

Sequence_id	Sequence	$\langle a \rangle$	$\langle b \rangle$	$\langle c \rangle$	$\langle d \rangle$	$\langle f \rangle$	$\langle aa \rangle$	...
10	$\langle a(abc)(ac)d(cf)\rangle$	2, 3, 6	4	5, 7	8	$	3, 6	...
20	$\langle (ad)c(bc)(ae)\rangle$	2	5	4, 6	3	$\emptyset$	7	...
30	$\langle (ef)(ab)(df)cb\rangle$	4	5	8	6	3, 7	$\emptyset$	...
40	$\langle eg(af)cbc\rangle$	4	6	6	$\emptyset$	5	$\emptyset$	...

if the pseudo-projection is used for disk-based accessing since random access disk space is costly. Based on this observation, the suggested approach is that if the original sequence database or the projected databases is too big to fit in memory, the physical projection should be applied, however, the execution should be swapped to pseudo-projection once the projected databases can fit in memory. This methodology is adopted in our PrefixSpan implementation.

Notice that the pseudo-projection works efficiently for PrefixSpan but not so for FreeSpan. This is because for PrefixSpan, an offset position clearly identifies the suffix and thus the projected subsequence. However, for FreeSpan, since the next step pattern-growth can be in both forward and backward directions, one needs to register more information on the possible extension positions in order to identify the remainder of the projected subsequences. Therefore, we only explore the pseudo-projection technique for PrefixSpan.

4 CloSpan: Mining Closed Frequent Sequential Patterns

The sequential pattern mining algorithms developed so far have good performance in databases consisting of short frequent sequences. Unfortunately, when mining long frequent sequences, or when using very low support thresholds, the performance of such algorithms often degrades dramatically. This is not surprising: Assume the database contains only one long frequent sequence $\langle (a_1)(a_2) \ldots (a_{100}) \rangle$, it will generate $2^{100} - 1$ frequent subsequences if the minimum support is 1, although all of them except the longest one are redundant because they have the same support as that of $\langle (a_1)(a_2) \ldots (a_{100}) \rangle$.

We propose an alternative but equally powerful solution: instead of mining the complete set of frequent subsequences, we mine frequent *closed subsequences* only, i.e., those containing no super-sequence with the same support. We develop CloSpan [35] (Closed Sequential pattern mining) to mine these patterns. CloSpan can produce a significantly less number of sequences than the traditional (i.e., full-set) methods while preserving the same expressive power since the whole set of frequent subsequences, together with their supports, can be derived easily from our mining results.

CloSpan first mines a closed sequence candidate set which contains all frequent closed sequences. The candidate set may contain some non-closed sequences. Thus, CloSpan needs a post-pruning step to filter out non-closed sequences. In order to efficiently mine the candidate set, we introduce a search space pruning condition: Whenever we find two exactly same prefix-based project databases, we can stop growing one prefix.

Let $\mathcal{I}(S)$ represent the total number of items in S, defined by

$$\mathcal{I}(S) = \sum_{\alpha \in S} l(\alpha) \, ,$$

where $l(\alpha)$ is α's length. We call $\mathcal{I}(S)$ the *the database size*. For the sample dataset in Table 1, $\mathcal{I}(S) = 31$.

Theorem 1 (Equivalence of Projected Databases). *Given two sequences,* $\alpha \sqsubseteq \beta$. *then*

$$S|_\alpha = S|_\beta \Leftrightarrow \mathcal{I}(S|_\alpha) = \mathcal{I}(S|_\beta) \tag{1}$$

Proof. It is obvious that $S|_\alpha = S|_\beta \Rightarrow \mathcal{I}(S|_\alpha) = \mathcal{I}(S|_\beta)$. Now we prove the sufficient condition. Since $\alpha \sqsubseteq \beta$, then $\mathcal{I}(S|_\alpha) \leqslant \mathcal{I}(S|_\beta)$. The equality between $\mathcal{I}(S|_\alpha)$ and $\mathcal{I}(S|_\beta)$ holds only if $\forall \gamma \in S|_\beta$, $\gamma \in S|_\alpha$, and vice vera. Therefore, $S|_\alpha = S|_\beta$. $\qquad\square$

For the sample database in Table 1, $S|_{\langle ac \rangle} = S|_{\langle c \rangle} = \{\langle (ac)d(cf) \rangle, \langle (bc)$ $(ae) \rangle, \langle b \rangle, \langle bc \rangle\}$ and $\mathcal{I}(S|_{\langle ac \rangle}) = \mathcal{I}(S|_{\langle c \rangle}) = 12$. According to Theorem 1, the search space can be pruned as follows.

Lemma 2 (Early Termination by Equivalence). *Given two sequences,* $\alpha \sqsubseteq \beta$. *if* $\mathcal{I}(S|_\alpha) = \mathcal{I}(S|_\beta)$, *then* $\forall \gamma, support(\alpha \diamond \gamma) = support(\beta \diamond \gamma)$, *where* $\alpha \diamond \gamma$ *and* $\beta \diamond \gamma$ *means* γ *is assembled to the last itemset of* α *and* β *(or appended to them), respectively.*

Considering the previous example, we have $\mathcal{I}(S|_{\langle ac \rangle}) = \mathcal{I}(S|_{\langle c \rangle})$. Based on Lemma 2, without calculating the supports of $\langle acb \rangle$ and $\langle cb \rangle$, we can conclude that they are the same.

The search space of PrefixSpan, when mining frequent sequences in Table 1, is depicted in Fig. 3. Each node in the figure represents one frequent sequence. PrefixSpan performs depth-first search by assembling one item to the last itemset of the current frequent sequence or appending one item to it. In Fig. 3, we use subscript "i" to denote the assembling extension, and "s" to denote the appending extension.

According to Lemma 2, it is recognized that if α and all of its descendants $(\alpha \diamond \gamma)$ in the prefix search tree have been discovered, it is unnecessary to search the branch under β. The reason is α and β share the exactly same descendants in the prefix search tree. So we can directly transplant the branch

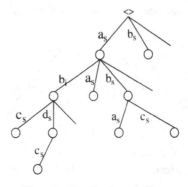

Fig. 3. Prefix Search Tree

under α to β. The power of such transplanting is that only two operations needed to detect such condition: first, containment between α and β; second, comparison between $\mathcal{I}(S|_\alpha)$ and $\mathcal{I}(S|_\beta)$. Since $\mathcal{I}(S|_\alpha)$ is just a number and can be produced as a side-product when we project the database, the computation cost introduced by Lemma 2 is nearly negligible. We define *projected database closed set*, $LS = \{\alpha \mid support(s) \geqslant min\_support$ and $\nexists \beta, s.t. \alpha \sqsubseteq \beta$ and $\mathcal{I}(S|_\alpha) = \mathcal{I}(S|_\beta)\}$. Obviously, LS is a superset of closed frequent sequences. In CloSpan, instead of mining closed frequent sequences directly, it first produces the complete set of LS and then applies the non-closed sequence elimination in LS to generate the accurate set of closed frequent sequences.

Corollary 1 (Backward Sub-Pattern). *If sequence α is discovered before β, $\alpha \sqsupseteq \beta$, and the condition $\mathcal{I}(S|_\alpha) = \mathcal{I}(S|_\beta)$ holds, it is sufficient to stop searching any descendant of β in the prefix search tree.*

We call β a *backward sub-pattern* of α if α is discovered before β and $\alpha \sqsupseteq \beta$. For the sample database in Table 1, if we know $\mathcal{I}(S|_{\langle ac \rangle}) = \mathcal{I}(S|_{\langle c \rangle})$, we can conclude that $S|_{\langle ac \rangle} = S|_{\langle c \rangle}$. We even need not compare the sequences in $S|_{\langle ac \rangle}$ and $S|_{\langle c \rangle}$ one by one to determine whether they are the same. This is the advantage of only comparing their size. Just as proved in Theorem 1, if their size is equal, we can conclude $S|_{\langle c \rangle} = S|_{\langle ac \rangle}$. We need not grow $\langle c \rangle$ anymore since all the children of $\langle c \rangle$ are the same as that of $\langle ac \rangle$ and vice versa under the condition of $S|_{\langle c \rangle} = S|_{\langle ac \rangle}$. Moreover, their supports are the same. Therefore, any sequence beginning with $\langle c \rangle$ **is absorbed** by the sequences beginning with $\langle ac \rangle$. Figure 4(a) shows that their subtrees (descendant branches) can be merged into **one** without mining the subtree under $\langle c \rangle$.

Corollary 2 (Backward Super-Pattern). *If a sequence α is discovered before β, $\alpha \sqsubseteq \beta$, and the condition $\mathcal{I}(S|_\alpha) = \mathcal{I}(S|_\beta)$ holds, it is sufficient to transplanting the descendants of α to β instead of searching any descendant of β in the prefix search tree.*

We call β a *backward super-pattern* of α if α is discovered before β and $\alpha \sqsubseteq \beta$). For example, if we know $\mathcal{I}(S|_{\langle b \rangle}) = \mathcal{I}(S|_{\langle eb \rangle})$, we can conclude that $S|_{\langle eb \rangle} = S|_{\langle b \rangle}$. There is no need to grow $\langle eb \rangle$ since all the children of $\langle b \rangle$ are

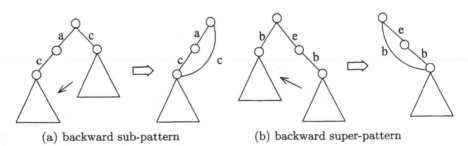

(a) backward sub-pattern   (b) backward super-pattern

Fig. 4. Backward Sub-Pattern and Super-Pattern

the same as those of $\langle eb \rangle$ and vice versa. Furthermore, they have the same support. Therefore, the sequences beginning with eb **can absorb** any sequence beginning with b. Figure 4(b) shows that their subtrees can be merged into **one** without discovering the subtree under $\langle eb \rangle$.

Based on the above discussion, we formulate the algorithm of CloSpan as follows.

Algorithm 2
(CloSpan) Closed frequent sequential pattern mining.

Input: A sequence database S, and the minimum support threshold *min_support*.
Output: The candidate set of closed sequential patterns.
Method: Call CloSpan($\langle \rangle, 0, S, L$).

> **Subroutine** CloSpan($\alpha, l, S|_\alpha, L$)
> The parameters are (1) α is a sequential pattern; (2) l is the length of α; and (3) $S|_\alpha$ is the α-projected database if $\alpha \neq \langle \rangle$, otherwise, it is the sequence database S. (4) L is the closed frequent sequence candidate set.
> **Method:**
> 1. Check whether a discovered sequence β exists s.t. either $\alpha \sqsubseteq \beta$ or $\beta \sqsubseteq \alpha$, and $\mathcal{I}(S|_\alpha) = \mathcal{I}(S|_\beta)$; If such pattern exists, then apply Corollary 1 or 2 and return;
> 2. Insert α into L;
> 3. Scan $S|_\alpha$ once, find each frequent item, b, such that
>   (a) b can be assembled to the last element of α to form a new sequential pattern; or
>   (b) $\langle b \rangle$ can be appended to α to form a new sequential pattern;
> 4. For each new sequential pattern α', if α' is frequent, construct α'-projected database $S|_{\alpha'}$, and call CloSpan($\alpha', l + 1, S|_{\alpha'}, L$).

After we perform CloSpan($\alpha, l, S|_\alpha, L$), we get a closed frequent sequence candidate set. L. A post-processing step is required in order to delete non-closed sequential patterns existing in L.

5 Experimental Results and Performance Analysis

Since GSP [30] and SPADE [36] are the two most influential sequential pattern mining algorithms, we conduct an extensive performance study to compare PrefixSpan with them. In this section, we first report our experimental results on the performance of PrefixSpan in comparison with GSP and SPADE and then present our performance results of CloSpan in comparison with PrefixSpan.

5.1 Performance Comparison
Among PrefixSpan, FreeSpan, GSP and SPADE

To evaluate the effectiveness and efficiency of the PrefixSpan algorithm, we performed an extensive performance study of four algorithms: PrefixSpan, FreeSpan, GSP and SPADE, on both real and synthetic data sets, with various kinds of sizes and data distributions.

All experiments were conducted on a 750 MHz AMD PC with 512 megabytes main memory, running Microsoft Windows-2000 Server. Three algorithms, GSP, FreeSpan, and PrefixSpan, were implemented by us using Microsoft Visual C++ 6.0. The implementation of the fourth algorithm, SPADE, is obtained directly from the author of the algorithm [36].

For real data set, we obtained the *Gazelle* data set from Blue Martini Software. This data set is used in KDD-CUP'2000. It contains customers' web click-stream data from Gazelle.com, a legwear and legcare web retailer. For each customer, there are several sessions of webpage click-stream and each session can have multiple webpage views. Because each session is associated with both starting and ending date/time, for each customer we can sort its sessions of click-stream into a sequence of page views according to the viewing date/time. This dataset contains 29369 sequences (i.e., customers), 35722 sessions (i.e., transactions or events), and 87546 page views (i.e., products or items). There are in total 1423 distinct page views. More detailed information about this data set can be found in [16].

For synthetic data sets, we have also used a large set of synthetic sequence data generated by a data generator similar in spirit to the IBM data generator [2] designed for testing sequential pattern mining algorithms. Various kinds of sizes and data distributions of data sets are generated and tested in this performance study. The convention for the data sets is as follows: $C200T2.5S10I1.25$ means that the data set contains 200 k customers (i.e., sequences) and the number of items is 10000. The average number of items in a transaction (i.e., event) is 2.5 and the average number of transactions in a sequence is 10. On average, a frequent sequential pattern consists of 4 transactions, and each transaction is composed of 1.25 items.

To make our experiments fair to all the algorithms, our synthetic test data sets are similar to that used in the performance study in [36]. Additional data sets are used for scalability study and for testing the algorithm behavior with varied (and sometimes very low) support thresholds.

The first test of the four algorithms is on the data set $C10T8S8I8$, which contains 10 k customers (i.e., sequences) and the number of items is 1000. Both the average number of items in a transaction (i.e., event) and the average number of transactions in a sequence are set to 8. On average, a frequent sequential pattern consists of 4 transactions, and each transaction is composed of 8 items. Figure 5 shows the distribution of frequent sequences of data set $C10T8S8I8$, from which one can see that when *min_support* is no less than 1%, the length of frequent sequences is very short (only

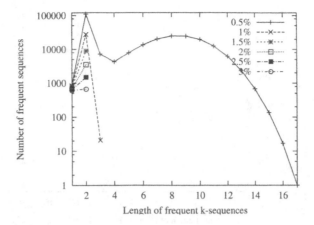

Fig. 5. Distribution of frequent sequences of data set $C10T8S8I8$

2–3), and the maximum number of frequent patterns in total is less than 10,000. Figure 6 shows the processing time of the four algorithms at different support thresholds. The processing times are sorted in time ascending order as "PrefixSpan < SPADE < FreeSpan < GSP". When $min\_support = 1\%$, PrefixSpan (runtime = 6.8 seconds) is about two orders of magnitude faster than GSP (runtime = 772.72 seconds). When $min\_support$ is reduced to 0.5%, the data set contains a large number of frequent sequences, PrefixSpan takes 32.56 seconds, which is more than 3.5 times faster than SPADE (116.35 seconds), while GSP never terminates on our machine.

The performance study on the real data set *Gazelle* is reported as follows. Figure 7 shows the distribution of frequent sequences of Gazelle dataset for different support thresholds. We can see that this dataset is a very sparse

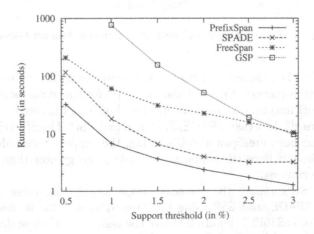

Fig. 6. Performance of the four algorithms on data set $C10T8S8I8$

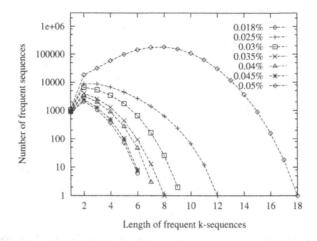

Fig. 7. Distribution of frequent sequences of data set Gazelle

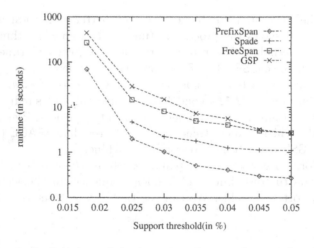

Fig. 8. Performance of the four algorithms on data set Gazelle

dataset: only when the support threshold is lower than 0.05% are there some long frequent sequences. Figure 8 shows the performance comparison among the four algorithms for Gazelle dataset. From Fig. 8 we can see that PrefixSpan is much more efficient than SPADE, FreeSpan and GSP. The SPADE algorithm is faster than both FreeSpan and GSP when the support threshold is no less than 0.025%, but once the support threshold is no greater than 0.018%, it cannot stop running.

Finally, we compare the memory usage among the three algorithms, PrefixSpan, SPADE, and GSP using both real data set Gazelle and synthetic data set C200T5S10I2.5. Figure 9 shows the results for Gazelle dataset, from which we can see that PrefixSpan is efficient in memory usage. It consumes

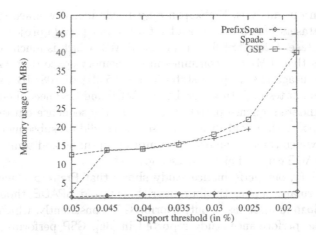

Fig. 9. Memory usage comparison among PrefixSpan, SPADE, and GSP for data set Gazelle

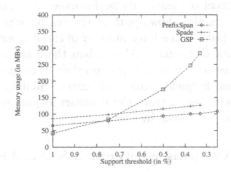

Fig. 10. Memory usage: PrefixSpan. SPADE, and GSP for synthetic data set C200T5S10I2.5

almost one order of magnitude less memory than both SPADE and GSP. For example, at support 0.018%, GSP consumes about 40 MB memory and SPADE just cannot stop running after it has used more than 22 MB memory while PrefixSpan only uses about 2.7 MB memory.

Figure 10 demonstrates the memory usage for dataset C200T5S10I2.5, from which we can see that PrefixSpan is not only more efficient but also more stable in memory usage than both SPADE and GSP. At support 0.25%, GSP cannot stop running after it has consumed about 362 MB memory and SPADE reported an error message *"memory:: Array: Not enough memory"* when it tried to allocate another bulk of memory after it has used about 262 MB memory, while PrefixSpan only uses 108 MB memory. This also explains why in several cases in our previous experiments when the support threshold becomes really low, only PrefixSpan can finish running.

Based on our analysis, PrefixSpan only needs memory space to hold the sequence datasets plus a set of header tables and pseudo-projection tables. Since the dataset C200T5S10I2.5 is about 46 MB, which is much bigger than Gazelle (less than 1 MB), it consumes more memory space than Gazelle but the memory usage is still quite stable (from 65 MB to 108 MB for different thresholds in our testing). However, both SPADE and GSP need memory space to hold candidate sequence patterns as well as the sequence datasets. When the *min_support* threshold drops, the set of candidate subsequences grows up quickly, which causes memory consumption upsurge, and sometimes both GSP and SPADE cannot finish processing.

In summary, our performance study shows that PrefixSpan has the best overall performance among the four algorithms tested. SPADE, though weaker than PrefixSpan in most cases, outperforms GSP consistently, which is consistent with the performance study reported in [36]. GSP performs fairly well only when *min_support* is rather high, with good scalability, which is consistent with the performance study reported in [30]. However, when there are a large number of frequent sequences, its performance starts deteriorating. Our memory usage analysis also shows part of the reason why some algorithms becomes really slow because the huge number of candidate sets may consume a tremendous amount of memory. Also, when there are a large number of frequent subsequences, all the algorithms run slow. This problem can be partially solved by closed frequent sequential pattern mining. In the remaining of this section, we will demonstrate the compactness of closed patterns and the better performance achieved by CloSpan.

5.2 Performance Comparison between CloSpan and PrefixSpan

The performance comparison between CloSpan and PrefixSpan are conducted in a different programming environment. All the experiments are done on a 1.7 GHZ Intel Pentium-4 PC with 1 GB main memory, running Windows XP Professional. All two algorithms are written in C++ with STL library support and compiled by g++ in cygwin environment with −O3 optimization.

The first experiment was conducted on the dataset $C10T2.5S6I2.5$ with 10 k items. Figure 11 shows the run time of both PrefixSpan and CloSpan with different support threshold. PrefixSpan cannot complete the task below support threshold 0.001 due to too long runtime. Figure 12 shows the distribution of discovered frequent closed sequences in terms of length. With the decreasing minimum support, the maximum length of frequent closed sequences grows larger. Figure 13 shows the number of frequent sequences which are discovered and checked in order to generate the frequent closed sequence set. This number is roughly equal to how many times the procedure, CloSpan, is called and how many times projected databases are generated. Surprisingly, this number accurately predicates the total running time as the great similarity exists between Fig. 11 and Fig. 12. Therefore, for the same dataset, the number of checked frequent sequences approximately determines the performance.

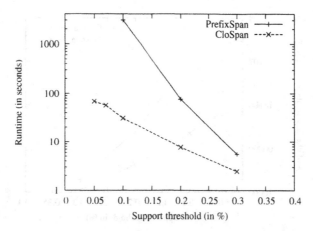

Fig. 11. The performance comparison between CloSpan and PrefixSpan on the dataset C10T2.5S6I2.5

We then test the performance of these two algorithms as some major parameters in the synthetic data generator are varied. The impact of different parameters is presented on the runtime of each algorithm. We select the following parameters as varied ones: the number of sequences in the dataset, the average number of transactions per sequence, and the average number of items per transaction. For each experiment, only one parameter varies with the others fixed. The experimental results are shown in Figs. 14 to 16. We also discovered in other experiments, the speed-up decreases when the number of distinct items in the dataset goes down. However, it is still faster than PrefixSpan.

Direct mining of closed patterns leads to much fewer patterns, especially when the patterns are long or when the minimum support threshold is low.

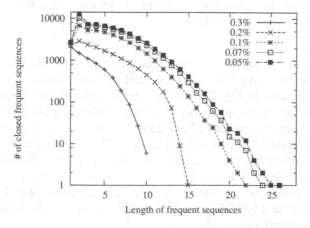

Fig. 12. The distribution of discovered frequent closed sequences in terms of length

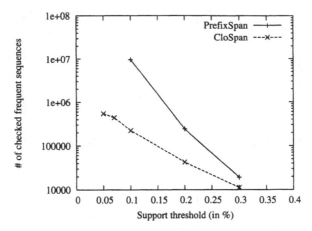

Fig. 13. The Number of frequent sequences checked

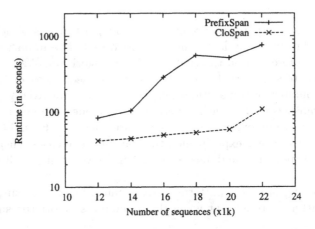

Fig. 14. Performance comparison vs. varying parameters: the number of sequences (C?T15S20I20, support threshold 0.8%

According to our analysis, the sets of patterns derived from CloSpan have the same expressive power as the traditional sequential pattern mining algorithms. As indicated in Fig. 13, CloSpan checks less frequent sequences than PrefixSpan and generates less number of projected databases. CloSpan clearly shows better performance than PrefixSpan in these cases. Our experimental results demonstrated this point since CloSpan often leads to savings in computation time over one order of magnitude in comparison with PrefixSpan. Based on the performance curves reported in Sect. 5.1 and the explosive number of subsequences generated for long sequences, it is expected that CloSpan will outperform GSP and SPADE as well when the patterns to be mined are long or when the support thresholds are low.

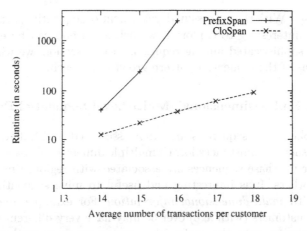

Fig. 15. Performance vs. varying parameters: Average number of transactions per sequence (C15T20S?I15, support threshold 0.75%)

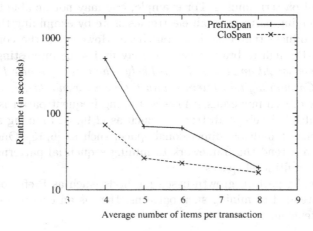

Fig. 16. Performance vs. varying parameters: Average number of items per transaction (C15T?S20I15, support threshold 1.2%)

CloSpan has been successfully used to improve the performance of storage systems by mining correlated access patterns [19]. A further study is in progress. Preliminary result shows that intelligent disk data layout using correlated access patterns can improve the average I/O response time by up to 25%.

6 Extensions of Sequential Pattern Growth Approach

Comparing with mining (unordered) frequent patterns, mining sequential patterns is one step towards mining more sophisticated frequent patterns in large databases. With the successful development of sequential pattern-growth method, it is interesting to explore how such a method can be extended to handle more sophisticated mining requests. In this section, we will discuss a few extensions of the sequential pattern growth approach.

6.1 Mining Multi-Dimensional, Multi-Level Sequential Patterns

In many applications, sequences are often associated with different circumstances, and such circumstances form a multiple dimensional space. For example, customer purchase sequences are associated with region, time, customer group, and others. It is interesting and useful to mine sequential patterns associated with *multi-dimensional information*. For example, one may find that retired customers (with age) over 60 may have very different patterns in shopping sequences from the professional customers younger than 40. Similarly, items in the sequences may also be associated with *different levels of abstraction*, and such multiple abstraction levels will form a multi-level space for sequential pattern mining. For example, one may not be able to find any interesting buying patterns in an electronics store by examining the concrete models of products that customers purchase. However, if the concept level is raised a little high to brand-level, one may find some interesting patterns, such as *"if one bought an IBM PC, it is likely s/he will buy a new IBM Laptop and then a Cannon digital camera within the next six months."*

There have been numerous studies at mining frequent patterns or associations at multiple levels of abstraction, such as [2,9], and mining association or correlations at multiple dimensional space, such as [6, 15]. One may like to see how to extend the framework to mining sequential patterns in multi-dimensional, multi-level spaces.

Interestingly, pattern growth-based methods, such as PrefixSpan, can be naturally extended to mining such patterns. Here is an example illustrating one such extension.

Example 8 (Mining multi-dimensional, multi-level sequential patterns). Consider a sequence database *SDB* in Table 4, where each sequence is associated with certain multi-dimensional, multi-level information. For example, it may contain multi-dimensional circumstance information, such as *cust-grp = business*, *city = Boston*, and *age-grp = middle_aged*. Also, each item may be associated with multiple-level information, such as item *b* being *IBM Laptop Thinkpad_X30*.

PrefixSpan can be extended to mining sequential patterns efficiently in such a multi-dimensional, multi-level environment. One such solution which we call *uniform sequential* (or *Uni-Seq*) [28] is outlined as follows. For each sequence, a set of multi-dimensional circumstance values can be treated as one added transaction in the sequence. For example, for *cid = 10*, (*business*, *Boston*, *middle_aged*) can be added into the sequence as one additional

Table 4. A multi-dimensional sequence database

cid	Cust-grp	City	Age-grp	Sequence
10	business	Boston	middle_aged	$\langle(bd)cba\rangle$
20	professional	Chicago	young	$\langle(bf)(ce)(fg)\rangle$
30	business	Chicago	middle_aged	$\langle(ah)abf\rangle$
40	education	New York	retired	$\langle(be)(ce)\rangle$

transaction. Similarly, for each item b, its associated multi-level information can be added as additional items into the same transaction that b resides. Thus the first sequence can be transformed into a sequence cid_{10} as, $cid_{10} : \langle$ (business, Boston, middle_aged), (($IBM, Laptop, Thinkpad_X$30), ($Dell, PC, Precision_3$30)) ($Canon, digital_camera, CD$420), ($IBM, Laptop, Thinkpad_X$30), ($Microsoft, RDBMS, SQLServer_$2000)$\rangle$. With such transformation, the database becomes a typical single-dimensional, single-level sequence database, and the PrefixSpan algorithm can be applied to efficient mining of multi-dimensional, multi-level sequential patterns. □

The proposed embedding of multi-dimensional, multi-level information into a transformed sequence database, and then extension of PrefixSpan to mining sequential patterns, as shown in Example 8, has been studied and implemented in [28]. In the study, we propose a few alternative methods, which integrate some efficient cubing algorithms, such as BUC [5] and H-cubing [10], with PrefixSpan. A detailed performance study in [28] shows that the *Uni-Seq* is an efficient algorithm. Another interesting algorithm, called *Seq-Dim*, which first mines sequential patterns, and then for each sequential pattern, forms projected multi-dimensional database and finds multi-dimensional patterns within the projected databases, also shows high performance in some situations. In both cases, PrefixSpan forms the kernel of the algorithm for efficient mining of multi-dimensional, multi-level sequential patterns.

6.2 Constraint-Based Mining of Sequential Patterns

For many sequential pattern mining applications, instead of finding all the possible sequential patterns in a database, a user may often like to enforce certain constraints to find desired patterns. The mining process which incorporates user-specified constraints to reduce search space and derive only the user-interested patterns is called *constraint-based mining*.

Constraint-based mining has been studied extensively in frequent pattern mining, such as [3, 23, 25]. In general, constraints can be characterized based on the notion of monotonicity, anti-monotonicity, succinctness, as well as convertible and inconvertible constraints respectively, depending on whether a constraint can be transformed into one of these categories if it does not naturally belong to one of them [25]. This has become a classical framework for constraint-based frequent pattern mining.

Interestingly, such a constraint-based mining framework can be extended to sequential pattern mining. Moreover, with pattern-growth framework, some previously not-so-easy-to-push constraints, such as regular expression constraints [7] can be handled elegantly. Let's examine one such example.

Example 9 (Constraint-based sequential pattern mining). Suppose our task is to mine sequential patterns with a regular expression constraint $C = \langle a *$ $\{bb|(bc)d|dd\}\rangle$ with *min_support* $= 2$, in a sequence database S (Table 1).

Since a regular expression constraint, like C, is neither anti-monotone, nor monotone, nor succinct, the classical constraint-pushing framework [23] cannot push it deep. To overcome this difficulty, Reference [7] develop a set of four SPIRIT algorithms, each pushing a stronger relaxation of regular expression constraint $\mathcal{R}$ than its predecessor in the pattern mining loop. However, the basic evaluation framework for sequential patterns is still based on GSP [30], a typical candidate generation-and-test approach.

With the development of the pattern-growth methodology, such kinds of constraints can be pushed deep easily and elegantly into the sequential pattern mining process [27]. This is because in the context of PrefixSpan a regular expression constraint has a nice property called *growth-based anti-monotonic*. A constraint is *growth-based anti-monotonic* if it has the following property: *if a sequence α satisfies the constraint, α must be reachable by growing from any component which matches part of the regular expression.*

The constraint $C = \langle a * \{bb|(bc)d|dd\}\rangle$ can be integrated with the pattern-growth mining process as follows. First, only the $\langle a\rangle$-projected database needs to be mined since the regular expression constraint C starting with a, and only the sequences which contain frequent single item within the set of $\{b, c, d\}$ should retain in the $\langle a\rangle$-projected database. Second, the remaining mining can proceed from the suffix, which is essentially "*Suffix-Span*", an algorithm symmetric to PrefixSpan by growing suffixes from the end of the sequence forward. The growth should match the suffix constraint "$\langle\{bb|(bc)d|dd\}\rangle$". For the projected databases which matches these suffixes, one can grow sequential patterns either in prefix- or suffix- expansion manner to find all the remaining sequential patterns.   □

Notice that the regular expression constraint C given in Example 9 is in a special form "$\langle prefix * suffix\rangle$" out of many possible general regular expressions. In this special case, an integration of PrefixSpan and *Suffix-Span* may achieve the best performance. In general, a regular expression could be of the form "$\langle * \alpha_1 * \alpha_2 * \alpha_3 * \rangle$", where α_i is a set of instantiated regular expressions. In this case, FreeSpan should be applied to push the instantiated items by expansion first from the instantiated items. A detailed discussion of constraint-based sequential pattern mining is in [27].

6.3 Mining Top-k Closed Sequential Patterns

Mining closed patterns may significantly reduce the number of patterns generated and is *information lossless* because it can be used to derive the complete set of sequential patterns. However, setting min_support is a subtle task: *A too small value may lead to the generation of thousands of patterns, whereas a too big one may lead to no answer found.* To come up with an appropriate min_support, one needs prior knowledge about the mining query and the task-specific data, and be able to estimate beforehand how many patterns will be generated with a particular threshold.

As proposed in [13], a desirable solution is to change the task of mining frequent patterns to *mining top-k frequent closed patterns of minimum length min_ℓ*, where k is the number of closed patterns to be mined, top-k refers to the k most frequent patterns, and *min_ℓ* is the minimum length of the closed patterns. We develop TSP [31] to discover top-k closed sequences. TSP is a multi-pass search space traversal algorithm that finds the most frequent patterns early in the mining process and allows dynamic raising of min_support which is then used to prune unpromising branches in the search space. Also, TSP devises an efficient closed pattern verification method which guarantees that during the mining process the candidate result set consists of the desired number of closed sequential patterns. The efficiency of TSP is further improved by applying the minimum length constraint in the mining and by employing the early termination conditions developed in CloSpan [35].

6.4 Mining Approximate Consensus Sequential Patterns

As we discussed before, conventional sequential pattern mining methods may meet inherent difficulties in mining databases with long sequences and noise. They may generate a huge number of short and trivial patterns but fail to find interesting patterns approximately shared by many sequences. In many applications, it is necessary to mine sequential patterns approximately shared by many sequences.

To attack these problems, in [17], we propose the theme of *approximate sequential pattern mining* roughly defined as *identifying patterns approximately shared by many sequences*. We present an efficient and effective algorithm, *ApproxMap* (for APPROXimate Multiple Alignment Pattern mining), to mine consensus patterns from large sequence databases. The method works in two steps. First, the sequences are clustered by similarity. Then, the consensus patterns are mined directly from each cluster through multiple alignments. A novel structure called weighted sequence is used to compress the alignment result. For each cluster, the longest consensus pattern best representing the cluster is generated from its weighted sequence.

Our extensive experimental results on both synthetic and real data sets show that *ApproxMap* is robust to noise and is both effective and efficient in mining approximate sequential patterns from noisy sequence databases with lengthy sequences. In particular, we report a successful case of mining a real data set which triggered important investigations in welfare services.

6.5 Clustering Time Series Gene Expression

Clustering the time series gene expression data is an important task in bioinformatics research and biomedical applications. Time series gene expression data is also in the form of sequences. Recently, some clustering methods have been adapted or proposed. However, some problems still remain, such as the robustness of the mining methods, the quality and the interpretability of the mining results.

In [14], we tackle the problem of *effectively clustering time series gene expression data* by proposing algorithm *DHC*, a *density-based, hierarchical clustering method aiming at time series gene expression data*. We use a density-based approach to identifying the clusters such that the clustering results are with high quality and robustness. Moreover, The mining result is in the form of a *density tree*, which uncovers the embedded clusters in a data sets. The inner-structures, the borders and the outliers of the clusters can be further investigated using the *attraction tree*, which is a intermediate result of the mining. By these two trees, the internal structure of the data set can be visualized effectively. Our empirical evaluation using some real-world data sets show that the method is effective, robust and scalable. It matches the ground truth given by bioinformatics experts very well in the sample data sets.

6.6 Towards Mining More Complex Kinds of Structured Patterns

Besides mining sequential patterns, another important task is the mining of frequent sub-structures in a database composed of structured or semi-structured data sets. The substructures may consist of trees, directed-acyclic graphs (i.e., DAGs), or general graphs which may contain cycles. There are a lot of applications related to mining frequent substructures since most human activities and natural processes may contain certain structures, and a huge amount of such data has been collected in large data/information repositories, such as molecule or bio-chemical structures, Web connection structures, and so on. It is important to develop scalable and flexible methods for mining structured patterns in such databases. There have been some recent work on mining frequent subtrees, such as [38], and frequent subgraphs, such as [18,33] in structured databases, where [33] shows that the pattern growth approach has clear performance edge over a candidate generation-and-test approach. Furthermore, as discussed above, is it is more desirable to mine closed frequent subgraphs (a subgraph g is *closed* if there exists no super-graph of g carrying the same support as g) than mining explicitly the complete set of frequent subgraphs because a large graph inherently contains an exponential number of subgraphs. A recent study [34] has developed an efficient closed subgraph pattern method, called *CloseGraph*, which is also based on the pattern-growth framework and influenced by this approach.

7 Conclusions

We have introduced a *pattern-growth approach* for efficient and scalable mining of sequential patterns in large sequence databases. Instead of refinement of the Apriori-like, candidate generation-and-test approach, such as GSP [30] and SPADE [36], we promote a divide-and-conquer approach, called *pattern-growth approach*, which is an extension of FP-growth [12], an efficient pattern-growth algorithm for mining frequent patterns without candidate generation.

An efficient pattern-growth method is developed for mining frequent sequential patterns, represented by PrefixSpan, and mining closed sequential patterns, represented by CloSpan, are presented and studied in this paper.

PrefixSpan recursively projects a sequence database into a set of smaller projected sequence databases and grows sequential patterns in each projected database by exploring only locally frequent fragments. It mines the complete set of sequential patterns and substantially reduces the efforts of candidate subsequence generation. Since PrefixSpan explores ordered growth by prefix-ordered expansion, it results in less "growth points" and reduced projected databases in comparison with our previously proposed pattern-growth algorithm, FreeSpan. Furthermore, a *pseudo-projection* technique is proposed for PrefixSpan to reduce the number of physical projected databases to be generated.

CloSpan mines closed sequential patterns efficiently by discovery of sharing portions of the projected databases in the mining process and prune any redundant search space and therefore substantially enhanced the mining efficiency and reduces the redundant patterns.

Our comprehensive performance study shows that PrefixSpan outperforms the Apriori-based GSP algorithm, FreeSpan, and SPADE in most cases, and PrefixSpan integrated with pseudo-projection is the fastest among all the tested algorithms for mining the complete set of sequential patterns; whereas CloSpan may substantially improve the mining efficiency over PrefixSpan and returns a substantially smaller set of results while preserving the completeness of the answer sets.

Based on our view, the implication of this method is far beyond yet another efficient sequential pattern mining algorithm. It demonstrates the strength of the pattern-growth mining methodology since the methodology has achieved high performance in both frequent-pattern mining and sequential pattern mining. Moreover, our discussion shows that the methodology can be extended to mining multi-level, multi-dimensional sequential patterns, mining sequential patterns with user-specified constraints, and a few interesting applications. Therefore, it represents a promising approach for the applications that rely on the discovery of frequent patterns and/or sequential patterns.

There are many interesting issues that need to be studied further. Especially, the developments of specialized sequential pattern mining methods for particular applications, such as DNA sequence mining that may admit faults, such as allowing insertions, deletions and mutations in DNA sequences, and handling industry/engineering sequential process analysis are interesting issues for future research.

References

1. R. Agrawal and R. Srikant. Fast algorithms for mining association rules. In *Proc. 1994 Int. Conf. Very Large Data Bases (VLDB'94)*, pp. 487–499, Santiago, Chile, Sept. 1994.
2. R. Agrawal and R. Srikant. Mining sequential patterns. In *Proc. 1995 Int. Conf. Data Engineering (ICDE'95)*, pp. 3–14, Taipei, Taiwan, Mar. 1995.
3. R. J. Bayardo, R. Agrawal, and D. Gunopulos. Constraint-based rule mining on large, dense data sets. In *Proc. 1999 Int. Conf. Data Engineering (ICDE'99)*, pp. 188–197, Sydney, Australia, April 1999.
4. C. Bettini, X. S. Wang, and S. Jajodia. Mining temporal relationships with multiple granularities in time sequences. *Data Engineering Bulletin*, 21:32–38, 1998.
5. K. Beyer and R. Ramakrishnan. Bottom-up computation of sparse and iceberg cubes. In *Proc. 1999 ACM-SIGMOD Int. Conf. Management of Data (SIGMOD'99)*, pp. 359–370, Philadelphia, PA, June 1999.
6. G. Grahne, L. V. S. Lakshmanan, X. Wang, and M. H. Xie. On dual mining: From patterns to circumstances, and back. In *Proc. 2001 Int. Conf. Data Engineering (ICDE'01)*, pp. 195–204, Heidelberg, Germany, April 2001.
7. S. Guha, R. Rastogi, and K. Shim. ROCK: A robust clustering algorithm for categorical attributes. In *Proc. 1999 Int. Conf. Data Engineering (ICDE'99)*, pp. 512–521, Sydney, Australia, Mar. 1999.
8. J. Han, G. Dong, and Y. Yin. Efficient mining of partial periodic patterns in time series database. In *Proc. 1999 Int. Conf. Data Engineering (ICDE'99)*, pp. 106–115, Sydney, Australia, April 1999.
9. J. Han and Y. Fu. Discovery of multiple-level association rules from large databases. In *Proc. 1995 Int. Conf. Very Large Data Bases (VLDB'95)*, pp. 420–431, Zurich, Switzerland, Sept. 1995.
10. J. Han, J. Pei, G. Dong, and K. Wang. Efficient computation of iceberg cubes with complex measures. In *Proc. 2001 ACM-SIGMOD Int. Conf. Management of Data (SIGMOD'01)*, pp. 1–12, Santa Barbara, CA, May 2001.
11. J. Han, J. Pei, B. Mortazavi-Asl, Q. Chen, U. Dayal, and M.-C. Hsu. FreeSpan: Frequent pattern-projected sequential pattern mining. In *Proc. 2000 ACM SIGKDD Int. Conf. Knowledge Discovery in Databases (KDD'00)*, pp. 355–359, Boston, MA, Aug. 2000.
12. J. Han, J. Pei, and Y. Yin. Mining frequent patterns without candidate generation. In *Proc. 2000 ACM-SIGMOD Int. Conf. Management of Data (SIGMOD'00)*, pp. 1–12, Dallas, TX, May 2000.
13. J. Han, J. Wang, Y. Lu, and P. Tzvetkov. Mining top-k frequent closed patterns without minimum support. In *Proc. 2002 Int. Conf. on Data Mining (ICDM'02)*, pp. 211–218, Maebashi, Japan, Dec. 2002.

14. D. Jiang, J. Pei, and A. Zhang. DHC: A density-based hierarchical clustering method for gene expression data. In *Proc. 3rd IEEE Symp. Bio-informatics and Bio-engineering (BIB'03)*, Washington D.C., March 2003.

15. M. Kamber, J. Han, and J. Y. Chiang. Metarule-guided mining of multi-dimensional association rules using data cubes. In *Proc. 1997 Int. Conf. Knowledge Discovery and Data Mining (KDD'97)*, pp. 207–210, Newport Beach, CA, Aug. 1997.

16. R. Kohavi, C. Brodley, B. Frasca, L. Mason, and Z. Zheng. KDD-Cup 2000 organizers' report: Peeling the onion. *SIGKDD Explorations*, 2:86–98, 2000.

17. H. Kum, J. Pei, and W. Wang. Approxmap: Approximate mining of consensus sequential patterns. In *Proc. 2003 SIAM Int. Conf. on Data Mining (SDM '03)*, San Francisco, CA, May 2003.

18. M. Kuramochi and G. Karypis. Frequent subgraph discovery. In *Proc. 2001 Int. Conf. Data Mining (ICDM'01)*, pp. 313–320, San Jose, CA, Nov. 2001.

19. Z. Li, S. M. Srinivasan, Z. Chen, Y. Zhou, P. Tzvetkov, X. Yan, and J. Han. Using data mining for discovering patterns in autonomic storage systems. In *Proc. 2003 ACM Workshop on Algorithms and Architectures for Self-Managing Systems (AASMS'03)*, San Diego, CA, June 2003.

20. H. Lu, J. Han, and L. Feng. Stock movement and n-dimensional inter-transaction association rules. In *Proc. 1998 SIGMOD Workshop Research Issues on Data Mining and Knowledge Discovery (DMKD'98)*, pp. 12:1–12:7, Seattle, WA, June 1998.

21. H. Mannila, H. Toivonen, and A. I. Verkamo. Discovery of frequent episodes in event sequences. *Data Mining and Knowledge Discovery*, 1:259–289, 1997.

22. F. Masseglia, F. Cathala, and P. Poncelet. The psp approach for mining sequential patterns. In *Proc. 1998 European Symp. Principle of Data Mining and Knowledge Discovery (PKDD'98)*, pp. 176–184, Nantes, France, Sept. 1998.

23. R. Ng, L. V. S. Lakshmanan, J. Han, and A. Pang. Exploratory mining and pruning optimizations of constrained associations rules. In *Proc. 1998 ACM-SIGMOD Int. Conf. Management of Data (SIGMOD'98)*, pp. 13–24, Seattle, WA, June 1998.

24. B. Özden, S. Ramaswamy, and A. Silberschatz. Cyclic association rules. In *Proc. 1998 Int. Conf. Data Engineering (ICDE'98)*, pp. 412–421, Orlando, FL, Feb. 1998.

25. J. Pei, J. Han, and L. V. S. Lakshmanan. Mining frequent itemsets with convertible constraints. In *Proc. 2001 Int. Conf. Data Engineering (ICDE'01)*, pp. 433–332, Heidelberg, Germany, April 2001.

26. J. Pei, J. Han, B. Mortazavi-Asl, H. Pinto, Q. Chen, U. Dayal, and M.-C. Hsu. PrefixSpan: Mining sequential patterns efficiently by prefix-projected pattern growth. In *Proc. 2001 Int. Conf. Data Engineering (ICDE'01)*, pp. 215–224, Heidelberg, Germany, April 2001.

27. J. Pei, J. Han, and W. Wang. Constraint-based sequential pattern mining in large databases. In *Proc. 2002 Int. Conf. Information and Knowledge Management (CIKM'02)*, pp. 18–25, McLean, VA, Nov. 2002.

28. H. Pinto, J. Han. J. Pei, K. Wang, Q. Chen, and U. Dayal. Multi-dimensional sequential pattern mining. In *Proc. 2001 Int. Conf. Information and Knowledge Management (CIKM'01)*, pp. 81–88, Atlanta, GA, Nov. 2001.

29. S. Ramaswamy, S. Mahajan, and A. Silberschatz. On the discovery of interesting patterns in association rules. In *Proc. 1998 Int. Conf. Very Large Data Bases (VLDB'98)*, pp. 368–379, New York, NY, Aug. 1998.

30. R. Srikant and R. Agrawal. Mining sequential patterns: Generalizations and per-
    formance improvements. In *Proc. 5th Int. Conf. Extending Database Technology
    (EDBT'96)*, pp. 3–17, Avignon, France, Mar. 1996.
31. P. Tzvetkov, X. Yan, and J. Han. Tsp: Mining top-k closed sequential patterns.
    In *Proc. 2003 Int. Conf. Data Mining (ICDM'03)*, Melbourne, FL, Nov. 2003.
32. J. Wang, G. Chirn, T. Marr, B. Shapiro, D. Shasha, and K. Zhang. Combinatior-
    ial pattern discovery for scientific data: Some preliminary results. In *Proc. 1994
    ACM-SIGMOD Int. Conf. Management of Data (SIGMOD'94)*, pp. 115–125,
    Minneapolis, MN, May, 1994.
33. X. Yan and J. Han. gSpan: Graph-based substructure pattern mining. In *Proc.
    2002 Int. Conf. on Data Mining (ICDM'02)*, pp. 721–724, Maebashi, Japan,
    Dec. 2002.
34. X. Yan and J. Han. CloseGraph: Mining closed frequent graph patterns. In
    *Proc. 2003 ACM SIGKDD Int. Conf. Knowledge Discovery and Data Mining
    (KDD'03)*, Washington, D.C., Aug. 2003.
35. X. Yan, J. Han, and R. Afshar. CloSpan: Mining closed sequential patterns
    in large datasets. In *Proc. 2003 SIAM Int. Conf. Data Mining (SDM'03)*,
    pp. 166–177, San Fransisco, CA, May 2003.
36. M. Zaki. SPADE: An efficient algorithm for mining frequent sequences. *Machine
    Learning*, 40:31–60, 2001.
37. M. J. Zaki. Efficient enumeration of frequent sequences. In *Proc. 7th Int. Conf.
    Information and Knowledge Management (CIKM'98)*, pp. 68–75, Washington
    DC, Nov. 1998.
38. M. J. Zaki. Efficiently mining frequent trees in a forest. In *Proc. 2002 ACM
    SIGKDD Int. Conf. Knowledge Discovery in Databases (KDD'02)*, pp. 71–80,
    Edmonton, Canada, July 2002.
39. M. J. Zaki and C. J. Hsiao. CHARM: An efficient algorithm for closed itemset
    mining. In *Proc. 2002 SIAM Int. Conf. Data Mining (SDM'02)*, pp. 457–473,
    Arlington, VA, April 2002.

Web Page Classification*

B. Choi and Z. Yao

Computer Science, College of Engineering and Science, Louisiana Tech University,
Ruston, LA 71272, USA
pro@BenChoi.org, zya001@latech.edu

Abstract. This chapter describes systems that automatically classify web pages into meaningful categories. It first defines two types of web page classification: subject based and genre based classifications. It then describes the state of the art techniques and subsystems used to build automatic web page classification systems, including web page representations, dimensionality reductions, web page classifiers, and evaluation of web page classifiers. Such systems are essential tools for Web Mining and for the future of Semantic Web.

1 Introduction

1.1 Motivation

Over the past decade we have witnessed an explosive growth on the Internet, with millions of web pages on every topic easily accessible through the Web. The Internet is a powerful medium for communication between computers and for accessing online documents all over the world but it is not a tool for locating or organizing the mass of information. Tools like search engines assist users in locating information on the Internet. They perform excellently in locating but provide limited ability in organizing the web pages. Internet users are now confronted with thousands of web pages returned by a search engine using simple keyword search. Searching through those web pages is in itself becoming impossible for users. Thus it has been of more interest in tools that can help make a relevant and quick selection of information that we are seeking. This chapter presents one of the most promising approaches: web page classification, which can efficiently support diversified applications, such as Web Mining, automatic web page categorization [17], information filtering, search engine, and user profile mining. It describes the state of the art techniques and subsystems used to build automatic web page classification

---

*This research was supported in part by a grant from the Center for Entrepreneurship and Information Technology (CEnIT), Louisiana Tech University.

systems. It starts with a definition of web page classification and a description
of two types of classifications.

1.2 A Definition of Web Page Classification

Web page classification, also known as web page categorization, may be defined as the task of determining whether a web page belongs to a category or
categories. Formally, let $C = \{c_1, \ldots, c_K\}$ be a set of predefined categories,
$D = \{d_1, \ldots, d_N\}$ be a set of web pages to be classified, and $A = D \times C$ be a
decision matrix:

Web Pages	Categories				
	c_1	$\ldots$	c_j	$\ldots$	c_K
d_1	a_{11}	$\ldots$	a_{1j}	$\ldots$	a_{1K}
$\ldots$	$\ldots$	$\ldots$	$\ldots$	$\ldots$	$\ldots$
d_i	a_{i1}	$\ldots$	a_{ij}	$\ldots$	a_{iK}
$\ldots$	$\ldots$	$\ldots$	$\ldots$	$\ldots$	$\ldots$
d_N	a_{N1}	$\ldots$	a_{Nj}	$\ldots$	a_{NK}

where, each entry a_{ij} ($1 \leq i \leq N$, $1 \leq j \leq K$) represents whether web page
d_i belongs to category c_j or not. Each $a_{ij} \in \{0, 1\}$ where 1 indicates web page
d_i belongs to category c_j, and 0 for not belonging. A web page can belong to
more than one category. The task of web page classification is to approximate
the unknown assignment function $f : D \times C \rightarrow \{0, 1\}$ by means of a learned
function $f' : D \times C \rightarrow \{0, 1\}$, called a classifier, a model, or a hypothesis, such
that f' coincides to f as much as possible [98].

The function f' is usually obtained by machine learning over a set of training examples of web pages. Each training example is tagged with a category
label. The function f' is induced during the training phase and is then used
during the classification phase to assign web pages to categories.

Throughout this chapter, we use "document" to denote a text document
including web pages, and use "HTML page", "web document", or "web page"
to indicate a web page. We also emphasize the characteristics of web page
classification that are different from traditional text classification.

1.3 Two Basic Approaches

Web page classification is in the area of machine learning, where learning is
over web pages. Using machine learning techniques on text databases is referred to as text learning, which has been well studied during the last two
decades [71]. Machine learning on web pages is similar to text learning since

web pages can be treated as text documents. Nevertheless, it is clear in advance that learning over web pages has new characteristics. First, web pages are semi-structured text documents that are usually written in HTML. Secondly, web pages are connected to each other forming direct graphs via *hyperlinks*. Thirdly, web pages are often short and by using only text in those web pages may be insufficient to analyze them. Finally, the sources of web pages are numerous, nonhomogeneous, distributed, and dynamically changing. In order to classify such large and heterogeneous web domain, we present in this chapter two basic classification approaches: *subject-based classification* and *genrebased classification*.

In *subject-based classification (also called topic-based classification)*, web pages are classified based on their contents or subjects. This approach defines numerous topic categories. Some examples of categories, under the "Science" domain used in Yahoo.com, are "Agriculture", "Astronomy", "Biology", "Chemistry", "Cognitive Science", "Complex Systems", and "Computer Science". The subject-based classification can be applied to build topic hierarchies of web pages, and subsequently to perform contextbased searches for web pages relating to specific subjects.

In *genre-based classification (also called style-based classification)*, web pages are classified depending on functional or genre related factors. In a broad sense, the word "genre" is used here merely as a literary substitute for "a kind of text". Some examples of web page genres are "product catalogue", "online shopping", "advertisement", "call for paper", "frequently asked questions", "home page", and "bulletin board". This approach can help users find immediate interests. Although text genre has been studied for a long history in linguistic literature, automatically text genre classification does not share much literature and fewer sophisticated research work has been done on web page genre classification. One important reason is that, up to recently, the digitized collections have been for the most part generically homogeneous, such as collections of scientific abstracts and newspaper articles. Thus the problem of genre identification could be set aside [52]. This problem does not become salient until we are confronted with heterogeneous domain like the Web. In fact, the Web is so diverse that no topic taxonomy can hope to capture all topics in sufficiently detail. It is worth noticing that the genre-based approach is not to reduce the importance of the subject-based approach, but rather to add another dimension to web page classification as a whole.

1.4 Overview of this Chapter

This chapter addresses three main topics of web page classification: web page representations, dimensionality reductions, and web page classifiers. It then concludes with an evaluation of classification methods and a summary. Section 2 describes how to encode or represent web pages for facilitating machine learning and classification processes. Web pages are usually represented by multi-dimensional vectors [93]. each dimension of which encodes a feature of the web pages. If all features of web pages are used in the representations, the number of dimensions of the vectors will usually be very high (hundreds of thousands). To reduce both time and space for computation, various methods are introduced to reduce the dimensionality as described in Sect. 3. Machine learning methods for classifying web pages are then applied to induce the classification function f' as defined in Sect. 1.2 or to induce representations for categories from the representations of training web pages. Section 4 describes various classifiers that have been proven efficient for web page classification. When a web page needed to be classified, the classifiers use the learned function to assign the web page to categories. Some classifiers compare the similarity of the representation of the web page to the representations of the categories. The category having the highest similarity is usually considered as the most appropriate category for assigning the web page. Section 5 discusses how to evaluate web page classification methods and provides experimental results for comparing various classifiers. Finally, Sect. 6 provides a summary of this chapter.

2 Web Page Representations

The first step in web page classification is to transform a web page, which typically composes of strings of characters, hyperlinks, images, and HTML tags, into a feature vector. This procedure is used to remove less important information and to extract salient features from the web pages. Apparently, the subject-based classification prefers features representing contents or subjects of web pages and these features may not represent genres of the web pages. Here we present different web page representations for the two basic classification approaches.

2.1 Representations for Subject Based Classification

Most work for subject-based classifications believes the text source (e.g. words, phases, and sentences) represents the content of a web page. In order to retrieve important textual features, web pages are first preprocessed to discard the less important data. The preprocessing consists of the following steps:

- **Removing HTML tags:** HTML tags indicate the formats of web pages. For instance, the content within <title> and </title> pair is the title of a web page; the content enclosed by <table> and </table> pair is a table. These HTML tags may indicate the importance of their enclosed content and they can thus help weight their enclosed content. The tags themselves are removed after weighting their enclosed content.
- **Removing stop words:** stop words are frequent words that carry little information, such as prepositions, pronouns, and conjunctions. They are removed by comparing the input text with a "stop list" of words.

- **Removing rare words:** low frequency words are also removed based on Luhn's idea that rare words do not contribute significantly to the content of a text [64]. This is to be done by removing words whose number of occurrences in the text are less than a predefined threshold.
- **Performing word stemming:** this is done by grouping words that have the same stem or root, such as computer, compute, and computing. The Porter stemmer is a well-known algorithm for performing this task.

After the preprocessing, we select features to represent each web page. Before going into details of web page representations, we standardize to use M to denote the dimension of a vector (or the number of features in the training corpus), N for the total number of web pages in the training collection, i for the ith vector, and j for the jth feature of a vector.

Bag-of-terms Representation

In the simplest way, a web page is represented by a vector of M weighted index words (see Fig. 1). This is often referred to as *bag-of-words* representation. The basic hypothesis behind the bag-of-words representation is that each word in a document indicates a certain concept in the document. Formally, a web page is represented by a vector d_i with words $t_1, t_2, \ldots, t_M$ as the features, each of which associates with a weight w_{ij}. That is,

$$d_i = (w_{i1}, w_{i2}, w_{i3}, \ldots, w_{iM}) \tag{1}$$

where M is the number of indexing words and w_{ij} $(1 \leq j \leq M)$ is the importance (or weight) of word t_j in the web page d_i.

Since a phrase usually contain more information than a single word, the bag-of-words representation can be enriched by adding new features generated

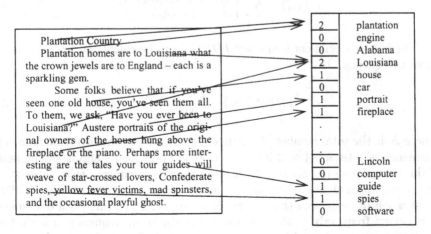

Fig. 1. Representing a web page in a vector space model. Each web page is converted into a vector of words in this case

from word sequences, also known as n-grams. For example, [71] generated features that compose up to five words (5-grams). By using n-grams we can capture some characteristic word combinations. Here we refer to the enriched bag-of-words representation as bag-of-terms, where a term can be a single word or any n-gram. The process of feature generation is performed in passes over documents, where i-grams are generated in the ith pass only from the features of length $i - 1$ generated in the previous pass [71]. Hereafter, we use "word" to denote a single word and use "term" to name an n-gram. Experiments showed that terms up to 3-grams are sufficient in most classification systems.

Deciding the weights of terms in a web page is a term-weighting problem. The simplest way to define the weight w_j of term t_j in a web page is to consider the binary occurrence as following:

$$w_j = \begin{cases} 1, & \text{if term } t_j \text{ is in the web page} \\ 0, & \text{otherwise} \end{cases} \tag{2}$$

Term Frequency Inverse Document Frequency

One of the most successful term weighting methods is TFIDF (Term Frequency Inverse Document Frequency) [94,95], which is obtained by the product of the local term importance (TF) and the global term importance (IDF):

$$w_{ij} = TF(t_j, d_i) \cdot IDF(t_j) \tag{3}$$

where term frequency $TF(t_j, d_i)$ is the number of times term t_j occurs in document d_i, and document frequency $DF(t_j)$ in (4) is the number of documents in which term t_j occurs at least once.

$$DF(t_j) = \sum_{i=1}^{N} \begin{cases} 1, & \text{if } d_i \text{ contains } t_j \\ 0, & \text{otherwise} \end{cases} \tag{4}$$

The inverse document frequency $IDF(t_j)$ can be calculated from the document frequency $DF(t_j)$:

$$IDF(t_j) = \log\left(\frac{N}{DF(t_j)}\right) \tag{5}$$

where N is the total number of documents. Intuitively, the inverse document frequency of a term is low if it occurs in many documents and is the highest if the term occurs in only one document. Many experiments have supported the discriminant characteristic of the inverse document frequency [48]. TFIDF term weighting expresses that a term is an important feature for a document if it occurs frequently in the document, i.e. the term frequency is high. On the other hand, a term becomes an unimportant feature for the document if the term occurs in many documents, i.e. the inverse document frequency is low [45]. Furthermore, the weight w_{ij} of term t_j in document d_i can be normalized by document length, e.g. Euclidian vector length (L^2-norm) of the document:

$$w_{ij} = \frac{TF(t_j, d_i)IDF(t_j)}{\sqrt{\sum_{j=1}^{M}(TF(t_j, d_i)IDF(t_j))^2}} \tag{6}$$

where M is the number of features (unique terms) in all training web pages. The normalization by document length can be considered as a part of term weighting.

Reference [45] proposed a modified version of the inverse document frequency. He modified the inverse document frequency as following:

$$DF'(t_j) = \sum_{i=1}^{N} \frac{TF(t_j, d_i)}{|d_i|} \tag{7}$$

$$IDF'(t_j) = sqrt\left(\frac{N}{DF'(t_j)}\right) \tag{8}$$

where the document length $|d_i|$ is the number of words in document d_i, and N is the total number of web pages in the training collection. As we can see, Joachims defined $DF'(t_j)$, document frequency of term t_j, as the sum of relative frequencies of term t_j in each web page. He argued that $IDF'(t_j)$ can make use of the frequency information instead of just considering binary occurrence information of $DF(t_j)$ (4). It can be noticed that the interpretations of $DF(t_j)$ and $DF'(t_j)$ are similar. The more often a term t_j occurs throughout the corpus, the higher $DF(t_j)$ and $DF'(t_j)$ will be. A difference is that the square root is used instead of the logarithm to dampen the effect of document frequency. Joachims then used $|d_i|$ for normalization instead of using Euclidian length. i.e.

$$w_{ij} = \frac{TF(t_j, d_i) \cdot IDF'(t_j)}{|d_i|} \tag{9}$$

The modified TFIDF term-weighting scheme applied in a probabilistic variant of Rocchio classifier showed performance improvement of up to 40% reduction of error rate in comparing to Rocchio classifier [45].

Balanced Term-weighting

The balanced term-weighting scheme [49] is optimized for similarity computation between documents, since similarity computation is often required in classification tasks. While the TFIDF term-weighting scheme only considers the terms that occur in documents, the basic premise under the balanced term-weighting scheme is that similarity between two documents is maximized if there are high number of matches not only in occurrence of terms but also in *absence* of terms. The main procedure of this scheme consists of the following four phases [49]:

1. Local term-weighting phase: assigns term frequency to the present terms, and assigns negative weights −1 for the absent terms. Use one vector to represent the positive weights, and another vector to represent the negative weights;
2. Global term-weighting phase: applies inverse document frequency for the positive and the negative weights;
3. Normalization phase: normalizes the positive weight vector and the negative weight vector independently by L^2-norm (Euclidean distance);
4. Merging phase: merges the positive and the negative weights into one vector.

By assigning negative weights to absence terms, the balanced termweighting scheme resolves the masking-by-zero problem in the dot product operation for calculating the similarity (or distance) of two vectors. Thus the similarity of two documents is increased if high number of same terms are absent in both documents. Experiments illustrated that the balanced term-weighting scheme achieved higher average precisions than other term-weighting schemes [49].

Web Structure Representation

The bag-of-terms web page representation does not exploit the structural information on the Web. There are at least two different kinds of structural information on the Web that could be used to enhance the performance of classification:

- The structure of an HTML representation which allows to easily identify important parts of a web page, such as its title and headings;
- The structure of the Web, where web pages are linked to each other via hyperlinks.

In this section, we will exploit different web page representation methods that are unique for the Web domain.

HTML Structure

For improving web page representation, exploiting HTML structure will help us identify where the more representative terms can be found. For example, we can consider that a term enclosed within the <title> and </title> tags is generally more representative for the topic of a web page than a term enclosed within the <body> and </body> tags. For instance, several sources (elements) for web page representation are:

- BODY, the body part of a web page;
- TITLE, the title of a web page;
- H1~H6, section headings;
- EM, emphasized content;

- URL, hyperlinks that may contain descriptor for the linked web pages. If keywords can be extracted from an URL, these keywords are relatively important;
- META, the meta-description of a web page. It is invisible to users of web browsing, but it may provide description, keywords, and date for a web page.

A Structure-oriented Weighting Technique (SWT) [89] can be used to weight the important features in web pages. The idea of SWT is to assign greater weights to terms that belong to the elements that are more suitable for representing web pages (such as terms enclosed in TITLE tags). SWT is defined by the function [89]:

$$\text{SWT}_w(t_j, d_i) = \sum_{e_k} (w(e_k) \cdot TF(t_j, e_k, d_i)) \tag{10}$$

where e_k is an HTML element, $w(e_k)$ denotes the weight assigned to the element e_k, and $TF(t_i, e_k, d_j)$ denotes the number of times term t_i is present in the element e_k of HTML page d_j. Reference [89] defined the function $w(e)$ as:

$$W(e) = \begin{cases} \alpha, & \text{if } e \text{ is META or TITLE} \\ 1, & \text{otherwise} \end{cases} \tag{11}$$

where, $\alpha = 2$, $\alpha = 3$, $\alpha = 4$, and $\alpha = 6$ were tested and compared with standard $TF(t_i, d_i)$. The experimental results showed that using SWT can improve classification accuracy.

A more sophisticated SWT was employed by [82]. They not only utilized the HTML elements but also the important sentences in a web page. The important sentences are identified by using Paice's scheme [81], such as,

- Sentences with most frequently used words;
- Sentences with words that occur within title or section headings;
- Sentences using cue words, such as "greatest" and "significant", or cue phrases, such as "the main aim of", "the paper describes" and "the purpose of";
- Sentences in the beginning or last part of the document; and
- Sentences at the beginning of each paragraph.

Reference [82] defined $w(e)$ function as shown in Table 1 and applied the function in (10) to weight the terms. Obviously, their local term-weighting scheme is more mature than the simple $TF(t_i, d_i)$ method. It is then applied together with global term-weight, the inverse document frequency, and normalization factor to weight the terms.

Hyper-textual Nature of Web Pages

Unlike the typical text database, web pages maybe contain no obvious textually clues as to their subjects. For example, the home page of *Microsoft.com*

Table 1. Weights assigned to different elements in a web page

$w(e)$	e (different elements)
2	First or last paragraph
3	The important sentences identified by cue words/phrases
4	<Title>
2	<H1>
2	<EM>
3	<Strong>
4	<Meta Name = "keywords" or "descriptions">
4	URL
1	Otherwise

does not explicitly mention the fact that Microsoft sells operating systems. Some web pages are very short providing few text information and some are non-text based, such as image, video, sound, or flash format. In addition to analyzing a web page itself, a feasible way to represent a web page is to use hyperlinks that link to other related web pages. The basic assumption made by link analysis is that a link is often created because of a subjective connection between the original web page and the linked web page.

Hyperlinks of web pages can be analyzed to extract additional information about the web pages. A hyperlink in web page A that points to or cite web page B is called an *in-link* (incoming link) of web page B; meanwhile, this hyperlink is also an *out-link* (outgoing link) of web page A. A hyperlink is associated with "anchortext" describing the link. For instance, a web page creator may create a link pointing to Google.com, and define the associated anchortext to be "My favorite search engine". Anchortext, since it is chosen by people who are interested in the cited web page, may better summarize the topic of the cited web page, such as the example from Yahoo.com (Fig. 2).

By making use of the information provided by hyperlinks, a target web page can be represented by not only its full text, but all terms in the web pages linking the target page (including in-links and out-links). However, the

---

<A HREF="http://www.fisita.com/">International Federation of Automotive Engineering Societies (FISITA)</A> - a global forum for transport engineers promoting advances in automotive engineering technology and working to create efficient, affordable, safe, and sustainable transportation worldwide.

Fig. 2. An example of hyperlink in web pages from Yahoo.com. The words between <A HREF = ...> and </A> are *anchortext*. The anchortext and the text nearby summarize the content of the cited web page

naïve combination of the local full text and the text in the linked web pages does not help the classification. Experiments conducted by [16] over the web pages of IBM patents and Yahoo corpus showed that the combined features increased the error rate of their system by 6% over the full text in the target page alone. Reference [77] also showed similar results that the performance of their classifier decreased by 24% when adding words in the linked web pages into the target web page in a collection of online encyclopedia articles. Hyperlinks clearly contain high quality semantic clues, but they are noisy. One reason is that those linked web pages go to a diverse set of topics and worse scenarios could be seen on the Web with completely unrelated topics.

Instead of adding all terms in the linked web pages into the target web page, a more appropriate way for web page representation is to retrieve only anchortext and text nearby anchortext in its in-linked web pages (the web pages pointing to the target page). This relaxes the relationship between the in-linked web pages and the target web page. For instance, some of the in-linked web pages may have different topics, but the anchortext in the in-linked web pages describes the hyperlinks and therefore the anchortext should help represent the target web page. Nevertheless, using anchortext alone to represent the target page is less powerful than using the full text in the target page [11, 39]. In many cases, the text nearby the anchortext is more discriminant. We can retrieve terms from the anchortext, headings preceding the anchortext, and paragraphs where the anchortext occurs in the in-linked web pages to represent a target page, instead of the local full text. This representation method has been shown to improve the accuracy by 20% compared to the performance of the same system when using full text in the target web page [36]. To make it simple, a window can be used to define the neighborhood of the anchortext, e.g., up to 25 words before and after the anchortext. We define the anchortext and its neighborhood as the extended anchortext [11, 39]. It has been reported that the best web page representation is the combination of the full text in a web page and the extended anchortext in its in-linked web pages, compared to the full text, anchortext, or extended anchortext alone [11, 39].

Whereas the work cited in this subsection provides insights in exploiting information in hyperlinks for web page classification, many questions still remain unanswered. For example, even though most classification tasks over the WebKB university corpus (see Sect. 5.2) reported good performance results, it is not clear whether the performances will increase over a more diverse web page corpus. Reference [114] showed that drawing general conclusions for using hyperlinks in web page classification without examinations over multiple datasets can be seriously misleading.

Link Based Representation

We believe that although hyperlinks are noisy, there are still some linked web pages that show the subjective connections with a target web page. Some linked web pages may share the same topic or subject as the target web page. Based on this assumption, the link based representation of web pages utilizes the neighborhood of a web page (created by its in-links and out-links) where some neighbors share a same topic. This representation method, making use of topics, differs from the methods in the above subsection where they make use of the text (e.g. anchortext) in the linked web pages.

For link based representation, the topics or class labels of the neighbors of a web page are used to represent the web page. Among all its neighbors, we should choose some of the neighbors who have the same topic as the target web page. One solution is to compute the probability of class i of some neighbors given all neighbors, $Pr(c_i|N_i)$, where c_i is the class i to which some neighbors belonged, and N_i represents the group of all the known class labels of the neighbors [16]. N_i includes the in-links I_i and the out-links O_i, $N_i = \{I_i, O_i\}$. The $Pr(c_i|N_i)$ can be computed according to Bayes theorem as following:

$$Pr(c_i|N_i) = \frac{Pr(N_i|c_i) \cdot Pr(c_i)}{Pr(N_i)} \qquad (12)$$

where $Pr(N_i)$ is no difference among classes of neighbors and thus can be removed, and $Pr(c_i)$ is the proportion of training examples that belongs to the class c_i. We assume all classes in N_i are independent, so that

$$Pr(N_i|c_i) = \prod_{\delta_j \in I_i} Pr(c_j|c_i) \cdot \prod_{\delta_k \in O_i} Pr(c_k|c_i) \qquad (13)$$

where δ_j is the neighbor in the in-links belonging to class c_j, and δ_k is the neighbor in the out-links belonging to class c_k [16]. The class which maximizes the $Pr(c_i|N_i)$ is selected as the class of the target web page. In order to get class labels for all neighbors, the neighbors should first be classified by employing a text based classifier. This procedure can be performed iteratively to assign the class labels to web pages and finally converge to a locally consistent assignment. This method significantly boosts classification accuracy by reducing error up to 70% from text based classifiers [16].

A similar approach using the neighborhood around a web page is to weight the categories (or classes) of its neighbors by the similarity between the categories of the neighbors and the category of the target web page [18]. The target web page and all its neighbors are first classified by a text based classifier. Then the similarities between the category of the target page and the categories of the neighbors are computed to weight each category of the neighbors. Some neighbors belonging to a same category are grouped. If the sum of similarities for a group exceeds a predefined threshold, the target web page is assigned to the category that the group belongs. This approach increases the accuracy by 35% over a local text based classifier when it was tested using Yahoo directory [18].

Word Sense Representation

The problem with the bag-of-words or bag-of-terms representation is that using word occurrence omits the fact that a word may have different word senses (or meanings) in different web pages or in the same web page. For instance, the word *"bank"* may have at least two different senses, as in the *"Bank"* of America or the *"bank"* of Mississippi river. However using a bag-of-words representation, these two *"bank"* are treated as a same feature. Rather than using a bag-of-words, using word senses to represent a web page can improve web page classification.

By using word senses to be features of a web page vector, a web page d_i is represented as:

$$d_i = (w_{i1}, w_{i2}, w_{i3}, \ldots, w_{iM}) \qquad (14)$$

where w_{ij} is the weight of *word sense* s_j $(1 \leq j \leq M)$ in web page d_i. Although this representation has the same format as (1), which is a vector for bag-of-words representation, the indexed features in (14) are word senses rather than words (character patterns) in (1).

Using word senses to be features of a web page, and further be features of each class, is more accurate than using words as features. However the problem is how to find the correct word sense that each word refers. This is also called *word sense disambiguation (WSD)*. Although solutions to WSD are provided in literature [57,68], semantic analysis has proved to be expensive to be implemented. No work in the literature of classification domain has used this method although initial attempt has been conducted by [19]. To reduce the complexity of semantic analysis. they use a thesaurus to relate a word to its synonyms. This approach does not address WSD; instead it expands word senses to groups of words having similar meanings. It extends the concept of term frequency (TF) to account for word senses. To determine TF of a word, the number of occurrences of its synonyms is also taken into account.

Employing word sense representation is still an open issue and it is not clear whether it will be found more successfully than the statistical approach, e.g. bag-of-terms representation which has been examined to be moderately successful.

2.2 Representations for Genre Based Classification

Web page representations for genre based classification is quite different from representations for subject based classification in that features representing the genre of a web page are different from features representing the subject of a web page. In order to retrieve important features representing the genre of a web page. here we first discuss genres of web pages.

Genre Taxonomy

Genre is necessarily a heterogeneous classification principle. While the concept of genre has a long tradition in rhetorical and literary analysis (Bakhtin 1986),

Table 2. Current contents in genre taxonomy

Sources	Genres
Widely recognized genres [117]	Business letter, memo, expense form, report, dialogue, proposal, announcement, thank you note, greeting card, face-to-face meeting system, frequently-asked-questions (FAQ), personal homepage, organizational homepage, bulletin board, hot-list, and intranet homepage
Acorn Project (Yates et al. 1999)	Official announcement, trip report, publication notice, release note, reference, lost and found system, team announcement, traditional memo, electronic memo, dialogue, solicitation, and team report
Online Process Handbook [65]	Welcome page, login page, introduction (user guide, reference), contents page, guide tour, search (search request, result), process knowledge viewer (process compass, process viewer, description, attributes list, tradeoff table, mail), discussion, and options
Web Page Types (Haas and Grams 1998)	Organizational pages (table of contents, site content pages, index pages), documentation (how-to pages, FAQ pages, descriptions of services or products), Text (article, scholarly paper, contract, bibliography, resume, biography, list, copyright), Homepage (organization homepage, personal homepages), Multimedia (non-textual documents), Tools (search tools, order forms, email or comment forms), and Database entry

a number of researches in cultural, rhetorical, and design studies, have recently begun using it to refer to *a typified social action* [6, 13, 69]. Reference [78] defined genre as "a distinctive type of communicative action, characterized by a socially recognized communicative purpose and common aspects of form", taking into account three main aspects of genre: content, form, and purpose. Reference [117] analyzed genres in terms of a number of dimensions of communications, such as What, When, Where, Why, Who, and How. Genre can be analyzed in terms of purposes, contents, participants, timings, places, forms, structural devices, and linguistic elements. How to define each genre of communicative actions is beyond the scope of this chapter. Here we show some genre taxonomy in Table 2, which may help us understand genre based web page classification.

As we can see from Table 2, genre is a subtle and difficult to define notion. One challenge is that it is difficult to know whether we have considered all possible purposes and forms. Without a strong foundational theory of genre to guide us, it is also problematic to construct a classification structure that will accommodate all genres.

The problem that we can solve now is to focus on some genres in which we are most interested and to identify salient features discriminating a genre from others. Once we have found all features for discriminating genres, we can represent a web page as a feature vector, as we have done for subject based classification. The features most useful for genre based classification can be broadly divided into two classes: *presentation features* and *lexical features* [4, 15, 27, 35, 50, 52, 56, 88, 101].

Presentation Features

There are many features that reflect the way in which a web is presented and these features are called as presentation features [27]. The presentation features can further be divided into syntactic features, text complexity features, character-level features, and layout features, as described in the following.

Syntactic Features

Syntactic features are a set of features directly encoding information about the syntactic structure of the text. Syntactic structure can be expected to be more revealing of genres than simple words counts, as argued in the work by [10]. The idea is that particular document genre will favor certain syntactic constructions. Typical syntactic features are passive count, nominalization count, and counts of the frequency of various syntactic categories (e.g. part of speech tags, such as noun, verb, and adjective) [52]. Previous work on syntactic features was based on automated parsing of the input documents. However, this is a time consuming and unreliable procedure. Whereas this set of features was used in some recent work [10, 50], others tried to avoid this set of features [52, 88].

Text Complexity Features

Text complexity features have been widely used due to their ease of computation and their discriminant ability. These features are based on text statistics, such as the average word length, the long word count, the average sentence length, the number of sentences and paragraphs, and the type-token ratio (the number of different words "type" divided by the number of word tokens). Text complexity correlates in certain ways with genre. For instance, long average word length may indicate documents containing technical terms. The complexity of certain constructs turns out to be captured quite well by these simple measures, such as the metrics for transforming all the counts into natural logarithms used by [52]. In addition to the mean measures, variation measures (e.g. the standard deviation in sentence length) can also be used [27, 52]. Furthermore, other derived measures, e.g. readability grade methods, such as Kincaid, Coleman-Liau, Wheeler-Smith Index, and Flesh Index, may be used as more condensed features. The readability measures are generated based on the above mentioned basic measures by using various transformations to obtain graded representations according to stylistic evaluation parameters.

Character Level Features

A wealth of genre information can be obtained from specific character counts, among which the most prominent features are punctuation marks [27,88,101]. For instances, exclamation marks seldom appear in scientific articles; high number of question marks may indicate interviews; high sentence lengths and high counts of commas or semicolon may indicate complicated sentence structures [88]. Example features include counts of quotation marks, exclamation marks, hyphens, periods, apostrophe, acronyms, slash marks, and various brackets.

Other specific characters, such as financial symbols, mathematical symbols, and copyright signs, can also be analyzed [88]. These specific characters hint special genre categories. For instance, the financial symbols, like $ and £, appear more often in the genre of online shopping, while equations frequently occur in technical reports.

Layout Features

Layout features are formed from mark-up tags that can be used to extract information from HTML documents. Mark-up tags provide information, such as the amount of images presented in a given web page, line spacing, and number of tables, equations, and links. Furthermore, mark-up tags can be used to retrieve common layouts for some specific genres like home pages and bulletin boards. Previous work exploiting mark-up tags for genre analysis includes [27,88], and others.

Lexical Features

Lexical features (or word features) are extracted from analyzing the use of words. This is further described in terms of stop words and keywords as following.

Stop Words

Contrary to the principle of subject based classification, a lot of genre information is conveyed by stop words, such as pronouns or adverbs [4,88,101]. A list of stop words is added as features such as *I, you, us, mine, yours, little, large, very, mostly, which, that, and where*. The rational behind their use is that the frequency of such words is presumably not driven by content and hence might be expected to remain invariant for a given genre over different topics. For instance, pronoun frequencies can be used to predict the formality or informality of a document; the word *"should"* may appear frequently in editorials; and the word *"today"* may appear frequently in news [4].

Keywords

For some genre analyses, keywords or key phrases may provide additional cues. For instance, keywords *"resume"*, *"education"*, and *"experience"* are usually expected to appear in *resume* genre. Keywords are retrieved in the same way as the subject based classification, and are primarily based on how discriminant the words or phases are among different genres. Selecting keywords as features is employed in recent work of [27, 35, 56].

Focusing on the goal of genre analysis, only a small subset of the available features may necessarily be selected for web page classification. How to select the features of high discriminant abilities will be addressed in the next section.

3 Dimensionality Reductions

In subject based classification, a web page may contain hundreds of unique terms and the whole collection of web pages may contain a total of hundreds of thousands of unique terms. If all the unique terms are included as features for representing web pages, the dimension of the feature vectors may be hundreds of thousands. Similarly, the total number of the presentation features and lexical features in the genre based classification may also be too large and superfluous for the genre classification. The high dimensionality of the feature space could be problematic and expensive for computation [102]. On the other hand, it is always expected that we can perform classification in a smaller feature space in order to reduce the computational complexity. Thus, techniques for dimensionality reduction should be employed, whose tasks are to reduce the dimensionality of feature vectors and also to insure that this process will not reduce the accuracy of classification.

Dimensionality reduction is also beneficial in that it tends to reduce the problem of over fitting. i.e. the phenomenon by which a classifier is tuned to the training data, rather than generalized from necessary characteristics of the training data [97]. Classifiers that over fit the training data tend to be very good at classifying the trained data, but are remarkably worse at classifying other data [70]. Some experiments suggested that to avoid over fitting. a number of training examples, roughly proportional to the number of features. is needed [97]. This means that, after dimensionality reduction is performed, over fitting may be avoided by using a smaller number of training examples.

Numerous dimensionality reduction functions, either from information theory or from linear algebra, have been proposed and their relative merits have been experimentally evaluated. These functions can be divided, based on what kinds of features are chosen. into *feature selection* and *feature extraction* [97], as described below.

- **Dimensionality Reduction by Feature Selection:** Feature selection selects a subset of the original feature space based on some criteria. Two broad approaches for feature selection have been presented in the literature: *the wrapper approach* and *the filter approach* [47]. *The wrapper approach* [47,53] employs a search through the space of feature subsets. It uses an estimated accuracy for a learning algorithm as the measure of goodness for a particular feature subset. Thus the feature selection is being "wrapped around" a learning algorithm. For example, for a neural network algorithm the wrapper approach selects an initial subset of features and measures the performance of the network; then it generates an "improved set of features" and measures the performance of the network. This process is repeated until it reaches a termination condition (either a minimal value of error or a number of iterations). While some wrapper based methods have encountered some success for classification tasks, they are often prohibitively expensive to run and can break down when a very large number of features are present [54]. For *the filter approach*, feature selection is performed as a preprocessing step before applying machine learning. Thus the method of feature selection is independent to the learning algorithm. The filter algorithm does not incur the high computational cost and is commonly used in classification systems even in a very high feature space. Concerning the advantages of the filter approach over the wrapper approach and the requirement of feature reduction in a very high feature space, the wrapper approaches are omitted in this chapter while several filter approaches are to be discussed and evaluated in Sect. 3.1.

- **Dimensionality Reduction by Feature Extraction:** For feature extraction, the resulting features are not necessary a subset of the original features. They are obtained by combinations or transformations of the original feature space, as discussed in Sect. 3.2.

3.1 Feature Selection

The principle under dimensionality reduction by feature selection is that the features that are more discriminant for some categories from others should be selected. The whole process of feature selection is simplified by the assumption of feature independence. All the features are independently evaluated based on some criteria and a score is assigned to each of them. Then a predefined threshold helps select the best features.

Many criteria have been employed to indicate the discriminant abilities of features. In a broad view, feature selection criteria can be divided into two sets: One set considers only the value denoting a feature occurred in an example, such as the feature selection by using *Document Frequency, Mutual Information, Cross Entropy, and Odds Ratio*, as described in the following subsections. The other set considers all possible values of a feature including both the present and the absent cases, such as the feature selection by using *Information Gain* and *Chi-square Statistic*, as described in the subsequent subsections. This section also describes special methods used for feature selection in a category hierarchy and concludes with the evaluations of the various feature selection techniques.

Document Frequency

A simple and surprising effective feature selection criterion is document frequency. Document frequency of a feature is the number of documents (or web pages) in which the feature occurs. The features whose document frequencies are less than some predetermined threshold are removed. The basic assumption is that rare features are either not informative for classification or not influential in performance [113]. An important characteristic of document frequency is that it does not require class labels of training examples.

This method has a time complexity of $O(n)$ where n is the number of training examples. Because of its simple computation and good performance, document frequency has been widely used in dimensionality reduction. Reference [113] has shown that by using document frequency as threshold, it is possible to reduce about 89% dimensionality of the feature space and results in either an improvement or no less in classification accuracy when tests were conducted over data from Reuters collection and OHSUMED collection (see Sect. 5.2). Similar good performance was also seen in [38].

Mutual Information

Mutual information [20, 113] is commonly used for word associations and can be used here for feature selection. Within a collection of web pages, the mutual information for each feature is computed and the features whose mutual information is less than some predetermined threshold are removed.

The mutual information between feature f and category (or class) c_i is defined to be

$$MI(f, c_i) = \log \frac{Pr(f \wedge c_i)}{Pr(f)Pr(c_i)} \tag{15}$$

where $Pr(f)$ is the proportion of examples containing feature f over all training examples, $Pr(c_i)$ is the proportion of examples in category c_i over all training examples, and $Pr(f \wedge c_i)$ is the joint probability of feature f and category c_i. which is equal to the proportion of examples in which feature f and category c_i both occur [113]. Equation (15) can be transformed to

$$MI(f, c_i) = \log \frac{Pr(c_i|f)}{Pr(c_i)} \quad \text{or} \quad MI(f, c_i) = \log \frac{Pr(f|c_i)}{Pr(f)} \tag{16}$$

where $Pr(c_i|f)$ is the conditional probability of category c_i given feature f, and $Pr(f|c_i)$ is the conditional probability of feature f given the category c_i. We can estimate $MI(f, c_i)$ by letting A be the number of times f and c_i co-occur, B be the number of times f occurs without c_i, C be the number of times c_i occurs without f, and N be the total number of examples [113]. Then $MI(f, c_i)$ can be estimated from (15) as:

$$MI(f, c_i) \approx \log \frac{AN}{(A+B)(A+C)} \tag{17}$$

The average and the maximum of mutual information of feature f are computed through all categories as:

$$MI_{\text{avg}}(f) = \sum_{i=1}^{K} Pr(c_i) MI(f, c_i) \tag{18}$$

$$MI_{\text{max}}(f) = \max_{i=1}^{K} \{MI(f, c_i)\} \tag{19}$$

where K is the total number of categories. The time complexity for computing mutual information is $O(MK)$, where M the size of feature space and K is the number of categories [113].

A weakness of mutual information is that favors rare features [113]. From (16), for features with an equal conditional probability $Pr(f|c_i)$, rare features will have a higher score than common features since rare features have small values of $Pr(f)$ at the denominator.

Cross Entropy

Cross entropy used in document classification [54,55,74] employs information theoretic measures [25] to determine a subset from an original feature space. It can also be used as a method for feature selection to remove redundant features. Formally, let μ and σ be two distributions over some probability space Ω. The cross entropy of μ and σ is defined as [54,55]:

$$D(\mu, \sigma) = \sum_{x \in \Omega} \mu(x) \cdot \log \frac{\mu(x)}{\sigma(x)} \tag{20}$$

It provides us with a notion of "distance" between μ and σ [54,55]. We can transform the above equation to be used for our feature selection as follows. For each feature f and a category set $C = \{c_1, c_2, \ldots, c_K\}$, $Pr(C|f)$ is substituted for μ and $Pr(C)$ for σ. Then, the expected cross entropy $CE(f)$ of feature f is determined as:

$$
\begin{aligned}
CE(f) &= Pr(f) \cdot D(Pr(C|f), Pr(C)) \\
&= Pr(f) \cdot \sum_{i=1}^{K} Pr(c_i|f) \cdot \log \frac{Pr(c_i|f)}{Pr(c_i)}
\end{aligned} \tag{21}
$$

where $Pr(f)$ is a normalization component [55,74]. The features whose $CE(f)$ are less than a certain predefined threshold are eliminated. The time complexity for computing cross entropy is $O(MK)$, which is same as that for mutual information. Cross entropy overcomes a weakness of mutual information by favoring common features instead of rare features.

Odd Ratio

Odds ratio is commonly used to indicate feature goodness in information retrieval, where the task is to rank documents according to their relevance [73, 74, 90, 92]. It is based on the idea that the distribution of features in the relevant documents is different from that in the non-relevant documents. Used for feature selection, the odds ratio of feature f and category c_i captures the difference between the distribution of feature f on its positive class c_i and the distribution of feature f on its negative class $\bar{c}_i$. It is defined as:

$$OR(f, c_i) = \log \frac{\text{odds}(f|c_i)}{\text{odds}(f|\bar{c}_i)} = \log \frac{Pr(f|c_i)(1 - Pr(f|\bar{c}_i))}{Pr(f|\bar{c}_i)(1 - Pr(f|c_i))} \tag{22}$$

where $Pr(f|c_i)$ is the conditional probability of feature f given category c_i, and $Pr(f|\bar{c}_i)$ is the conditional probability of feature f given categories $\bar{c}_i$ [72, 74]. From the definition, we can see that $OR(f, c_i)$ will have a high score if feature f appears frequently in the positive training example set c_i and infrequently in the negative training example set $\bar{c}_i$. The average and the maximum odds ratio of feature f are computed through all categories:

$$OR_{\text{avg}}(f) = \sum_{i=1}^{K} Pr(c_i) OR(f, c_i) \tag{23}$$

$$OR_{\max}(f) = \max_{i=1}^{K} \{OR(f, c_i)\} \tag{24}$$

The features whose odds ratios are less than a certain predefined threshold are removed. The time complexity for computing odds ratio is $O(MK)$, where M is the size of feature space and K is the number of categories.

Information Gain

Information gain is frequently employed as a feature goodness criterion in machine learning [70]. The information gain of a feature measures the expected reduction in entropy caused by partitioning the training examples according to the feature. Entropy characterizes the impurity of an arbitrary collection of training examples. Information gain is also called expected entropy loss [39]. More precisely, the information gain $IG(f)$ of a feature f is defined as:

$$IG(f) \equiv \text{Entropy}(D) - \sum_{v \in (f, \bar{f})} \frac{D_v}{|D|} \text{Entropy}(D_v) \tag{25}$$

where D is a collection of training examples, and D_v is a subset of D which is determined by binary feature value v [70]. For instance, D_f is a subset of D in which each example contains feature f, and $D_{\bar{f}}$ is a subset of D in which each example does not contain feature f. Entropy (D) is defined as

$$\text{Entropy}(D) \equiv -\sum_{i=1}^{K} Pr(c_i) \log Pr(c_i) \qquad (26)$$

where K is the total number of classes (or categories) in the collection D, and $Pr(c_i)$ is the proportion of examples in category c_i over total training examples [70]. By substituting (26) into (25), the information gain of feature f is

$$
\begin{aligned}
IG(f) = &-\sum_{i=1}^{K} Pr(c_i) \log Pr(c_i) \\
&+ Pr(f) \sum_{i=1}^{K} Pr(c_i|f) \log Pr(c_i|f) \\
&+ Pr(\bar{f}) \sum_{i=1}^{K} Pr(c_i|\bar{f}) \log Pr(c_i|\bar{f}) \qquad (27)
\end{aligned}
$$

which is equivalent to

$$
\begin{aligned}
IG(f) = &\, Pr(f) \cdot \sum_{i=1}^{K} Pr(c_i|f) \log \frac{Pr(c_i|f)}{Pr(c_i)} \\
&+ Pr(\bar{f}) \cdot \sum_{i=1}^{K} Pr(c_i|\bar{f}) \log \frac{Pr(c_i|\bar{f})}{Pr(c_i)} \qquad (28)
\end{aligned}
$$

where $Pr(f)$ is the proportion of examples in which feature f is present, $Pr(\bar{f})$ is the proportion of examples in which feature f is absent, $Pr(c_i|f)$ is the conditional probability of category c_i given feature f, and $Pr(c_i|\bar{f})$ is the conditional probability of category c_i given feature f is absent.

The information gain of each feature is computed and the features whose information gain is less than a predetermined threshold are removed. The computation includes the estimation of the conditional probabilities of a category given a feature and the entropy computations. The probability estimation has a time complexity of $O(N)$ where N is the number of training examples. The entropy computation has a time complexity of $O(MK)$ where M the size of feature space and K is the number of categories [113].

It is worthy to note the difference between information gain and cross entropy. As we can see from (21) and (28), the difference between the information gain $IG(f)$ and the cross entropy $CE(f)$ of feature f is that the former makes use of the feature presence and the feature absence, i.e. $IG(f) = CE(f) + CE(\bar{f})$. The similar difference exists between information gain and mutual information. From (28), Information Gain of feature f can be proven equivalent to:

$$IG(f) = \sum_{X \in \{f, \bar{f}\}} \sum_{Y \in \{c_i\}} Pr(X \wedge Y) \log \frac{Pr(X \wedge Y)}{Pr(X)Pr(Y)}$$

$$= \sum_{i=1}^{K} Pr(f \wedge c_i)MI(f, c_i) + \sum_{i=1}^{K} Pr(\bar{f} \wedge c_i)MI(\bar{f}, c_i) \quad (29)$$

This shows that information gain is the weighted average of the mutual information $MI(f, c)$ and $MI(\bar{f}, c)$ (see (15)). Thus information gain is also called average mutual information [113].

Chi-Square Statistic

The Chi-Square (χ^2) statistic measures the lack of independence between feature f and category c_i. The feature goodness metric by χ^2 statistic is defined as [38, 113]

$$\chi^2(f, c_i) = \frac{N[Pr(f \wedge c_i)Pr(\bar{f} \wedge \bar{c}_i) - Pr(f \wedge \bar{c}_i)Pr(\bar{f} \wedge c_i)]^2}{Pr(f)Pr(\bar{f})Pr(c_i)Pr(\bar{c}_i)} \quad (30)$$

where N is the total number of training examples. The features f with the high values of $\chi^2(f, c_i)$ are thus the more dependent (closely related) with category c_i. which are selected for the purpose of feature selection. We can estimate the value of $\chi^2(f, c_i)$ by letting A be the number of times f and c_i co-occur, B be the number of time f occurs without c_i, C be the number of times c_i occurs without f. and D be the number of times neither c_i nor f occurs. i.e.

$$\chi^2(f, c_i) = \frac{N(AD - BC)^2}{(A + B)(C + D)(A + C)(B + D)} \quad (31)$$

The average and the maximum χ^2 statistic values of feature f over all categories are then computed as following [113]:

$$\chi^2_{avg}(f) = \sum_{i=1}^{K} Pr(c_i)\chi^2(f, c_i) \quad (32)$$

$$\chi^2_{max}(f) = \max_{i=1}^{K}\{\chi^2(f, c_i)\} \quad (33)$$

The computation of χ^2 statistic has a time complexity of $O(MK)$, where M is the size of feature space and K is the number of categories. The features with low χ^2 statistic values are removed. Using $\chi^2_{max}(f)$ for feature selection outperformed using $\chi^2_{avg}(f)$ as reported in [113] and [38].

Rare features are emphasized by χ^2 in the form of $Pr(f)$ at the denominator. This is not what we expected based on the fact that rare features are not influential in performance as in the case of Document Frequency. To avoid emphasizing rare features. a simplified χ^2 statistics was proposed in [38]:

$$\chi^2(f, c_i) = Pr(f \wedge c_i)Pr(\bar{f} \wedge \bar{c}_i) - Pr(f \wedge \bar{c}_i)Pr(\bar{f} \wedge c) \quad (34)$$

It emphasizes the positive correlation between f and c_i (i.e. $Pr(f \wedge c_i)$ and $Pr(\bar{f} \wedge \bar{c_i})$) and de-emphasizes the negative correlation (i.e. $Pr(f \wedge \bar{c_i})$ and $Pr(\bar{f} \wedge c_i)$). Reference [38] showed that the simplified χ^2 outperformed the original χ^2 when feature reduction above 95% is required. However, when below 95% reduction is required, the simplified version is slight inferior to the original one.

Feature Selection in a Category Hierarchy

For a large corpus, we may have hundreds of categories and hundreds of thousands of features. Even by applying the above feature selection techniques, the computational cost for the remaining features may still poses significant limitations. Another approach is to organize categories in a hierarchy (e.g. Yahoo directory) and to divide the classification task into a set of smaller classification tasks, each of which corresponds to some splits in the classification hierarchy. The key insight is that within each of the smaller subtasks and their corresponding smaller subset of feature space, it is possible to use fewer features to differentiate among a smaller number of categories [55]. The approach requires creating a category hierarchy and dividing a classification task into smaller subtasks as described in following.

Creating a Category Hierarchy

The most representative example for web page category hierarchy is provided by Yahoo.com. The top level of the hierarchy consists of 14 main categories (see Fig. 3), such as "Business and Economy", "Computer and Internet", "Society and Culture", "Education", and "Reference". Each main category contains many sub-categories. For instance, "References" main category contains 129 sub-categories, "Education" contains 349 sub-categories, and "Computer and Internet" contains 2652 sub-categories. To represent a category, features are selected from the training web pages taken from the category. To represent the hierarchy, the features in the low level categories are added into the

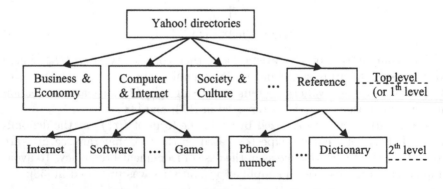

Fig. 3. Top two levels of Yahoo directory

top level categories along the paths in the category tree, which is based on the fact that the low level categories belong to its more general top level categories. When propagating the features upward from the lowest level categories among the paths in the category tree to the root, the weights of the features are reduced and normalized in proportional to the size of its category [73,82].

Dividing Classification Task into Smaller Subtasks

It is important to note that the key here is not only the use of feature selection, but also the integration of feature selection process within the hierarchical structure. As we can see from Fig. 3, focusing on a main category at the top level, we are only interested in those features that can discriminate its main category from the other 13 main categories. In other words, when selecting features for one category, we only consider features that are discriminant from its sibling categories, which share a same parent node. Among a category and its sibling categories, this smaller feature selection task can be done by applying any feature selection criterions discussed. For instance, Odds Ratio was used to select features in a hierarchy in [73]; Cross Entropy was employed in [55]; and Information Gain was used in [32].

Specially for selecting features among categories, [82] proposed to use a *Uniqueness* criterion, which scored feature f in category c_i (or node c_i) according to the uniqueness in comparing to its sibling categories:

$$U(f, C_i) = \frac{Pr(SubT_{c_i}|T)Pr(f|c_i)}{Pr(f|parent)} \qquad (35)$$

where *parent* is the parent node of node c_i, T represents the tree rooted at parent node, $SubT c_i$ is a sub-tree located underneath tree T, and $Pr(SubT c_i|T)$ is a weight factor assigned to node c_i which is equal to the proportion of the size of node c_i over the size of its parent node. The idea of the uniqueness criterion is that if a feature is unique in one node, it is the only source that can be propagated to the parent feature space. By regarding the parent category C as a whole domain $C = \{c_i, c_{\text{siblings}}\}$, the $U(f, c_i)$ can be shown equivalent to

$$U(f, c_i) = \frac{Pr(c_i|C)Pr(f|c_i)}{Pr(f|C)} = \frac{Pr(c_i)Pr(f|c_i)}{Pr(f)} = Pr(c_i|f) \qquad (36)$$

From the above equation, it can be noticed that the uniqueness criterion $U(f, c_i)$ is comparable to the mutual information $MI(f, c_i)$ (16); the main difference is that the uniqueness criterion removes the factor $Pr(c_i)$ at the denominator of $MI(f, c_i)$. Since [82] only selected features whose uniqueness score is 1 for a category c_i, the $Pr(c_i|f)$ alone contains enough information to be a criterion for their feature selection.

Evaluations of Feature Selection Techniques

While many feature selection techniques have been proposed, thorough evaluations have rarely carried out for classification in a large feature space. The most impressive work on evaluating some feature selection techniques can be found in [113] and [73, 74]. To assess the effectiveness of feature selection techniques, two classifiers, a k-nearest-neighbor (kNN) classifier and a Linear Least Squares Fit mapping (LLSF) classifier (described Sect. 4), were employed in [113]. The classifiers learned over two text collections, the Reuters collection and the OHSUMED collection.

Reference [113] found Information Gain and χ^2 statistics most effective in their experiments compared to Document Frequency, Mutual Information, and Term Strength, which is omitted in this chapter. Using Information Gain with a kNN classifier on the Reuters collection not only achieved up to 98% reduction of feature space but also yielded improvements in classification accuracy. They found strong correlation between Document Frequency, Information Gain, and χ^2 statistics. This suggests that Document Frequency, the simplest method with the lowest cost in computation, can be reliably used instead of Information Gain and χ^2 statistics when the computations of the later criterions are too expensive [113]. In contrast, Term Strength was not competitive at high percentage of feature reduction. Mutual Information had relatively poor performance due to its use of feature presence only and its bias toward favoring rare features. However, the effect of favoring rare features can be compensated by first removing those rare features and Mutual Information showed no significant performance difference among other feature selection techniques in [92].

Reference [73, 74] showed that the best performing feature selection methods were Odds Ratio among eleven feature selection methods tested. They employed a Naïve Bayes classifier for learning over web pages derived from Yahoo directory. The next group of methods that achieve good results favors common features (e.g. Cross Entropy). Mutual Information differs from Cross Entropy only in favoring rare features and achieved worse results than Cross Entropy. The worst feature selection method was Information Gain, which on the other hand achieved the best performance in experiments by [113].

The differences in evaluation results reflect the differences in classification algorithms and test domain used. We can observe that Information Gain makes use of feature presence and feature absence while Odds Ratio, Cross Entropy and Mutual Information only consider information of feature presence. In experiments by [73, 74], the data collection from Yahoo directory has unbalanced class distribution and highly unbalanced feature distribution. They observed that the prior probability of a feature in a web page, $Pr(f)$, is rather small. Most of the features selected by Information Gain are features having high absent feature value $Pr(\bar{f})$. If $Pr(\bar{f})$ is much larger than $Pr(f)$, the high value of Information Gain in most cases means that the second part $Pr(\bar{f}) \sum_{i=1}^{k} Pr(c_i|\bar{f}) \log \frac{Pr(c_i|\bar{f})}{c_i}$ of the Information Gain formula (28) is high.

In other words, knowing that feature f does not occur in a web page brings useful information about the category of the web page. The problem is that a classification relied mostly on the absence of features is usually more difficult and requires larger feature space than a classification relied on feature presence [73, 74]. In contrast, Odds Ratio (22) favors features from positive examples (high $Pr(f|c_i)$). Thus, Odds Ratio outperformed all other feature selection techniques in [73,74]. Another reason for the differences of evaluation results is that the classification algorithm used by [73, 74] is a Naïve Bayes classifier, which considers only features that occur in training examples. This means that the selected features should be the features that will probably occur in new web pages to be classified [73,74].

The common conclusions made by [113] and [73, 74] include the followings. When choosing a feature selection method, both classification algorithm and data domain should be taken into considerations. A rather small feature subset should be used since it gives either better or as good results as large feature space. A simple frequency count of features, such as document frequency, achieves very good results. Feature selection methods favoring frequent features achieve better results than methods favoring rare features. This indicates that frequent features are informative for classification.

One limitation of using feature selection techniques as described in this subsection is the inability to consider co-occurrence of features. Two or more features individually may not be useful, but when combined may become highly effective. This limitation is addressed by using feature extraction.

3.2 Feature Extraction

Feature extraction synthesizes a set of new features from the set of original features where the number of new features is much smaller than the number of original features. The rationale for using synthetic, rather than naturally occurring, features is that the original features may not form an optimal dimension for web page representation [97]. Methods for feature extraction aim at creating artificial features that do not suffer the problems of polysemy, homonymy, and synonymy present in the original features. Several approaches have been reported and successfully tested [96, 105, 108]. In the following, we describe two approaches: latent semantic indexing and word clustering.

Latent Semantic Indexing

Latent semantic indexing (LSI) is based on the assumption that there is an underlying or latent semantic structure in the pattern of features used across the web page corpus and some statistical techniques can be used to estimate the structure [9, 26, 96, 105, 108]. LSI uses singular value decomposition (SVD), which is a technique related to eigenvector decomposition and factor analysis.

The main idea in latent semantic indexing (see also [9]) is to explicitly model the interrelationships among features by using SVD and to exploit this to improve classification. The process begins by constructing a M features by N documents matrix called A, where each entry represents the weight of a feature in a document, i.e.,

$$A = (a_{ij}) \tag{37}$$

where, a_{ij} is the weight of feature i $(1 \le i \le M)$ in document j $(1 \le j \le N)$. Since not every feature normally appears in every document, the matrix A is usually sparse. The singular value decomposition (SVD) of matrix A is given by:

$$A_{M \times N} = U_{M \times R} \sum_{R \times R} V_{R \times N}^T \tag{38}$$

where R is the rank of A $(R \le \min(M, N))$; U and V have orthogonal unitlength columns $(U^T U = I; V^T V = I)$; and Σ is the diagonal matrix of singular values of A $(\Sigma = \mathrm{diag}(\sigma_1, \ldots, \sigma_R))$ which are the nonnegative square roots of the eigenvalues of AA^T. Table 3 outlines the definition of the terms.

Table 3. Interpretation of SVD components within LSI

A = matrix of M features × N documents	M = number of features
U = feature vectors	N = number of documents
Σ = singular values	R = rank of A
V = document vectors	k = number of factors
	(k highest singular values)

If the singular values in Σ are ordered by size, the k largest may be kept and the remaining smaller ones set to zero, i.e. $\Sigma = \mathrm{diag}(\sigma_1, \sigma_2, \ldots, \sigma_k, \ldots, \sigma_R)$, where $\sigma_i > 0$ for $1 \le i \le k$ and $\sigma_i = 0$ for $i > k$. The product of the resulting matrices is a matrix A_k which is an approximation to A with rank k, i.e.

$$A_k = U_k \sum_k V_k^T \tag{39}$$

where $k \ll M$, Σ_k is obtained by deleting the zero rows and columns of Σ, and U_k and V_k are obtained by deleting the corresponding rows and columns of U and V (showed in Fig. 4).

The resulting A_k captures most of the underlying structure in the association of features and documents in A. The three matrices U_k, Σ_k, and V_k reflect a breakdown of the original feature-document relationships into linearly-independent vectors or factors. The use of k factors or k-largest singular triplets is equivalent to approximate the original matrix. In addition, a new document d can be represented as a vector in k-dimensional space as:

$$\tilde{d} = d^T U_k \sum_k^{-1} \tag{40}$$

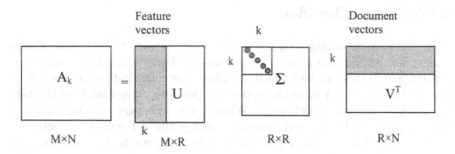

Fig. 4. Graphical interpretation of the matrix A_k

where $d^T U_k$ reflects the sum of k-dimensional feature vectors and Σ_k^{-1} weights the separate dimensions [9, 40].

It is difficult to interpret the new smaller k-dimensional space although it is assumed to work well in bringing out the latent semantic structure of feature-document matrix. An example provided in [9] may help us understand the new space: consider the words *car, automobile, driver* and *elephant*. The words *car* and *automobile* are synonyms, *driver* is a related concept and *elephant* is unrelated. The words *car* and *automobile* will occur with many of the same words, such as *motor, model, vehicle, chassis,* and *engine*. Thus, they will have similar representations in the k-dimensional space. The context for *driver* will overlap to a lesser extent, and those for *elephant* will be quite dissimilar. This relates to the fact that features which occur in similar documents will be near each other in the k-dimensional space even if these features do not co-occur in the same documents. This farther means that two documents may be similar even if they do not share same keywords.

Word Clustering

Word clustering aims at grouping words or phrases into clusters based on their semantic relatedness. The resulting clusters or their centroids are then used in place of the original groups of words or phases [97]. A word clustering method can also be interpreted as a method for converting the original representation of a document into a new representation that has a much smaller dimensionality than the original one.

One example of this approach is the work by [63], who view semantic relatedness between words in terms of their co-occurrence and co-absence within training documents. By using the criteria in the context of a hierarchical clustering algorithm, they witnessed only a marginal improvement, which may not by conclusive due to the small size of their experiments. Other works [5, 28, 100], such as distributional clustering of words, has achieved improvements over feature selection methods in terms of classification accuracy, especially at lower number of features. Additional related research could be found in [7, 23, 28, 31, 34, 44, 51, 66, 87, 99, 106, 116].

4 Web Page Classifiers

After features of training web pages have been selected to form concise representations of the web pages, various machine learning methods and classification algorithms can be applied to induce the classification function f' as defined in Sect. 1.2 or to induce representations for categories from the representations of training web pages. When a new web page is to be classified, the classifiers use the learned function to assign the web page to categories.

In what follows we discuss the state of the art classifiers in terms of web page classification. We partition classifiers appeared in literature into profile, rule learning, direct example, and parameter based classifiers, where the first three are called non-parametric approaches and the last one is called parametric approach [31,58]. Here first the definitions and some general notations are given. A web page is usually represented by a vector $d_i = \{w_1, w_2, \ldots, w_M\}$, where each w_i is the weight of a feature of the web page and M is the size of feature space. Predefined categories are denoted by a set $C = \{c_1, c_2, \ldots, c_K\}$, where each c_i is a category label and there are K categories. Training examples consists of N web pages represented by vectors $d_1, d_2, \ldots d_N$, which are tagged with true category labels $y_1, y_2, \ldots y_N$, respectively. Let N_j be the number of training web pages for which the true category label is c_j. In general, the classification process consists of a training phase and a testing phase: during the training phase, training examples are used to train the classifiers; during the testing phase, the classifiers are applied to classify web pages. Some rule learning classifiers also consist of a validation phase for optimizing the rules.

4.1 Profile Based Classifiers

For profile based classifiers, a profile (or a representation) for each category is extracted from a set of training web pages that has been predefined as examples of the category. After training all categories, the classifiers are used to classify new web pages. When a new web page is to be classified, it is first represented in the form of a feature vector. The feature vector is compared and scored with profiles of all the categories. In general, the new web page may be assigned to more than one category by thresholding on those webpage-category scores and the thresholding methods used can influence the classification results significantly [114]. In the case where a web page has one and only one category, the new web page is assigned to the category that has the highest resulting score. Examples of classifiers using this approach are Rocchio classifier, Support Vector Machine, Neural Network classifier, and Linear Least Square Fit classifier, each of which is reviewed in the followings.

Rocchio Classifier

Rocchio algorithm is a classic algorithm for document routing and filtering in information retrieval [14, 43, 91]. Rocchio algorithm employs TFIDF feature weighting method to create a feature vector for each document. Learning (or training) is achieved by combining feature vectors into a prototype vector $\vec{c}_j$ for each class c_j. The normalized feature vectors of the positive examples for class c_j and those of the negative examples are first summed up. Then, the prototype vector $\vec{c}_j$ is calculated as a weighted difference as:

$$\vec{c}_j = \alpha \frac{1}{|c_j|} \sum_{\vec{d} \in c_j} \frac{\vec{d}}{\|\vec{d}\|} - \beta \frac{1}{|C - c_j|} \sum_{\vec{d} \in C - c_j} \frac{\vec{d}}{\|\vec{d}\|} \tag{41}$$

where α and β are parameters that adjust the relative impact of positive and negative training examples, $|c_j|$ is the number of elements in set c_j and $\|\vec{d}\|$ is the length of vector $\vec{d}$ [45]. The resulting set of prototype vectors, one vector for each class, represents the learned model.

Using cosine as a similarity metric and letting $\alpha = \beta = 1$, Rocchio shows that each prototype vector maximizes the mean similarity of the positive training examples minus the mean similarity of the negative training examples [45], i.e.

$$\frac{1}{|c_j|} \sum_{\vec{d} \in c_j} \cos(\vec{c}_j, \vec{d}) - \frac{1}{|C - c_j|} \sum_{\vec{d} \in C - c_j} \cos(\vec{c}_j, \vec{d}) \tag{42}$$

After obtaining a prototype vector $\vec{c}_j$ for each of the predefined categories C, the classifier is then used to classify a new document d'. To classify document d' the cosine similarity measures of each prototype vectors $\vec{c}_j$ with $\vec{d'}$ are calculated. The document d' is assigned to the class with which $\vec{d'}$ has the highest cosine metric:

$$H(d') = \underset{\vec{c}_j \in C}{\arg\max}(\cos(\vec{c}_j, \vec{d'})) = \underset{\vec{c}_j \in C}{\arg\max} \frac{\vec{c}_j}{\|\vec{c}_j\|} \cdot \frac{\vec{d'}}{\|\vec{d'}\|} \tag{43}$$

where $\arg\max f(x)$ returns the argument x for which $f(x)$ is maximum and $H(d')$ is the category to which the algorithm assigns document d' [45].

Note that the cosine similarity measure is nonlinear. However, this model can be recast as linear classification by incorporating its length normalization into each of the elements of its weight vector:

$$\vec{c}_j \leftarrow \frac{\vec{c}_j}{\|\vec{c}_j\|} \quad \text{and} \quad \vec{d'} \leftarrow \frac{\vec{d'}}{\|\vec{d'}\|} \tag{44}$$

Thus $H(d')$ is transformed to be:

$$H(d') = \underset{\vec{c}_j \in C}{\arg\max} \vec{c}_j \cdot \vec{d'} \tag{45}$$

Previous work using the Rocchio algorithm in text classification could be found in [22, 61, 62, 86, 109]. More interesting, [45] proposed a probabilistic analysis of the Rocchio algorithm, which he called PrTFIDF classifier and showed improvement compared to the original Rocchio algorithm.

Support Vector Machine

Support Vector Machines (SVMs) have shown to yield good performance on a wide variety of classification problems, most recently on text classification [27, 33, 46, 62, 80, 103, 112]. They are based on Structural Risk Minimization principle from computational learning theory [24, 104]. The idea of structural risk minimization is to find a hypothesis h which is defined as the decision function with maximal margin between the vectors of positive examples and the vectors of negative examples [105], see Fig. 5. It was shown that if the training set is separated without errors by h, the expectation value of the probability of committing an error on a test example is bounded by a very small number, i.e. 0.03 [24].

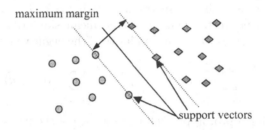

Fig. 5. Linear support vector machine. This figure shows an example of a simple two-dimensional problem that is linearly separable. The diamonds in the figure represent positive examples and the circles represent negative examples. SVM finds the hyperplane h (denoted by the *solid line*), which separates the positive and negative training examples with a maximum margin that is the distance between the two parallel *dashed lines*. The examples closest to the hyperplane are called Support Vectors (indicated in the figure with *arrows*). In other words, SVM finds h that maximizes distances to Support Vectors [104]

In its simplest linear form, an SVM is a hyperplane that separates a set of positive examples from a set of negative examples with a maximum margin (see Fig. 5). Let $D = \{(y_i, d_i)\}$ denote the training set, and $y_i \in \{+1, -1\}$ be the classification of a document vector $\vec{d}_i$, where $+1$ indicates a positive example and -1 indicates a negative example of a given category. SVM learns linear threshold functions [46] of the type:

$$h(\vec{d}) = \text{sign}(\vec{w} \cdot \vec{d} + b) = \begin{cases} +1, & \text{if } \vec{w} \cdot \vec{d} + b > 0 \\ -1, & \text{otherwise} \end{cases} \qquad (49)$$

where $h(\vec{d})$ represents a hypotheses given $\vec{d}$, and $\vec{w}$ represents a weight vector, while b is a scalar to be defined in (54). Finding the hyperplane having the maximum margin can be translated into the following optimization problem [104]:

$$\text{Minimizes}: \quad \|\vec{w}\| \tag{50}$$

so that:

$$\forall i : y_i[\vec{w} \cdot \vec{d}_i + b] \geq 1 \tag{51}$$

where $\|\vec{w}\|$ denotes the Enclidean length of a weight vector $\vec{w}$. The constraint expressed in (51) requires that all training examples are classified correctly. In order to solve the above optimization problem, Lagrange multipliers are used to translate the problem into an equivalent quadratic optimization problem [46, 104]:

$$\text{Minimize}: -\sum_{i=1}^{N} \alpha_i + \frac{1}{2} \sum_{i,j=1}^{N} \alpha_i \alpha_j y_i y_j \vec{d}_i \cdot \vec{d}_j \tag{52}$$

so that:

$$\sum_{i=1}^{N} \alpha_i y_i = 0 \quad \text{and} \quad \forall i : \alpha_i \geq 0 \tag{53}$$

For this quadratic optimization problem, efficient algorithms can be used to find the global optimum [80]. The result of the optimization process is a set of coefficients α_i^* for which (52) is minimized [46]. These coefficients can be used to construct the hyperplane as follows:

$$\vec{w} \cdot \vec{d} = \left(\sum_{i=1}^{N} \alpha_i^* y_i \vec{d}_i \right) \cdot \vec{d} \quad \text{and} \quad b = \frac{1}{2}(\vec{w} \cdot \vec{d}_+ + \vec{w} \cdot \vec{d}_-) \tag{54}$$

From the above equation [46], we can see that the resulting weight vector, $\vec{w}$, is constructed as a linear combination of the training examples. Only the training vectors, for which the coefficient α_i is greater than zero, contribute the combination. These vectors are called Support Vectors, as shown in Fig. 5. To calculate b, an arbitrary support vector $\vec{d}_+$ from positive examples and one $\vec{d}_-$ from negative examples are used [46].

Once the weight vector for each of the given categories is obtained, a new document d can be classified by computing $\vec{w} \cdot \vec{d} + b$ in (49), where $\vec{w}$ is the learned weight vector of a given category, and $\vec{d}$ is the vector representing the new document. If the value is larger than 0, then the new document is assigned to this category.

Neural Network Classifier

Neural network (NNet) approaches to text classification were evaluated by many researchers, such as [27, 52, 76, 96, 105]. Reference [105] employed a perceptron approach (without a hidden layer) and a three-layered neural network (with a hidden layer), while [76] evaluated only perceptrons. Since neural networks are among the top ranking classifiers [62, 109], Perceptrons and Least Mean Square rules will be briefly described in terms of web page classification.

Perceptrons

For web page classification, a perceptron is used for a given category. It takes a feature vector representing a web page as input, calculates a linear combination of the features of the input vector, then outputs a +1 if the result is greater than a threshold (that is automatically learned during the training process) or −1 otherwise, which indicates whether the web page belongs to the category or not, respectively. For illustration (see also [70]), we write a perceptron function as

$$o(\vec{d}) = \text{sgn}(\vec{w} \cdot \vec{d} + b) = \text{sgn}\left(\sum_{i=1}^{M} w_i d_i + b\right) \tag{55}$$

where $\vec{w} = (w_1, w_2, \ldots, w_M)$ is an M-dimensional weight vector and $\vec{d} = (f_1, f_2, \ldots, f_M)$ is an M-dimensional input vector representing a web page, in which each element f_i is the ith feature value, b is a threshold, and the function sgn is defined as

$$\text{sgn}(y) = \begin{cases} +1, & \text{if } y > 0 \\ -1, & \text{otherwise} \end{cases} \tag{56}$$

To simplify notation, we transform the vectors $\vec{w}$ and $\vec{d}$ to be $M + 1$ dimensional vectors by adding $w_0 = b$ and $f_0 = 1$. This allows us to rewrite the above (55) and (56) as

$$\text{sgn}(\vec{w} \cdot \vec{d}) = \begin{cases} +1, & \text{if } \vec{w} \cdot \vec{d} > 0 \\ -1, & \text{otherwise} \end{cases} \tag{57}$$

We can view the perceptron as a hyperplane decision surface in an $M + 1$-dimensional space [70]. The perceptron outputs +1's for all positive examples on one side of the hyperplane and outputs −1's for negative examples on the other side. The equation for this decision hyperplane is $\vec{w} \cdot \vec{d} = 0$. Sets of web pages that can be separated by the hyperplane are called linearly separable. We can also notice that the hyperplane produced by the perceptron does not require the maximum margin between the vectors of the two classes which is required by the SVM.

When training a perceptron, the weights $w_1, w_2, \ldots, w_M$ for the perceptron are adjusted based on the training examples. The learning process begins with a setup of random weights, then iteratively applies the perceptron to each training example, and modifies the weights whenever the perceptron misclassifies an example. The weights are modified at each step according to a training rule, which revises the weights w_i in associated with the inputs f_i according to the following learning rule:

$$w_i \leftarrow w_i + \eta(y_i - o_i)f_i \tag{58}$$

where η is the learning rate, y_i is the actual class label ($+1$ or -1), o_i is the output generated by the perceptron [70]. This learning process iterates through the training examples as many times as needed until the perceptron classifies most training examples correctly as a result of converging toward a set of weights. The learned weight vector is then used to classify new web pages.

Least Mean Square Rule

Least Mean Square (LMS) training rule, also known as Widrow-Hoff rule, was employed and showed good performance for text classification [61, 109]. Although the perceptron rule finds a successful weight vector when the training examples are linearly separable, it may fail to converge if the examples are not linearly separable. LMS training rule is designed to overcome this difficulty.

LMS training rule is best understood by considering the task of training a perceptron without the threshold (see also [70]); that is, a linear unit (without threshold) for which the output o is given by

$$o(\vec{d}) = \vec{w} \cdot \vec{d} \tag{59}$$

The basic principle under LMS rule is to minimize the error function $E_i(\vec{w})$ defined for each individual training example d_i:

$$E_i(\vec{w}) = \frac{1}{2}(y_i - o_i)^2 \tag{60}$$

The negative of the gradient of E with respect to the weight vector $\vec{w}$ gives the direction of steepest decrease [70]. Thus the weight update rule for incremental gradient descent is:

$$\Delta \vec{w}_i = -\eta \frac{\partial E}{\partial w_i} = \eta(y_i - o_i) \cdot f_i \tag{61}$$

It can be seen that the expression of LMS rule appears to be identical with the perceptron weight update rule (58). These two training rules are different in terms of that for LMS rule the output o refers to the linear unit output $o(\vec{d}) = \vec{w} \cdot \vec{d}$ whereas for perceptron rule the output o refers to the threshold output $o(\vec{d}) = \mathrm{sgn}(\vec{w} \cdot \vec{d})$. Similar to a perceptron, the learned weight vectors can then be used to classify new web pages.

Linear Least Squares Fit Classifier

Linear Least Squares Fit (LLSF) is a successful classifier for document classification [62, 108, 109, 111, 112, 119]. It is associated with a linear regression model [110, 111]. The training data are represented using two matrices D and C, where D is a document matrix and C is a category matrix. An element of matrix D is the weight of a feature in a corresponding document, and an element of matrix C is the weight ($+1$ or -1) of a category in a corresponding document, where $+1$ indicates belonging to the category and -1 for not belonging to the category. The LLSF problem is defined as finding a matrix W that minimizes the squared error of the Frobenius matrix norm of $WD - C$ [108]; in other words, the objective is to find W that minimizes the squared error in the mapping from training documents to their categories. The solution of W is then used to transform an arbitrary document, represented by a feature vector $\vec{d}$, to a category vector $\vec{c}$ by computing $W\vec{d} = \vec{c}$ [110]. The elements of vector $\vec{c}$ are interpreted as the relevance scores of categories with respect to document $\vec{d}$. By sorting the scores, the most relevant categories are obtained, which are the output of the LLSF mapping [110].

A conventional method for solving a LLSF problem employs a Singular Value Decomposition (SVD) of the input matrix D as a part of the computation [40, 110]. Compute an SVD of D, yielding $D = USV^T$ (see Sect. 3.2). Then compute the mapping function $W = CVS^{-1}U^T$ [110]. Because of the high time complexity of computing SVD, dimensionality of the document vectors must be reduced before LLSF is employed.

It is worth noting that the linear classifiers (such as Perceptrons, LMS rule, LLSF, and SVM) do not explicitly construct feature combinations but use the context implicitly [110, 111]. For instance, the classification function in LLSF is sensitive to weighted linear combinations of features that co-occur in training examples [109]. This is a fundamental difference from the classification methods based on the assumption of feature independence, such as Naïve Bayes classifier (see Sect. 4.4).

4.2 Rule Learning Based Classifiers

The one of the most expressive and human readable representations for learned hypotheses is sets of if-then rules. The most important property of rule induction algorithms is that they allow the interrelationships of features to influence the outcome of the classification, whereas some other classification schemes, e.g. Naïve Bayes, assume the features as independent components. Hence, rule learning based classifiers are context sensitive classifiers [3, 22].

In general, for rule learning based classifiers, the training web pages for a category are used to induce a set of rules for describing the category. A web page to be classified is used to match the conditions of the rules. The matched rules predict the class for the web page based on the consequents of the rules. In this section we discuss three successful representatives of the rule learning based classifiers: Disjunctive Normal Form (DNF) rule, Association rule, and Decision tree. Each of them uses a different rule induction algorithm but they are theoretically equivalent because each learned model is a disjunction of conjunction rules.

Disjunctive Normal Form Rule

An example of a classifier using Disjunctive Normal Form (DNF) rules is RIPPER [21, 22], which performs quite well for document classification [109]. DNF classifiers can be interpreted as learning a disjunction of contexts, each of which defines a conjunction of simple features. For instance, the context of a feature f_1 in document d_i is a conjunction of the form:

$$f_1 \text{ and } f_2 \text{ and } f_3 \ldots \text{ and } f_k \qquad (62)$$

where $f_j \in d_i$ for $j = 1$ to k. This means that the context of feature f_1 consists of a number of other feature $f_2, f_3, \ldots f_k$ that must co-occur with f_1. These features may occur in any order and in any location in the document. The classifier constructed by RIPPER is a set of rules that can be interpreted as a disjunction of conjunctions. For instance,

Document d_i belongs to category "*Louisiana*" IF AND ONLY IF
("*Louisiana*" appears in d_i AND "*Cajun*" appears in d_i) OR ("*New Orleans*" appears in d_i AND "*French Quarter*" appeared in d_i)

The classification of new documents is based on the learned rules that test for the simultaneous presence or absence of features. The rule learning process consists of two main stages. The first stage is a greedy process that constructs an initial rule set. The second stage is an optimization process that attempts to further improve the compactness and accuracy of the rule set. These two main stages are briefly discussed as follows.

The first stage is based on a variant of the rule learning algorithm called incremental reduced error pruning (IREP) [37]. To construct a rule (see also [22]), the uncovered examples are randomly partitioned into two subsets, a growing set containing two-third of the examples and a pruning set containing the remaining one-third. The algorithm will first grow a rule by repeatedly adding conditions and then prune the rule. In the procedure of growing a rule, at each step i, a single condition is added to the rule r_i, producing a more specialized rule r_{i+1}. The added condition is the one that yields the largest information gain [85] for r_{i+1} relative to r_i, which is defined as

$$\text{Gain}(r_{i+1}, r_i) = T_{i+1}^+ \cdot \left(\log_2 \frac{T_{i+1}^+}{T_{i+1}^+ + T_{i+1}^-} - \log_2 \frac{T^+}{T_i^+ + T_i^-} \right) \qquad (63)$$

where T_i^+ is the number of positive examples and T_i^- is the number of negative examples in the growing set covered by rule r_i [22]. The addition of new conditions continues until the rule covers no negative examples in the growing set or until no condition results in a positive information gain.

After growing a rule, as described in [22], the rule is then pruned or simplified. At each step, the algorithm considers deleting a condition from a rule. It chooses a condition for deletion that maximizes the function

$$f(r_i) = \frac{U_{i+1}^+ - U_{i+1}^-}{U_{i+1}^+ + U_{i+1}^-} \tag{64}$$

where U_{i+1}^+ is the number of positive examples and U_{i+1}^- is the number negative examples in the pruning set covered by the pruned rule [85]. After pruning conditions, the rule is added into the initial rule set and the examples covered by the condition are removed.

The second stage by RIPPER is an optimization procedure in which it optimizes each rule in the current rule set in order to avoid the over fitting problem. For each rule r, two additional rules are constructed: a revised rule r' and a new replacement rule r''. The revised rule r' is constructed by greedily adding literals to r. The resulting r' is then pruned to minimize error of the current rule set. The new replacement rule r'' is constructed from an empty rule by growing. After growing, r'' is also pruned to minimize error of the current rule set. The final step is to select r, r' or r''. This selection is based on minimum description length principle. The rule that has the smallest description length and has no less descriptive power than the other rules is selected [22].

After the optimization, the current rule set may not cover all positive examples. For uncovered positive examples, additional rules are constructed and added to cover them. The optimization step is then repeated, occasionally resulting in further improvements in the current rule set. Previous experiments show that two rounds of the optimization are usually sufficient [22].

Association Rule

An example of a classifier that uses association rules is proposed by [118], who applied association rule mining in building a document categorization system. Association rules mining [2, 42] is a data mining task aiming at discovering relationships between items in a dataset. This approach has the advantage of fast training and has the performance comparable to most well known document classifiers.

For document classification, as described in [118], a set of rules can be used to encode each category. A rule is represented in "$F => c$" formal, where F is a conjunction (such as $f_1 \wedge f_2 \wedge f_3$) of features extracted from a set of training documents, which are taken from the same category c. Once a set of rules is discovered for each category, the rule set can be used to classifier new documents.

The process for discovering a set of rules for category c begins by counting the number of occurrences (or frequency) of each feature in the set D of training documents for the category. It then selects features that have concurrent frequency larger than a threshold called support. The selected features are paired to form 2-item features. Then, the process counts frequency of each of the 2-item features, and selects the 2-item features that have concurrent frequency larger the support threshold. The resulting 2-item features are combined with 1-item features to form 3-item features, and so on. This process is repeated until k-item features are selected, where k is a predefined number. The set of k-item features are transformed to a set of rules, each of which is in form of $(f_1 \wedge f_2 \wedge \ldots f_k) \Rightarrow c$. Only some of the rules are used for classification. These rules are selected based on a confidence criterion. The confident of a rule $(f_1 \wedge f_2 \wedge \ldots f_k) \Rightarrow c$ can be defined here as the proportion of documents which belong to category c over those documents each of which contains features f_1 and f_2 and $\ldots f_k$. A threshold of 70% for the confidence criterion was used in [118].

Decision Tree

Classifiers using decision tree are considered as rule learning based classifiers since a decision tree can be converted into a disjunction of conjunction rules (same as DNF). Decision trees have been employed for document classifications [27, 35, 60, 75, 108]. However, this approach contains no special mechanism to handle the large feature sets encountered in document classification and probably accounts for its relatively poor performance in experiments by [60, 75, 108].

To classify a document d using a decision tree, the document vector $\vec{d}$ is matched against the decision tree to determine to which category the document d belongs [1, 12, 70, 83, 84]. The decision tree is constructed from training documents. A popular approach is CART algorithm [12].

CART approach, as described in [1], constructs a binary decision tree (e.g. Fig. 6) from a set of training documents that are represented as feature

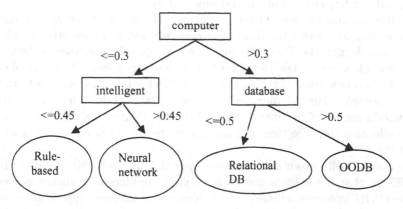

Fig. 6. A decision tree. Each node, except the leaf nodes, in the tree represents a feature, each branch descending from the node corresponds to one of the possible feature values (e.g. the TFIDF value of a term), and each leaf node is a class label. A new test document is classified by traversing it through the tree to the appropriate leaf node, and then returned the category associated with the leaf node

vectors. At each step, a feature is selected from the set of feature vectors and is used to split the set of feature vectors into two subsets. A measure called *diversity* is used to determine which feature to select. The best selection is done by maximizing:

$$diversity(before\_split) - [diversity(left\_child) + diversity(right\_child) \quad (65)$$

One of the commonly used *diversity* measures is entropy (see (26)) and another one is Gini Entropy (*GE*) which is used in CART. The Gini Entropy of a node t is defined as

$$GE(t) = 1 - \sum_{j=1}^{K} Pr(c_j|t) \quad (66)$$

where K is number of categories, and $Pr(c_j|t)$ is the probability of a training example being in class c_j that falls into node t. $Pr(c_j|t)$ can be estimated by

$$Pr(c_j|t) = \frac{N_j(t)}{N(t)} \quad (67)$$

where $N_j(t)$ is the number of training examples of class c_j at node t and $N(t)$ is the total number of training examples at node t.

To select a feature for a node (e.g. Fig. 6), each feature in all training feature vectors is evaluated using (65) and (66). The feature resulting in the maximum value in (65) is selected and used to split the set of training feature vectors. This process is repeated until no training feature vectors can be partitioned any further. Each training feature vector can then be used to traverse the resulting binary tree from root to a leaf node. The resulting leaf node is assigned a category label of the training feature vector.

After building the initial binary tree using the above algorithm, the resulting tree usually overfits the training documents and is not effective for classifying new documents. Thus, the initial tree is pruned to remove the branches that provide the least additional predictive power per leaf. The result of each pruning is a new tree. The final task is to select a tree that will best classify new documents. For this purpose, a new set of labeled documents are used as the validation set of documents. Each of the candidate trees is used to classify the validation set. The tree that has the highest accuracy is selected as the final tree.

Another well known decision tree algorithm is C4.5 [84]. It differs from CART in that it produces tree with varying numbers of branches per node while CART produces a binary tree. It also uses a different approach to prune the tree by converting the learned tree into an equivalent set of rules, which are the result of creating one rule for each path from the root to a leaf node. Each feature along the path becomes a condition while the category label at the leaf node becomes the consequent. For instance, the leftmost path of the tree in Fig. 6 is translated into the rule: *IF (computer <=0.3) and (intelligent*

<=0.45)THEN class label is "Rule-based". Next, each such rule is pruned by removing any condition, whose removal does not worsen the estimated accuracy (see also [70]).

4.3 Direct Example Based Classifiers

For a direct example based classifier, a web page to be classified is used as a query directly against a set of examples that identify categories. The web page is assigned to the category whose set of examples has the highest similarity with the web page. These classifiers are called lazy learning systems. K-nearest-neighbors classifier is a representative.

K-Nearest-Neighbors Classifier

In contrast to "eager learning" classifiers (e.g. Rocchio classifier) that have an explicit training phase before classify new documents, K-nearest-neighbors (KNN) [30, 70] is a lazy learning method that delays the learning process until a new document must be classified. KNN classifier has been successfully applied for document classification [62, 67, 107, 109, 111, 112].

KNN classifier compares a new document directly with the given training documents. It uses cosine metric to compute the similarity between two document vectors. It ranks, in a descend order, the training documents based on their similarities with the new document. The top k training documents are k-nearest neighbors of the new document and the k-nearest neighbors are used to predict the categories of the new document. Reference [109] showed that the performance of kNN is relatively stable for a large range of k, and $k = 30$, 45 or 65 were tested in their experiments.

The similarity score of each neighbor document is used as a weight for the associated category. To classify a new document, the likelihood score of a category can be calculated as [112]

$$y(\vec{d'}.c_j) = \sum_{\vec{d_i} \in KNN} \text{sim}(\vec{d'}, \vec{d_i}) y(\vec{d_i}, c_j) - b_j \qquad (68)$$

where $\vec{d_i}$ is one of k-nearest neighbors of the new document $\vec{d'}$, $y(\vec{d_i}, c_j) \in \{0, 1\}$, is the classification for the neighbor $\vec{d_i}$ with respect to category c_j ($y = 1$ for yes; and $y = 0$ for no), and $\text{sim}(\vec{d'}, \vec{d_i})$ is the similarity (e.g. cosine similarity) between the new document d' and its neighbor d_i, and b_j is the category specific threshold. The category specific threshold b_j is automatically learned by using a validation set of documents. That is, KNN algorithm learns the optimal threshold b_j for category c_j in that it yields the best performance on the validation documents [112]. The new document is assigned to those categories having likelihood scores larger than a predefined threshold.

4.4 Parameter Based Classifiers

For parameter based classifiers, training examples are used to estimate parameters of a probability distribution. Probabilistic Naïve Bayes classifier is an example.

Naïve Bayes Classifiers

The naïve Bayes classifier, as described in [45, 70], is constructed by using training examples to estimate the probability of each category given a new document d', which is written as $P(c_j|d')$:

$$Pr(c_j|d') = \frac{Pr(c_j) \cdot Pr(d'|c_j)}{Pr(d')} \qquad (69)$$

The denominator in the above equation does not differ between categories and can be left out. The naïve Bayes classifier makes an assumption of feature independence in order to estimate $P(d'|c_j)$ as

$$Pr(c_j|d') = Pr(c_j) \prod_{f_i \in d'} Pr(f_i|c_j) \qquad (70)$$

where $Pr(c_j)$ is the proportion of training examples in category c_j, and f_i is a feature (or term) found in document d'. An estimate for $Pr(f_i|c_j)$ is given by [1, 70]:

$$\tilde{Pr}(f_i|c_j) = \frac{1 + N_{ij}}{M + \sum_{k=1}^{M} N_{kj}} \qquad (71)$$

where N_{ij} is the number of times feature f_i occurring within documents from category c_j, and M is the total number of features in the training set. The category with the maximum value of $P(c_j|d')$ is the desired category for document d'. The work of the Naïve Bayes classifier applied in text classification could be found in [46, 58, 59, 63, 73, 74, 109]. Because the fact that the assumption of feature independence is general not true for documents, the Naïve Bayes classifier showed relatively worse performance in [109] and [63].

5 Evaluation of Web Page Classifiers

This section describes how to evaluate web page classifiers and reports experimental results. While the above section concerns more about the training phase for building a classifier, this section discusses how to evaluate a classifier in the testing phase. The testing phase performs on the testing examples, which are a part of all available examples. The other part of available examples consists of training examples used in the training phase and/or validation examples used for optimizing the model generated from the training phase. Criteria for performance measures are first described in the following. Results of experiments are then provided.

5.1 Performance Measures

The experimental evaluation of classifiers tries to evaluate the effectiveness of the learned model, i.e. its capability of making the right classification decision. The most frequently used measures of classification effectiveness are presented as follows.

A classifier can be evaluated in terms of precision, recall, accuracy, and error. Precision may be viewed as the degree of soundness of the classifier, while recall may be viewed as its degree of completeness. The precision and recall can be estimated in terms of the contingency table for category c_i on a given test set (see Table 4) [1,97].

Table 4. The contingency table for category c_i

a: the number of testing examples correctly assigned to this category
b: the number of testing examples incorrectly assigned to this category
c: the number of testing examples incorrectly rejected from this category
d: the number of testing examples correctly rejected from this category

From the quantities in Table 4, precision and recall for a category are defined as:

$$\text{precision} = \frac{a}{a+b} \tag{72}$$

$$\text{recall} = \frac{a}{a+c} \tag{73}$$

Precision has similar meaning as classification accuracy. But they are difference in that precision considers only examples assigned to the category, while accuracy considers both assigned and rejected cases. Accuracy and error for a category are defined as:

$$\text{accuracy} = \frac{a+d}{a+b+c+d} \tag{74}$$

$$\text{error} = \frac{b+c}{a+b+c+d} \tag{75}$$

The above definitions are applicable for each category. To obtain measures relating to all categories, two methods may be adopted: microaveraging and macro-averaging [1,97,109]:

- **Micro-averaging**: the performance measures are obtained by globally summing over all individual decisions, i.e.

$$\text{precision} = \frac{\sum_{i=1}^{K} a_i}{\sum_{i=1}^{K} (a_i + b_i)} \tag{76}$$

where K is the number of categories, a_i is the number of testing examples correctly assigned to category i, and b_i is the number of testing examples incorrectly assigned to category i.

- **Macro-averaging**: the performance measures are first evaluated "locally" for each category, and then "globally" by averaging over the results of the different categories, i.e.

$$\text{precision} = \frac{\sum_{i=1}^{K} \text{precision}_i}{K} \tag{77}$$

Recall, accuracy, and error for all categories can be computed similarly. It is important to recognize that these two methods may give quite different results, especially if the categories are unevenly populated.

Precision or recall may be misleading when considered alone since they are interdependent. Thus, a combined measure is considered [1, 97]. The effectiveness of a classifier is expressed as a function F_α in terms of both precision and recall as follow:

$$F_\alpha = \frac{1}{\alpha \cdot \frac{1}{\text{precision}} + (1 - \alpha) \cdot \frac{1}{\text{recall}}} \tag{78}$$

where $0 \le \alpha \le 1$. A value of $\alpha = 0.5$ is usually used, which attributes equal importance to precision and recall and is usually referred to as F_1 measure.

5.2 Experimental Settings and Results

A large number and diversity of document classifiers have been proposed. Comparing the effectiveness of these classifiers has shown to be difficult. Many of the classifiers were tested with different data sets and under different experimental settings. However, only the evaluations of different classifiers under same experimental setting and using same data set are comparable. The following first outlines commonly used data sets and then reports attempts to compare classifiers under same experimental settings.

Data Sets

For testing classifiers, standard text collections can be found in public domain. Typical examples include [109, 119]:

- The *Reuters*-21578 corpus: The documents are newswire stories covering the period between 1987 and 1991.
- The *OHSUMED* corpus: The documents are title and abstract from medical journals.
- The *Yahoo* corpus: It is provided by Yahoo.com consisting of a directory of manually organized web pages.
- The 20 *Newsgroups* collection: The documents are messages posted to Usenet newsgroups, and the categories are the newsgroups themselves.

- The *Hoovers* corpora of company Web pages: It contains detailed information about a large number of companies and is a reliable source of corporate information.
- The *WebKB* collection: The documents are web pages that were assembled for training an intelligent web crawler.
- The *Brown* text document collection: This dataset is used for genrebased classification.

Experimental Comparison of Classifiers

An evaluation of fourteen classifiers for the subject-based classification was reported in [109]. Here we cite the experimental results in [109] to show the performance difference among classifiers (see Table 5). The results indicate that kNN classifier has the best performance. Among the top classifiers are LLSF, NNet, and WH. The next group is the rule induction algorithms, such as SWAP-1, RIPPER and CHARADE, showing a similar performance. Rocchio and Naïve Bayes had relatively worse performance.

Reference [112] re-examined five document classifiers, SVM, kNN, NNet, LLSF and Naïve Bayes and focused on the robustness of these classifiers in dealing with a skewed category distribution. The experimental results showed that SVM, kNN, and LLSF significantly outperformed NNet and Naïve Bayes when the number of positive training examples per category is less than ten.

Reference [112] evaluated eight classifiers, including SVM, linear regression (LLSF), logistic regression (LR), NNet, Rocchio, Prototypes, kNN, and Naïve Bayes. They used a loss function to analyze the optimization criterion of each classifier. The reason for using a loss function is that the optimization of a classifier is not only driven by the training set error but also driven by the model complexity. The loss function of a classifier L_c is defined as *the training set loss + the complexity penalty*. Balancing between the two criteria has been referred as the regularization of a classifier. The degree of regularization is controlled by a parameter in the classifier. They showed that regularization made significant improvement on the performances of LLSF and NNet, and also showed that the performances of regularized LLSF, NNet, LR and SVM were quite competitive. kNN was among the top classifiers that include SVM, regularized LR, NNet and LLSF. Rocchio was second. Naïve Bayes and Prototype were the last.

However, no such experiments have been conducted for the genre-based classifiers. Most work for genre-based classification employed neural network, decision tree, or rule-based learning methods [27,35]. No such thorough evaluation of these classifiers in terms of genre-based classification has been reported in the literature.

Table 5. Test results for comparing fourteen classifiers [109]

	Reuters Apte BrkEvn	Reuters PARC BrkEvn	OHSUMED Full Range $F(\beta=1)$	OHSUMED HD Big $F(\beta=1)$	Reuters Lewis BrkEvn	Reuters CONS. BrkEvn
kNN(N)	0.85*	0.82*	0.51*	0.56	0.69	–
LLSF(L)	0.85*	0.81* (−1%)	–	–	–	–
NNets(N)	–	0.82*	–	–	–	–
WH	–	–	–	0.59* (+5%)	–	–
EG(L)	–	–	–	0.54 (−4%)	–	–
RIPPER(N)	0.80 (−6%)	–	–	–	0.72	–
DTree (N)	[0.79]	–	–	–	0.67	–
SWAP–1 (N)	0.79 (−7%)	–	–	–	–	–
CHARADE (N)	0.78 (−8%)	–	–	–	–	–
EXPERTS (N)	0.76 (−11%)	–	–	–	0.75*	–
Rocchio (L)	0.75 (−12%)	–	–	0.46 (−18%)	0.66	–
NaiveBayes (L)	0.71 (−16%)	–	–	–	0.65	–
CONSTRUE	–	–	–	–	–	0.90*
WORD	0.29 (−66%)	0.25 (−69%)	0.27 (−47%)	0.44 (−21%)	0.15	–

L a linear method, N a non-linear model, *the local optimal on a fixed collection, (x%) the performance improvement relative to kNN, [x] a F_1 measure. kNN k-nearestneighbor algorithm, WH Widrow-Hoff learning algorithm (also known as Least Mean Square), EG Exponential Gradient algorithm, DTree decision tree learning algorithm, SWAP-1 an rule-based learning algorithm, CHARADE an expert system consisting of manually developed categorization rules, EXPERTS Sleeping Experts, CONSTRUE a rule learning system, WORD a non-learning algorithm as a baseline of the comparison.

6 Summary

Web page classification is the assignment of web pages to one or more predefined categories based on their subjects or genres. A general inductive process automatically builds a model by learning over a set of previously classified web pages. This model is then used to classify new web pages.

A typical web page classification process consists of the following steps: extracting salient features from training web pages, creating a feature vector for each web page, reducing dimensionality of the feature vectors, creating a classifier by learning over the training feature vectors, and then classifying new web pages using the classifier.

Web pages can be classified in terms of subjects or genres. For subject-based classification, the salient features of web pages are the text contents, such as words, phases, and sentences. For genre-based classification, the salient features are the genre attributes, such as presentation characteristics and functional words.

Dimensionality reduction methods, including feature selections and extractions, are used to reduce the feature space of the training web pages. Feature selection techniques are used to select subset from the original feature space based on criterions, such as Information Gain, Cross Entropy, and Odds Ratio. Feature extraction techniques, such as Latent Semantic Indexing and Word Clustering, are used to transform the original large feature space into a smaller feature space having possibly new set of features.

Numerous classifiers proposed and used for machine learning can be applied for web page classification. Various classifiers that have been applied and to some extent proven efficient for web page classification are described in this chapter.

References

1. Aas K, Eikvil L (1999) Text categorization: a survey. Report NR 941, Norwegian Computing Center
2. Agrawal R, Srikant R (1994) Fast algorithm for mining association rules. In: Proceedings of 1994 Int. Conf. Very Large Data Bases. Santiago, Chile, pp. 487–499
3. Apte C, Damerau F, Weiss S (1994) Towards language independent automated learning of text categorization models. In: Proceedings of SIGIR '94: 17th ACM international conference on research and development in information retrieval. Dublin, Ireland, pp. 24–30
4. Argamon S, Koppel M, Avneri G (1998) Routing documents according to style. In: First International Workshop on Innovative Information Systems. Pisa, Italy
5. Baker LD, McCallum A (1998) Distributional clustering of words for text classification. In: Proceedings of SIGIR '98: 21st ACM international conference on research and development in information retrieval. Melbourne, Australia, pp. 96–103
6. Bazerman C (1998) Shaping written knowledge: the genre and activity of the experimental article in science. The University of Wisconsin Press, Madison, WI, USA
7. Berkhin P (2000) Survey of clustering data mining techniques. http://www.accrue.com/products/researchpapers.html
8. Berry MW (1992) Large-scale sparse singular value computations. The International J. of Supercomputer Applications 6: 13–49

9. Berry MW, Dumais ST, O'Brien GW (1995) Using linear algebra for intelligent information retrieval. SIAM Review 37: 573–595

10. Biber, D (1995) Dimensions of register variation: a cross-linguistic comparison. Cambridge University Press, Cambridge, England

11. Blum A, Mitchell T (1998) Combining labeled and unlabeled data with co-training. In: COLT: Proceedings of the workshop on computational learning theory. Morgan Kaufmann Publishers Inc., San Francisco, CA

12. Breiman L, Friedman JH, Olshen RA, Stone CJ (1984) Classification and regression trees. Belmont, CA: Wadsworth

13. Brown JS (1994) Borderline issues: social and material aspects of design. Human Computer Interactions 9: 3–36

14. Buckley C, Salton G, Allan J (1994) The effect of adding relevance information in a relevance feedback environment. In: Proceedings of SIGIR '94: 17th ACM international conference on research and development in information retrieval. Dublin, Ireland, pp. 292–300

15. Burrows J (1992) Not unless you ask nicely: the interpretative nexus between analysis and information. Literary and Linguistic Computing 7: 91–109

16. Chakrabati S, Dom BE, Indyk P (1998) Enhanced hypertext categorization using hyperlinks. In: Proceedings of SIGMOD-98: ACM International conference on management of data. Seattle, WA, USA, pp. 307–318

17. Choi B (2001) Making sense of search results by automatic web-page classification. In: WebNet 2001. Orlando, Florida, USA, pp. 184–186

18. Choi B, Guo Q (2003) Applying semantic links for classifying web pages. Lecture Notes in Artificial Intelligence 2718: 148–153

19. Choi B, Li B (2002) Abstracting keywords from hypertext documents. In: International conference on information and knowledge engineering. Las Vegas, Nevada, USA, pp. 137–142

20. Church KW, Hanks P (1989) Word association norms, mutual information and lexicography. In: Proceedings of ACL 27. Vancouver, Canada, pp. 76–83

21. Cohen WW (1995) Fast effective rule induction. In: Proceedings of the 12th international conference on machine learning. Lake Tahoe, CA, USA, pp 115–123

22. Cohen WW, Singer Y (1996) Context-sensitive learning methods for text categorization. In: Proceedings of SIGIR '96: 19th ACM international conference on research and development in information systems. Zurich, CH, pp. 307–315

23. Cormen TH, Leiserson CE and Rivest RL (1990) Introduction to algorithms. Cambridge, MA: MIT Press

24. Cortes C, Vapnik V (1995) Support vector networks. Machine learning 20: 273–297

25. Cover TM, Thomas JA (1991) Elements of information theory. Wiley, New York

26. Deerwester S, Dumais ST, Furnas GW, Landauer TK, Harshman R (1990) Indexing by latent semantic analysis. J. Amer. Soc. Inf. Sci. 41: 391–407

27. Dewdney N, VanEss-Dykema C, MacMillan R (2001) The form is the substance: classification of genres in text. In: ACL workshop on human language technology and knowledge management. Toulouse, France

28. Dillon IS, Mallela S, Kumar R (2002) Enhanced word clustering for hierarchical text classification. In: Proceedings of the 8th ACM SIGKDD. Edmonton, Canada, pp. 191–200

29. Dongarra JJ, Moler CB, Bunch JR, Stewart GW (1979) LINPACK users' guide. Philadelphia, PA: SIAM
30. Duda RO, Hart PE (1973) Pattern Classification and Scene Analysis. John Wiley and Sons, New York
31. Duda RO, Hart PE, Stork DG (2001). Pattern classification, 2nd edn. Wiley, New York
32. Dumais S, Chen H (2000) Hierarchical classification of web content. In: Proceedings of the 23rd ACM international conference on research and development in information retrieval (ACM SIGIR). Athens, Greece, pp 256–263
33. Dumais S, Platt J, Hecherman D, Sahami M (1998) Inductive learning algorithm and representations for text categorization. In: Proceedings of CIKM-98: 7th ACM international conference on information and knowledge management. Bethesda, MD, USA, pp. 148–155
34. Everitt BS, Landua S, Leese M (2001) Cluster Analysis. Arnold, London, UK
35. Finn A, Kushmerick N (2003) Learning to classify documents according to genre. In: IJCAI-2003 workshop on computational approaches to text style and synthesis. Acapulco, Mexico
36. Furnkranz J (1999) Exploiting structural information for text classification on the WWW. In: Proceedings of IDA 99: 3rd Symposium on intelligent data analysis. Amsterdam, NL, pp. 487–497
37. Furnkranz J, Widmer G (1994) Incremental reduced error pruning. In: Proceedings of the 11th annual conference on machine learning. Morgan Kaufmann Publishers Inc., San Francisco, CA
38. Galavotti L, Sebastiani F, Simi M (2000) Experiments on the use of feature selection and negative evidence in automated text categorization. In: Proceedings of ECDL-00: 4th European conference on research and advanced technology for digital libraries. Lisbon, PT, pp. 59–68
39. Glover EJ, Tsioutsiouliklis K, Lawrence S, Pennock DM, Flake GW (2002) Using web structure for classifying and describing web pages. In: Proceedings of international world wide web conference. Honolulu, Hawaii, pp. 562–569
40. Golub B, Loan CV (1989) Matrix computations (2nd Ed). Johns-Hopkins, Baltimore
41. Haas SW, Grams ES (1998) Page and link classifications: connecting diverse resources. In: Proceedings of the 3rd ACM conference on digital libraries. Pittsburgh, PA, USA, pp. 99–107
42. Han J, Pei J, Yin Y (2000) Mining frequent patterns without candidate generation. In: ACM-SIGMOD international conference on management of data. Dallas, TX, USA, pp. 1–12
43. Ittner DJ, Lewis DD, Ahn DD (1995) Text categorization of low quality images. In: Symposium on document analysis and information retrieval. Las Vegas, NV, USA, pp. 301–315
44. Jain AK, Murty M N, Flynn PJ (1999) Data clustering: a review. ACM computing surveys 31: 255–323
45. Joachims T (1997) A probabilistic analysis of the Rocchio algorithm with TFIDF for text categorization. In: International conference on machine learning (ICML). Nashville, TN, USA, pp. 143–151
46. Joachims T (1998) Text categorization with support vector machines: learning with many relevant features. In: European conference on machine learning. Berlin, German, pp. 137–142

47. John G, Kohavi R, Pfleger K (1994) Irrelevant features and the subset selection problem. In: Machine learning: proceedings of the 11th international conference. Morgan Kaufman Publishers Inc., San Francisco, CA, pp. 121–129

48. Jones KS (1972) A statistical interpretation of term specificity and its application in retrieval. Journal of Documentation 28: 11–20

49. Jung Y, Park H, Du D (2000) An effective term-weighting scheme for information retrieval. Computer science technical report TR008. Department of computer science, University of Minnesota, Minneapolis, Minnesota

50. Karlgren J, Cutting D (1994) Recognizing text genres with simple matrices using discriminant analysis. In: Proceeding of the 15th international conference on computational linguistics. Kyoto, Japan, pp. 1071–1075

51. Kaufman L and Rousseeuw PJ (1990) Finding groups in data. New York: Wiley

52. Kessler B, Nunberg G, Schutze H (1997) Automatic detection of text genre. In: Proceeding of 35th annual meeting of the association for computational linguistics. Madrid, Spain, pp. 32–38

53. Kohavi R, John GH (1997) Wrappers for feature subset selection. Artificial Intelligence Journal 97(1–2): 273–324

54. Koller D, Sahami M (1996) Toward optimal feature selection. In: Lorenza Saitta (eds) Machine learning: proceedings of the 13th international conference. Morgan Kaufmann Publishers Inc., San Francisco, CA, pp. 294–292

55. Koller D, Sahami M (1997) Hierarchically classifying documents using very few words. In: Proc. of the 14th international conference on machine learning (ICML97). Nashville, TN, USA, pp. 170–178

56. Lee Y, Myaeng SH (2002) Text genre classification with genre-revealing and subject-revealing features. In: Proceedings of SIGIR '02: 25th ACM international conference on research and development in information retrieval. Tampere, Finland, pp. 145–150

57. Lesk ME (1986) Automatic word sense disambiguation: how to tell a pine cone from an ice cream cone. In: DeBuys V (eds) Proceedings of SIGDOC-86: 5th ACM international conference on systems documentation. New York, US, pp. 24–26

58. Lewis DD (1992) An evaluation of phrasal and clustered representation on a text categorization task. In: Proceedings of SIGIR-92: 15th ACM international conference on research and development in information retrieval. Kobenhavn, DK, pp. 37–50

59. Lewis DD (1998) Naïve (Bayes) at forty: the independence assumption in information retrieval. In: Proceedings of ECML-98: 10th European conference on machine learning. Chemnitz, DE, pp. 4–15

60. Lewis DD, Ringuette M (1994) Comparison of two learning algorithms for text categorization. In: Proceedings of the 3rd annual symposium on document analysis and information retrieval (SDAIR '94). Las Vegas, Nevada, USA, pp. 81–93

61. Lewis DD, Schapire RE, Callan JP, Papka R (1996) Training algorithms for linear text classifiers. In: Proceedings of SIGIR-96: 19th ACM international conference on research and development in information retrieval. Zurich, CH, pp. 298–306

62. Li F, Yang Y (2003) A loss function analysis for classification methods in text categorization. In: Proceeding of the twentieth international conference on machine learning (ICML 2003). Washington, DC, USA, pp. 472–479

63. Li YH, Jain AK (1998) Classification of text documents. The Computer Journal 41: 537–546
64. Luhn HP (1958) The automatic creation of literature abstracts. IBM Journal of Research and Development 2: 159–165
65. Malone TW et al. (1999) Tools for inventing organizations: toward a handbook of organizational process. Management Science 45: 425–443
66. Manning and Schutze (2001) Foundations of statistical natural language processing. The MIT Press, Cambridge Massachusetts, London England
67. Masand B, Linoff G, Waltz D (1992) Classifying news stories using memory based reasoning. In: 15th Ann. Int. ACM SIGIR conference on research and development in information retrieval (SIGIR '92). Kobenhavn, DK, pp. 59–64
68. Mihalcea R, Moldovan D (2001) A highly accurate bootstrapping algorithm for word sense disambiguation. International Journal on Artificial Intelligence Tools 10: 5–21.
69. Miller CR (1984) Genre as social action. Quarterly Journal of Speech 70: 151–167
70. Mitchell T (1997) Machine Learning. McGraw Hill, New York
71. Mladenic D (1998) Feature subset selection in text-learning. In: Proceeding of the 10th European conference on machine learning (ECML98). Chemnitz, Germany, pp. 95–100
72. Mladenic D (1998) Machine learning on non-homogeneous, distributed text data. PhD thesis, University of Ljubljana, Slovenia
73. Mladenic D, Grobelnik M (1998) Feature selection for classification based on text hierarchy. In: Working notes of learning from text and the web: conference on automated learning and discovery. Carnegie Mellon University, Pittsburgh
74. Mladenic D, Grobelnik M (1999) Feature selection for unbalanced class distribution and Naïve Bayes. In: 16th International conference on machine learning. Bled, Slovenia, pp. 258–267
75. Moulinier I (1997) Is learning bias an issue on the text categorization problem? Technical report. LAFORIA-LIP6, University Paris, VI
76. Ng HT, Goh WB, Low KL (1997) Feature selection, perceptron learning, and a usability case study for text categorization. In: 20th Ann. Int. ACM SIGIR conference on research and development in information retrieval (SIGIR '97). Philadelphia, PA, USA, pp. 67–73
77. Oh H, Myaeng SH, Lee M (2000) A practical hypertext categorization method using links and incrementally available class information. In: Proceedings of SIGIR-00: 23rd ACM international conference on research and development in information retrieval. Athens, GR. pp. 264–271
78. Orlikowski WJ, Yates J (1994) Genre repertoire: examining the structuring of communicative practices in organizations. Administrative Science Quarterly 39: 541–574
79. Orlikowski WJ, Yates J (1998) Genre systems as communicative norms for structuring interaction in groupware. Working paper #205, MIT Center for Coordination Science
80. Osuna E, Freund R, Girosi F (1997) Support vector machines: training and applications. A. I. memo No. 1602, A. I. Lab, Massachusetts Institute of Technology
81. Paice CD (1990) Constructing literature abstracts by computer: techniques and prospects. Information Processing and Management 26: 171–186

82. Peng X, Choi B (2002) Automatic web page classification in a dynamic and hierarchical way. In: IEEE international conference on data mining. pp. 386–393
83. Quinlan JR (1986) Induction of decision trees. Machine Learning 1: 81–106
84. Quinlan JR (1993) C4.5: programming for machine learning. Morgan Kaumann publishers, San Francisco, CA
85. Quinlan JR (1995) MDL and categorical theories (continued). In: Proceedings of the 12th international conference on machine learning. Lake Tahoe, CA, pp. 464–470
86. Ragas H, Koster CH (1998) Four text classification algorithms compared on a dutch corpus. In: Proceedings of SIGIR-98: 21st ACM international conference on research and development in information retrieval. Melbourne, AU, pp. 369–370
87. Rasmussen E (1992) Clustering algorithms. In: Willian B. Frakes and Ricardo Baeza-Yates (eds.), information retrieval. Englewood Cliffs, NJ: Prentice Hall, pp. 419–442
88. Rauber A, Muller-Kogler A (2001) Integrating automatic genre analysis into digital libraries. In: 1st ACM-IEEE joint conference on digital libraries. Roanoke, VA, USA, pp. 1–10
89. Riboni D (2002) Feature selection for web page classification. In: 2002 proceedings of the workshop of first EurAsian conference on advances in information and communication technology. Shiraz, Iran
90. Rijsbergen V, Harper CJ, Porter DJ (1981) The selection of good search terms. Information Processing and Management 17: 77–91
91. Rocchio J (1971) Relevance feedback in information retrieval. In: Salton (ed) The SMART retrieval system: experiments in automatic document processing. Prentice-Hall, New Jersey, pp. 313–323
92. Ruiz ME, Srinivasan P (2002) Hierarchical text categorization using neural networks. Information Retrieval 5: 87–118
93. Salton G (1989) Automatic text processing: the transformation, analysis, and retrieval of information by computer. Addison-Wesley, New York
94. Salton G (1991) Developments in automatic text retrieval. Science 253: 974–979
95. Salton G, Buckley C (1988) Term weighting approaches in automatic text retrieval. Information Processing and Management 24: 513–523
96. Schutze H, Hull DA, Pedersen JO (1995) A comparison of classifiers and document representations for the routing problem. In: 18th Ann. Int. ACM SIGIR conference on research and development in information retrieval. Seattle, WA, USA, pp. 229–237
97. Sebastiani F (1999) A tutorial on automated text categorization. In: Proceedings of ASAI-99: 1st Argentinean symposium on artificial intelligence. Buenos Aires, AR, pp. 7–35
98. Sebastiani F (2002) Machine learning in automated text categorization. ACM Computing Surveys 34: 1–47
99. Silverstein C and Pedersen JO (1997) Almost-constant-time clustering of arbitrary corpus subsets. In: Proceedings of the 20th annual international ACM SIGIR conference on research and development in information retrieval (SIGIR '97). Philadelphia, PA, pp. 60–66

100. Slonim N, Tishby N (2001) The power of word clusters for text classification. In: Proc. 23rd European colloquium on information retrieval research (ECIR). Darmstadt, DE.
101. Stamatatos E, Fakotakis N, Kokkinakis G (2000) Text genre detection using common word frequencies. In: Proceeding of the 18th international conference on computational linguistics (COLING2000). Luxembourg, pp. 808–814
102. Strehl A (2002) Relationship-based clustering and cluster ensembles for high-dimensional data mining. Dissertation, the University of Texas at Austin
103. Sun A, Lim E, Ng W (2002) Web classification using support vector machine. Proceedings of the 4th international workshop on Web information and data management. McLean, Virginia, USA, pp. 96–99
104. Vapnik V (1995) The nature of statistical learning theory. Springer, New York
105. Wiener E, Pederson JO, Weigend AS (1995) A neural network approach to topic spotting. In: Proceedings of the 4th annual symposium on document analysis and information retrieval (SDAIR '95). Las Vegas, US, pp. 317–332
106. Willett P (1988) Recent trends in hierarchic document clustering: a critical review. Information Processing & Management 24: 577–597
107. Yang Y (1994) Expert network: effective and efficient learning for human decision in text categorization and retrieval. In: Proc. of the 7th annual international ACM-SIGIR conference on research and development in information retrieval. Dublin, Ireland, pp. 13–22
108. Yang Y (1995) Noise reduction in a statistical approach to text categorization. In: Proceedings of the 18th Ann. Int. ACM SIGIR conference on research and development in information retrieval (SIGIR '95). Seattle, WA, USA, pp. 256–263
109. Yang Y (1999) An evaluation of statistical approaches to text categorization. Information Retrieval 1: 69–90
110. Yang Y, Chute CG (1992) A linear least squares fit mapping method for information retrieval from natural language texts. In: Proceedings of the 14th international conference on computational linguistics (COLING 92). Nantes, France, pp. 447–453
111. Yang Y, Chute CG (1994) An example-based mapping method for text categorization and retrieval. ACM Transaction and Information System 12: 253–277
112. Yang Y, Liu X (1999) A re-examination of text categorization methods. In: Proceedings of SIGIR-99: 22nd ACM international conference on research and development in information retrieval. Berkeley, CA, US, pp. 42–49
113. Yang Y, Pederson JO (1997) A comparative study on feature selection in text categorization. In: Proc. of the 14th international conference on machine learning (ICML97). Morgan Kaufmann Publishers Inc, San Francisco, USA, pp. 412–420
114. Yang Y (2001) A study on thresholding strategies for text categorization. In: Proceedings of the 24th annual international ACM SIGIR conference on research and development in information retrieval (SIGIR 2001). New Orleans, LA, USA, pp. 137–145
115. Yang Y, Slattery S, Ghani R (2001) A study of approaches to hypertext categorization. Journal of Intelligent Information Systems 18(2–3): 219–241
116. Yao Z, Choi B (2003) Bidirectional hierarchical clustering for web mining. In: Proc. of the 2003 IEEE/WIC international conference on web intelligence. Halifax, Canada, pp. 620–624

117. Yoshioka T, Herman G (1999) Genre taxonomy: a knowledge repository of communicative actions. Working paper #209, MIT Center for Coordination Science
118. Zaiane OR, Antonie ML (2002) Classifying text documents by associating terms with text categories. In: Proceedings of the 13th Australasian conference on database technologies, Melbourne, Australia, pp. 215–222
119. Zhang T, Oles FJ (2001) Text categorization based on regularized linear classification methods. Information Retrieval 4: 5–31

Web Mining – Concepts, Applications and Research Directions

J. Srivastava, P. Desikan, and V. Kumar

Department of Computer Science,
University of Minnesota
{srivasta, desikan, kumar}@cs.umn.edu

Abstract. From its very beginning, the potential of extracting valuable knowledge from the Web has been quite evident. Web mining, i.e. the application of data mining techniques to extract knowledge from Web content, structure, and usage, is the collection of technologies to fulfill this potential. Interest in Web mining has grown rapidly in its short history, both in the research and practitioner communities. This paper provides a brief overview of the accomplishments of the field, both in terms of technologies and applications, and outlines key future research directions.

1 Introduction

Web mining is the application of data mining techniques to extract knowledge from Web data, including Web documents, hyperlinks between documents, usage logs of web sites, etc. A panel organized at ICTAI 1997 [91] asked the question "Is there anything distinct about Web mining (compared to data mining in general)?" While no definitive conclusions were reached then, the tremendous attention on Web mining in the past five years, and a number of significant ideas that have been developed, have answered this question in the affirmative in a big way. In addition, a fairly stable community of researchers interested in the area has been formed, largely through the successful series of WebKDD workshops, which have been held annually in conjunction with the ACM SIGKDD Conference since 1999 [53,63,64,81], and the Web Analytics workshops, which have been held in conjunction with the SIAM data mining conference [46,47]. A good survey of the research in the field till the end of 1999 is provided in [82] and [62].

Two different approaches were taken in initially defining Web mining. First was a "process-centric view", which defined Web mining as a sequence of tasks [71]. Second was a "data-centric view", which defined Web mining in terms of the types of Web data that was being used in the mining process [29]. The second definition has become more acceptable, as is evident from the approach adopted in most recent papers [45, 62, 82] that have addressed the issue. In this paper we follow the data-centric view of Web mining which is defined as,

> **Web mining** is the application of data mining techniques to extract knowledge from Web data, i.e. Web Content, Web Structure and Web Usage data.

The attention paid to Web mining, in research, software industry, and Web-based organizations, has led to the accumulation of a lot of experiences. It is our attempt in this paper to capture them in a systematic manner, and identify directions for future research.

The rest of this paper is organized as follows: In Sect. 2 we provide a taxonomy of Web mining, in Sect. 3 we summarize some of the key concepts in the field, and in Sect. 4 we describe successful applications of Web mining techniques. In 5 we present some directions for future research, and in Sect. 6 we conclude the paper.

2 Web Mining Taxonomy

Web Mining can be broadly divided into three distinct categories, according to the kinds of data to be mined. We provide a brief overview of the three categories and a figure depicting the taxonomy is shown in Fig. 1:

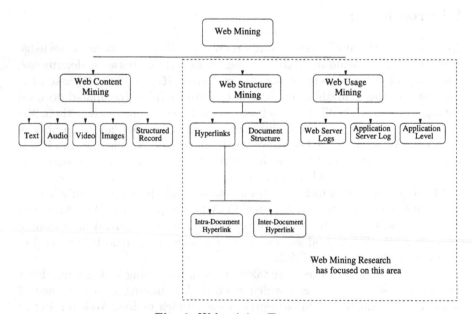

Fig. 1. Web mining Taxonomy

1. **Web Content Mining:** Web Content Mining is the process of extracting useful information from the contents of Web documents. Content data corresponds to the collection of facts a Web page was designed to convey to the users. It may consist of text, images, audio, video, or structured records such as lists and tables. Application of text mining to Web content has been the most widely researched. Issues addressed in text mining are, topic discovery, extracting association patterns, clustering of web documents and classification of Web Pages. Research activities on this topic have drawn heavily on techniques developed in other disciplines such as Information Retrieval (IR) and Natural Language Processing (NLP). While there exists a significant body of work in extracting knowledge from images, in the fields of image processing and computer vision, the application of these techniques to Web content mining has been limited.

2. **Web Structure Mining:** The structure of a typical Web graph consists of Web pages as nodes. and hyperlinks as edges connecting related pages. Web Structure Mining is the process of discovering structure information from the Web. This can be further divided into two kinds based on the kind of structure information used.

   - *Hyperlinks*: A Hyperlink is a structural unit that connects a location in a Web page to different location, either within the same Web page or on a different Web page. A hyperlink that connects to a different part of the same page is called an *Intra-Document Hyperlink*, and a hyperlink that connects two different pages is called an *Inter-Document Hyperlink*. There has been a significant body of work on hyperlink analysis, of which [75] provides an up-to-date survey.

   - *Document Structure*: In addition, the content within a Web page can also be organized in a tree-structured format, based on the various HTML and XML tags within the page. Mining efforts here have focused on automatically extracting document object model (DOM) structures out of documents [22. 55].

3. **Web Usage Mining:** Web Usage Mining is the application of data mining techniques to discover interesting usage patterns from Web data, in order to understand and better serve the needs of Web-based applications [49]. Usage data captures the identity or origin of Web users along with their browsing behavior at a Web site. Web usage mining itself can be classified further depending on the kind of usage data considered:

   - Web Server Data: The user logs are collected by Web server. Typical data includes IP address, page reference and access time.

   - Application Server Data: Commercial application servers, e.g. Weblogic [6. 11]. StoryServer [94], etc. have significant features in the framework to enable E-commerce applications to be built on top of them with little effort. A key feature is the ability to track various kinds of business events and log them in application server logs.

- **Application Level Data:** Finally, new kinds of events can always be defined in an application, and logging can be turned on for them – generating histories of these specially defined events.

The usage data can also be split into three different kinds on the basis of the source of its collection: on the server side, the client side, and the proxy side. The key issue is that on the server side there is an aggregate picture of the usage of a service by all users, while on the client side there is complete picture of usage of all services by a particular client, with the proxy side being somewhere in the middle [49].

3 Key Concepts

In this section we briefly describe the key new concepts introduced by the Web mining research community.

3.1 Ranking Metrics – For Page Quality and Relevance

Searching the Web involves two main steps: *Extracting the relevant pages to a query* and *ranking them according to their quality*. Ranking is important as it helps the user look for "quality" pages that are relevant to the query. Different metrics have been proposed to rank Web pages according to their quality. We briefly discuss two of the prominent metrics.

- **PageRank:** PageRank is a metric for ranking hypertext documents that determines the quality of these documents. Reference [61] developed this metric for the popular search engine Google [41,83]. The key idea is that a page has a high rank if it is pointed to by many highly ranked pages. So, the rank of a page depends upon the ranks of the pages pointing to it. This process is done iteratively till the rank of all the pages is determined. The rank of a page p can thus be written as:

$$PR(p) = d/n + (1-d) \sum_{(q,p) \in G} \left(\frac{PR(q)}{Outdegree(q)} \right)$$

Here, n is the number of nodes in the graph and $OutDegree(q)$ is the number of hyperlinks on page q. Intuitively, the approach can be viewed as a stochastic analysis of a random walk on the Web graph. The first term in the right hand side of the equation corresponds to the probability that a random Web surfer arrives at a page p by typing the URL or from a bookmark, or may have a particular page as his/her homepage. Here, d is the probability that a random surfer chooses a URL directly, rather than traversing a link[1] and $1-d$ is the probability that a person arrives at a

---

[1]The parameter d, called the dampening factor, is usually set between 0.1 and 0.2 [83]

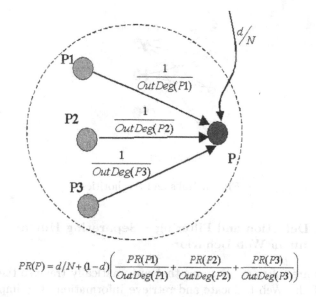

$$PR(P) = d/N + (1-d)\left(\frac{PR(P1)}{OutDeg(P1)} + \frac{PR(P2)}{OutDeg(P2)} + \frac{PR(P3)}{OutDeg(P3)}\right)$$

Fig. 2. PageRank -Markov Model for Random Web Surfer

page by traversing a link. The second term in the right hand side of the equation corresponds to the probability of arriving at a page by traversing a link. Figure 2 illustrates this concept, by showing how the PageRank of the page p is calculated.

• **Hubs and Authorities**: Hubs and Authorities can be viewed as "fans" and "centers" in a bipartite core of a Web graph. This is depicted in Fig. 3, where the nodes on the left represent the hubs and the nodes on the right represent the authorities. The hub and authority scores computed for each Web page indicate the extent to which the Web page serves as a "hub" pointing to good "authority" pages or as an "authority" on a topic pointed to by good hubs. The hub and authority scores are computed for a set of pages related to a topic using an iterative procedure called HITS [52]. First a query is submitted to a search engine and a set of relevant documents is retrieved. This set, called the "root set", is then expanded by including Web pages that point to those in the "root set" and are pointed by those in the "root set". This new set is called the "Base Set". An adjacency matrix, A is formed such that if there exists at least one hyperlink from page i to page j, then $A_{i,j} = 1$, otherwise $A_{i,j} = 0$. HITS algorithm is then used to compute the "hub" and "authority" scores for these set of pages.

There have been modifications and improvements to the basic *PageRank* and *Hubs and Authorities* approaches such as SALSA [59], Topic Sensitive PageRank [42] and Web page Reputations [65]. These different hyperlink based metrics have been discussed in [75].

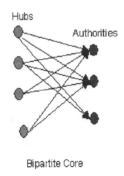

Fig. 3. Hubs and Authorities

3.2 Robot Detection and Filtering – Separating Human and Non-Human Web Behavior

Web robots are software programs that automatically traverse the hyperlink structure of the Web to locate and retrieve information. The importance of separating robot behavior from human behavior prior to building user behavior models has been illustrated by [80]. First of all, e-commerce retailers are particularly concerned about the unauthorized deployment of robots for gathering business intelligence at their Web sites. In addition, Web robots tend to consume considerable network bandwidth at the expense of other users. Sessions due to Web robots also make it difficult to perform click-stream analysis effectively on the Web data. Conventional techniques for detecting Web robots are often based on identifying the IP address and user agent of the Web clients. While these techniques are applicable to many well-known robots, they are not sufficient to detect camouflaged and previously unknown robots. Reference [93] proposed an approach that uses the navigational patterns in click-stream data to determine if it is due to a robot. Experimental results have shown that highly accurate classification models can be built using this approach. Furthermore, these models are able to discover many camouflaged and previously unidentified robots.

3.3 Information Scent – Applying Foraging Theory to Browsing Behavior

Information scent is a concept that uses the snippets and information presented around the links in a page as a "scent" to evaluate the quality of content of the page it points to, and the cost of accessing such a page [21]. The key idea is to model a user at a given page as "foraging" for information, and following a link with a stronger "scent". The "scent" of a path depends on how likely it is to lead the user to relevant information, and is determined by a network flow algorithm called spreading activation. The snippets, graphics, and other information around a link are called "proximal cues". The user's

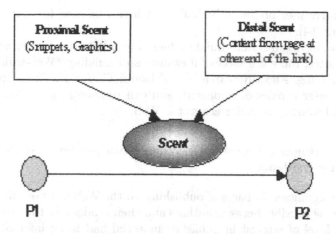

Fig. 4. Information Scent

desired information need is expressed as a weighted keyword vector. The similarity between the proximal cues and the user's information need is computed as "Proximal Scent". With the proximal cues from all the links and the user's information need vector, a "Proximal Scent Matrix" is generated. Each element in the matrix reflects the extent of similarity between the link's proximal cues and the user's information need. If enough information is not available around the link, a "Distal Scent" is computed with the information about the link described by the contents of the pages it points to. The "Proximal Scent" and the "Distal Scent" are then combined to give the "Scent" Matrix. The probability that a user would follow a link is decided by the "scent" or the value of the element in the "Scent" matrix. Figure 4 depicts a high level view of this model. Reference [21] proposed two new algorithms called Web User Flow by Information Scent (WUFIS) and Inferring User Need by Information Scent (IUNIS) using the theory of information scent based on Information foraging concepts. WUFIS predicts user actions based on user needs, and IUNIS infers user needs based on user actions. The concept is illustrated in 4.

3.4 User Profiles – Understanding How Users Behave

The Web has taken user profiling to completely new levels. For example, in a "brick-and-mortar" store, data collection happens only at the checkout counter, usually called the "point-of-sale". This provides information only about the final outcome of a complex human decision making process, with no direct information about the process itself. In an on-line store, the complete click-stream is recorded, which provides a detailed record of every single action taken by the user, providing a much more detailed insight into the decision making process. Adding such behavioral information to other kinds of information about users, e.g. demographic, psychographic, etc., allows a comprehensive user profile to be built, which can be used for many different applications [64].

While most organizations build profiles of user behavior limited to visits to their own sites, there are successful examples of building "Web-wide" behavioral profiles, e.g. Alexa Research [2] and DoubleClick [31]. These approaches require browser cookies of some sort, and can provide a fairly detailed view of a user's browsing behavior across the Web.

3.5 Interestingness Measures – When Multiple Sources Provide Conflicting Evidence

One of the significant impacts of publishing on the Web has been the close interaction now possible between authors and their readers. In the pre-Web era, a reader's level of interest in published material had to be inferred from indirect measures such as buying/borrowing, library checkout/renewal, opinion surveys, and in rare cases feedback on the content. For material published on the Web it is possible to track the precise click-stream of a reader to observe the exact path taken through on-line published material. We can measure exact times spent on each page, the specific link taken to arrive at a page and to leave it, etc. Much more accurate inferences about readers' interest in content can be drawn from these observations. Mining the user click-stream for user behavior, and using it to adapt the "look-and-feel" of a site to a reader's needs was first proposed in [69].

While the usage data of any portion of a Web site can be analyzed, the most significant, and thus "interesting", is the one where the usage pattern differs significantly from the link structure. This is interesting because the readers' behavior, reflected by Web usage, is very different from what the author would like it to be – reflected by the structure created by the author. Treating knowledge extracted from structure data and usage data as evidence from independent sources, and combining them in an evidential reasoning framework to develop measures for interestingness has been proposed in [9,28].

3.6 Pre-Processing – Making Web Data Suitable for Mining

In the panel discussion referred to earlier [91], pre-processing of Web data to make it suitable for mining was identified as one of the key issues for Web mining. A significant amount of work has been done in this area for Web usage data, including user identification [79], session creation [79], robot detection and filtering [93], extracting usage path patterns [89], etc. Cooley's Ph.D. thesis [28] provides a comprehensive overview of the work in Web usage data preprocessing.

Preprocessing of Web structure data, especially link information, has been carried out for some applications, the most notable being Google style Web search [83]. An up-to-date survey of structure preprocessing is provided in [75].

3.7 Topic Distillation

Topic Distillation is the identification of a set of documents or parts of document that are most relevant to a given topic. It has been defined [15] as

> "the process of finding authoritative Web pages and comprehensive "hubs" which reciprocally endorse each other and are relevant to a given query."

Kleinberg's HITS approach [52] was one of early link based approach that addressed the issue of identifying Web pages related to a specific topic. Bharath and Henzinger [8] and Chakrabarti et al [13, 24] used hyperlink analysis to automatically identify the set of documents relevant to a given topic. Reference [95] used a three step approach – (i) Document Keyword Extraction, (ii) Keyword propagation across pages connected by links, and (iii) keyword propagation through category tree structure – to automatically distill topics from the set of documents belonging to a category or to extract documents related to certain topics. The FOCUS project [17,18,37] concentrates on building portals pertaining to a topic automatically. A "fine-grained model" based on the Document Object Model (DOM) of a page and the hyperlink structure of hubs and authorities related to a topic has also been developed [16]. This approach reduces topic drift and helps in identifying parts of a Web page relevant to a query.

In recent work on identifying topics, [65] define a new measure called "reputation" of a page and compute the set of topics for which a page will be rated high. Reference [42] proposed a "Topic-Sensitive PageRank", which pre-computes a set of PageRank vectors corresponding to different topics.

3.8 Web Page Categorization

Web page categorization determines the category or class a Web page belongs to, from a pre-determined set of categories or classes. Topic Distillation is similar but in Web page categorization, the categories can be based on topics or other functionalities. e.g. home pages, content pages, research papers, etc, whereas Topic Distillation is concerned mainly with content-oriented topics. Reference [76] defined a set of 8 categories for nodes representing Web pages and identified 7 different features based on which a Web page could be categorized into these 8 categories. Reference [15] use a relaxation labeling technique to model a class-conditional probability distribution for assigning a category by looking at the neighboring documents that link to the given document or linked by the given the document. Reference [5] proposed an automatic method of classifying Web pages based on the link and context. The idea is that if a page is pointed to by another page, the link would carry certain context weight since it induces someone to read the given page from the page that is referring to it. Reference [60] treat documents and links as entities in an Entity-Relationship model and use a Probabilistic Relational Model to specify the probability distribution over the document-link database, and classify the documents using belief propagation methods. Reference [14] describe how topic taxonomies and automatic classifiers can be used to estimate the distribution of broad topics in the whole Web.

3.9 Identifying Web Communities of Information Sources

The Web has had tremendous success in building communities of users and information sources. Identifying such communities is useful for many purposes. We discuss here a few significant efforts in this direction.

Reference [35] identified Web communities as *"a core of central 'authoritative' pages linked together by 'hub' pages"*. Their approach was extended by Ravi Kumar et al. in [54] to discover emerging Web communities while crawling. A different approach to this problem was taken by [38] who applied the "maximum-flow minimum cut model" [48] to the Web graph for identifying "Web communities". This principle is illustrated in Fig. 5. Reference [70] compare the HITS and the maximum flow approaches and discuss

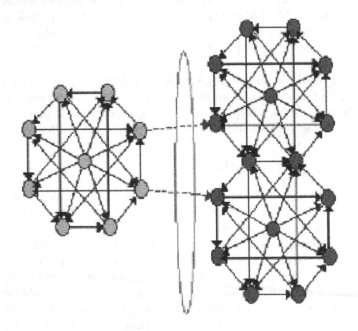

Minimum Cut Edges

Fig. 5. Maximal Flow Model for Web Communities

the the strengths and weakness of the two methods. Reference [77] propose a dense bipartite graph method, a relaxation to the complete bipartite method followed by HITS approach, to find Web communities. A related concept of "Friends and Neighbors" was introduced by Adamic and Adar in [56]. They identified a group of individuals with similar interests, who in the cyber-world would form a "community". Two people are termed "friends" if the similarity between their Web pages is high. The similarity is measured using the features:*text, out-links, in-Links and mailing lists.*

3.10 Online Bibiliometrics

With the Web having become the fastest growing and most upto date source of information, the research community has found it extremely useful to have online repository of publications. Lawerence et al. have observed in [86] that having articles online makes them more easily accessible and hence more often cited than articles that were offline. Such online repositories not only keep the researchers updated on work carried out at different centers, but also makes the interaction and exchange of information much easier.The concept is illustrated in Fig. 6

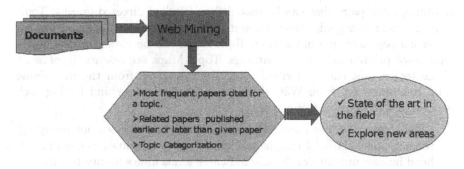

Fig. 6. Information Extraction in Online Bibiliometrics

With such information stored in the Web, it becomes easier to point to the most frequent papers that are cited for a topic and also related papers that have been published earlier or later than a given paper. This helps in understanding the "state of the art" in a particular field, helping researchers to explore new areas. Fundamental Web mining techniques are applied to improve the search and categorization of research papers, and citing related articles. Some of the prominent digital libraries are SCI [85], ACM portal [1], CiteSeer [23] and DBLP [30].

3.11 Semantic Web Mining

The data in the World Wide Web is largely in an unstructured format which makes information retrieval and navigation on the Web a difficult task. Automatic retrieval of information is an even more daunting task as the large amount of data available in the web is written mostly for human interpretation with no semantic structure attached. With the amount of information overload on the internet, it is necessary to develop automatic agents that can perform the challenging task of extracting information from the web. Existing search engines apply their own heuristics to arrive at the most relevant web pages for a query. Though they are found to be very useful, there is lack of preciseness as the search engines are not able to identify the exact semantics of the documents. Hence, there is a need for a more structured semantic document that would help in better retrieval and exchange of information.

At the highest level Semantic Web can be thought as adding certain semantic structures to the existing Web data. Semantic Web is also closely related to other areas such as Semantic Networks [98] and Conceptual graphs [88] that have been extensively studied and have been adopted to the web domain. In Semantic Networks, the edges of such a graph represent the semantic relationship between the vertices. Among other techniques used for Semantic Web is the RDF and XML Topic Maps. RDF data consists of nodes and attached attribute/value pairs that can be modeled as labelled directed graphs. Topic maps are used to organise large amount of information in an optimal way for better management and navigation. Topic maps can be viewed as the online versions of printed indices and catalogs. Topic Maps are essentially network of the topics that can be formed using the semantics from the underlying data. Reference [58] from W3C describes best the idea behind having such structured well-defined documents as:

> "The concept of machine-understandable documents does not imply some magical artificial intelligence which allows machines to comprehend human mumblings. It only indicates a machine's ability to solve a well-defined problem by performing well-defined operations on existing well-defined data. Instead of asking machines to understand people's language, it involves asking people to make the extra effort."

Web Mining techniques can be applied to learn ontologies for the vast source of unstructured web data available. Doing this manually for the whole web is definitely not scalable or practical. Conversely, defining ontologies for existing and future documents will help in faster and more accurate retrieval of documents. Reference [7] discuss in more detail about the integration of the two topics – "Semantic Web" and "Web Mining".

3.12 Visualization of the World Wide Web

Mining Web data provides a lot of information, which can be better understood with visualization tools. This makes concepts clearer than is possible

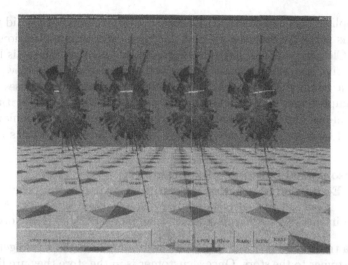

Fig. 7. Time Tube consisting of four disk trees representing evolution of Web Ecology. Figure taken from [20]

with pure textual representation. Hence, there is a need to develop tools that provide a graphical interface that aids in visualizing results of Web mining.

Analyzing the web log data with visualization tools has evoked a lot of interest in the research community. Reference [20] developed a Web Ecology and Evolution Visualization (WEEV) tool to understand the relationship between Web content, Web structure and Web Usage over a period of time. The site hierarchy is represented in a circular form called the "Disk Tree" and the evolution of the Web is viewed as a "Time Tube". Reference [12] present a tool called WebCANVAS that displays clusters of users with similar navigation behavior. Reference [10] introduce Ñaviz, a interactive web log visualization tool that is designed to display the user browsing pattern on the web site at a global level and then display each browsing path on the pattern displayed earlier in an incremental manner. The support of each traversal is represented by the thickness of the edge between the pages. The user browsing path that is of interest can be displayed by specifying the pages, or the number of intermediate nodes that have been traversed to reach a page. Such a tool is very useful in analyzing user behavior and improving web sites.

4 Prominent Applications

An outcome of the excitement about the Web in the past few years has been that Web applications have been developed at a much faster rate in the industry than research in Web related technologies. Many of these are based on the use of Web mining concepts. even though the organizations that developed these applications, and invented the corresponding technologies, did not consider it as such. We describe some of the most successful applications in this section. Clearly, realizing that these applications use Web mining is largely a retrospective exercise. For each application category discussed below, we have selected a prominent representative, purely for exemplary purposes. This in no way implies that all the techniques described were developed by that organization alone. On the contrary, in most cases the successful techniques were developed by a rapid "copy and improve" approach to each other's ideas.

4.1 Personalized Customer Experience in B2C E-commerce – Amazon.com

Early on in the life of Amazon.com, its visionary CEO Jeff Bezos observed,

> "In a traditional (brick-and-mortar) store, the main effort is in getting a customer to the store. Once a customer is in the store they are likely to make a purchase – since the cost of going to another store is high – and thus the marketing budget (focused on getting the customer to the store) is in general much higher than the in-store customer experience budget (which keeps the customer in the store). In the case of an online store, getting in or out requires exactly one click, and thus the main focus must be on customer experience in the store."[2]

This fundamental observation has been the driving force behind Amazon's comprehensive approach to personalized customer experience, based on the mantra "a personalized store for every customer" [68]. A host of Web mining techniques, e.g. associations between pages visited, click-path analysis, etc., are used to improve the customer's experience during a "store visit". Knowledge gained from Web mining is the key intelligence behind Amazon's features such as "instant recommendations", "purchase circles", "wish-lists", etc. [3].

4.2 Web Search – Google

Google [41] is one of the most popular and widely used search engines. It provides users access to information from over 2 billion web pages that it has indexed on its server. The quality and quickness of the search facility, makes it the most successful search engine. Earlier search engines concentrated on Web content alone to return the relevant pages to a query. Google was the first to introduce the importance of the link structure in mining the information from the web. PageRank, that measures the importance of a page, is the underlying

---

[2]The truth of this fundamental insight has been borne out by the phenomenon of "shopping cart abandonment", which happens frequently in on-line stores, but practically never in a brick-and-mortar one.

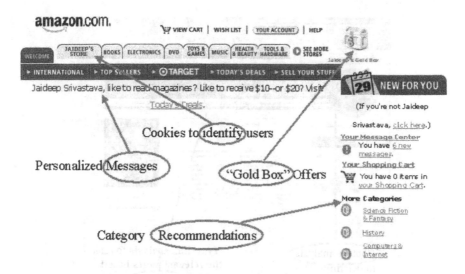

Fig. 8. Amazon.com's personalized Web page

technology in all Google search products, and uses structural information of the Web graph to return high quality results.

The "Google Toolbar" is another service provided by Google that seeks to make search easier and informative by providing additional features such as highlighting the query words on the returned web pages. The full version of the toolbar, if installed, also sends the click-stream information of the user to Google. The usage statistics thus obtained is used by Google to enhance the quality of its results. Google also provides advanced search capabilities to search images and find pages that have been updated within a specific date range. Built on top of Netscape's Open Directory project, Google's web directory provides a fast and easy way to search within a certain topic or related topics. The Advertising Programs introduced by Google targets users by providing advertisements that are relevant to a search query. This does not bother users with irrelevant ads and has increased the clicks for the advertising companies by four or five times. According to BtoB, a leading national marketing publication, Google was named a top 10 advertising property in the Media Power 50 that recognizes the most powerful and targeted business-to-business advertising outlets [39].

One of the latest services offered by Google is, "Google News" [40]. It integrates news from the online versions of all newspapers and organizes them categorically to make it easier for users to read "the most relevant news". It seeks to provide latest information by constantly retrieving pages from news site worldwide that are being updated on a regular basis. The key feature of this news page, like any other Google service, is that it integrates information from various Web news sources through purely algorithmic means, and thus does not introduce any human bias or effort. However, the publishing industry

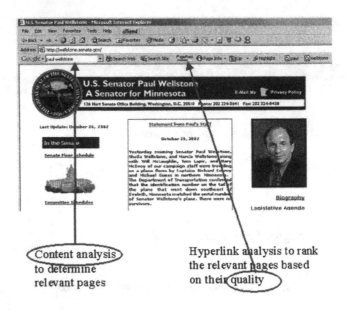

Fig. 9. Web page returned by Google for query "Paul Wellstone"

is not very convinced about a fully automated approach to news distillations
[90].

4.3 Web-Wide Tracking – DoubleClick

"Web-wide tracking", i.e. tracking an individual across all sites he visits is one
of the most intriguing and controversial technologies. It can provide an understanding of an individual's lifestyle and habits to a level that is unprecedented,
which is clearly of tremendous interest to marketers. A successful example of
this is DoubleClick Inc.'s DART ad management technology [31]. DoubleClick
serves advertisements, which can be targeted on demographic or behavioral
attributes, to end-user on behalf of the client, i.e. the Web site using Dou-
bliClick's service. Sites that use DoubleClick's service are part of "The DoubleClick Network" and the browsing behavior of a user can be tracked across
all sites in the network, using a cookie. This makes DoubleClick's ad targeting
to be based on very sophisticated criteria. Alexa Research [2] has recruited a
panel of more than 500,000 users, who have voluntarily agreed to have their
every click tracked, in return for some freebies. This is achieved through having
a browser bar that can be downloaded by the panelist from Alexa's website,
which gets attached to the browser and sends Alexa a complete click-stream
of the panelist's Web usage. Alexa was purchased by Amazon for its tracking
technology.

Clearly Web-wide tracking is a very powerful idea. However, the invasion
of privacy it causes has not gone unnoticed, and both Alexa/Amazon and

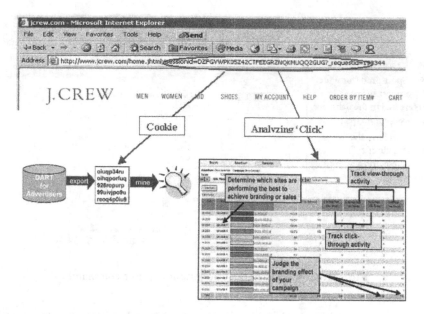

Fig. 10. DART system for Advertisers, DoubleClick

DoubleClick have faced very visible lawsuits [32, 34]. Microsoft's "Passport" technology also falls into this category [66]. The value of this technology in applications such a cyber-threat analysis and homeland defense is quite clear, and it might be only a matter of time before these organizations are asked to provide this information to law enforcement agencies.

4.4 Understanding Web Communities – AOL

One of the biggest successes of America Online (AOL) has been its sizeable and loyal customer base [4]. A large portion of this customer base participates in various "AOL communities", which are collections of users with similar interests. In addition to providing a forum for each such community to interact amongst themselves, AOL provides them with useful information and services. Over time these communities have grown to be well-visited "waterholes" for AOL users with shared interests. Applying Web mining to the data collected from community interactions provides AOL with a very good understanding of its communities, which it has used for targeted marketing through ads and e-mail solicitation. Recently, it has started the concept of "community sponsorship", whereby an organization, say Nike, may sponsor a community called "Young Athletic TwentySomethings". In return, consumer survey and new product development experts of the sponsoring organization get to participate in the community, perhaps without the knowledge of other participants. The idea is to treat the community as a highly specialized focus

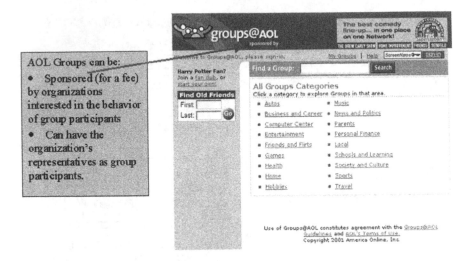

Fig. 11. Groups at AOL: Understanding user community

group, understand its needs and opinions on new and existing products; and also test strategies for influencing opinions.

4.5 Understanding Auction Behavior – eBay

As individuals in a society where we have many more things than we need, the allure of exchanging our "useless stuff" for some cash, no matter how small, is quite powerful. This is evident from the success of flea markets, garage sales and estate sales. The genius of eBay's founders was to create an infrastructure that gave this urge a global reach, with the convenience of doing it from one's home PC [36]. In addition, it popularized auctions as a product selling/buying mechanism, which provides the thrill of gambling without the trouble of having to go to Las Vegas. All of this has made eBay as one of the most successful businesses of the Internet era. Unfortunately, the anonymity of the Web has also created a significant problem for eBay auctions, as it is impossible to distinguish real bids from fake ones. eBay is now using Web mining techniques to analyze bidding behavior to determine if a bid is fraudulent [25]. Recent efforts are towards understanding participants' bidding behaviors/patterns to create a more efficient auction market.

4.6 Personalized Portal for the Web – MyYahoo

Yahoo [99] was the first to introduce the concept of a "personalized portal", i.e. a Web site designed to have the look-and-feel and content personalized to the needs of an individual end-user. This has been an extremely popular concept and has led to the creation of other personalized portals, e.g. Yodlee [100] for private information, e.g bank and brokerage accounts.

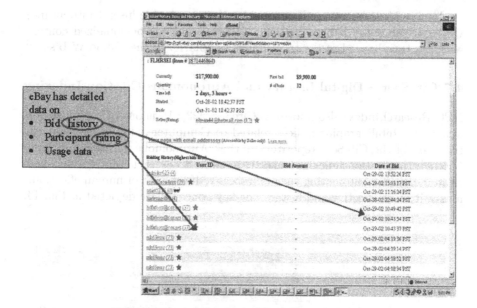

eBay has detailed data on
- Bid history
- Participant rating
- Usage data

Fig. 12. E-Bay: Understanding Auction Behavior

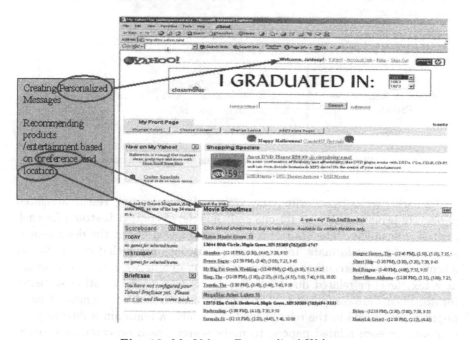

Creating Personalized Messages

Recommending products /entertainment based on preference and location

Fig. 13. My Yahoo: Personalized Webpage

Mining MyYahoo usage logs provides Yahoo valuable insight into an individual's Web usage habits, enabling Yahoo to provide personalized content, which in turn has led to the tremendous popularity of the Yahoo Web site.[3]

4.7 CiteSeer – Digital Library and Autonomous Citation Indexing

NEC ResearchIndex, also known as CiteSeer [23,51], is one of the most popular online bibiliographic indices related to Computer Science. The key contribution of the CiteSeer repository is the "Autonomous Citation Indexing" (ACI) [87]. Citation indexingmakes it possible to extract information about related articles. Automating such a process reduces a lot of human effort, and makes it more effective and faster. The key concepts are depicted in Fig. 14.

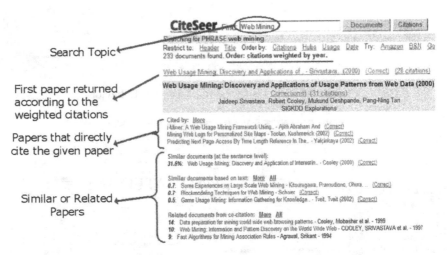

Fig. 14. CiteSeer – Autonomous Citation Indexing

CiteSeer works by crawling the Web and downloading research related papers. Information about citations and the related context is stored for each of these documents. The entire text and information about the document is stored in different formats. Information about documents that are similar at a sentence level (percentage of sentences that match between the documents), at a text level or related due to co-citation is also given. Citation statistics for documents are computed that enable the user to look at the most cited or popular documents in the related field. They also a maintain a directory for computer science related papers, to make search based on categories easier. These documents are ordered by the number of citations.

---

[3]Yahoo has been consistently ranked as one of the top Web property for a number of years [67].

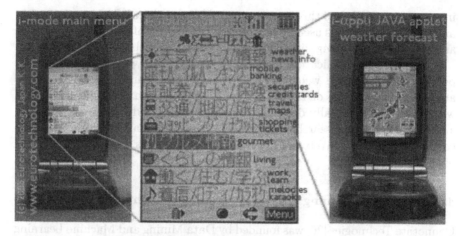

Fig. 15. i-MODE: NTT DoCoMo's mobile internet acccess system. The figure is taken from the The Eurotechnology Website [43]

4.8 i-MODE: Accessing the Web through Cell Phones

i-MODE is a cell-phone service from NTT DoCoMo, Japan [43]. It has about 40 million users who access the internet from their cell phones. The internet connections are continuous and the customers can access the specially tailored web sites as long as they are in the area that receives the i-mode the signal. Unlike the "circuit-switched" based systems that require dial-up, i-mode is "packet-switched" and hence continuous. This enables the download of information from the web sites to the cell phones faster, and without having to worry about the connection time. Users can receive and send email, do online shopping or banking, stock trading, receive traffic news and weather forecasts, and search for restaurants and other local places.

The usual speed for i-mode download ranges from 28.8 kbit/sec for top range models to the order of 200 kbit/sec for FOMA (3rd Generation) services. As a markup language. I-mode uses cHTML (compact HTML), which is in an extended subset of ordinary HTML that concentrates on text and simple graphics. The i-mode markup language also has certain special i-mode only tags and image characters that are used as emoticon symbols. The size of an i-mode page is limited to 5 kbytes. These set of web pages open a new domain of information. they have their own structure, semantics and usage. The content of these pages are alo restricted depending on the needs of the end users. Such a domain provides usage data based on an individual and this can be very useful to identify interesting user behavior patterns.

4.9 OWL: Web Ontology Language

The OWL Web Ontology Language [72] is designed for automatic processing of web content information to derive meaningful contexts without any human intervention. OWL has three expressive sublanguages: OWL Lite, OWL DL, and OWL Full. OWL is used to express the ontological information – meanings and relationships among different words – from the Web content.

One succesful application of such a language is a web portal. The web portal has an a single web page that is used as a starting point to search among the listed topics on the page. The list of topics and the topics associated with a web page is usally done manually and submitted. However, there has also been extensive research on automatic "topic distillation" and "web page categorization". OntoWeb and Open Directory Projects are typical examples of such portals.

4.10 vTag Web Mining Server – Connotate Technologies

Connotate Technogies [26] was founded by Data Mining and Machine Learning Scientists at Rutger's Univeristy. They are the developers of Web Services products that help users to browse and convert information from unstructured documents on the Web to a more structured format like XML. This conversion helps in providing better insight for personalizations, business intelligence and other enterprise solutions. The overall architecture of the the vTag Web Mining Server can be seen in Fig. 16.

The Web Mining Server supports information agents that monitor, extract and summarise the information from the various web sources. These information agents are easy and quick to set up using a a graphical user interface. The user can set it up according to the fatures they they think are essential to keep track of. There is no special need for programmers. The automation of this process helps busineeses and enterprises to better track the neccesary information from the large amount of web pages and summarise them for further analysis and action. Information agents are also capable of converting

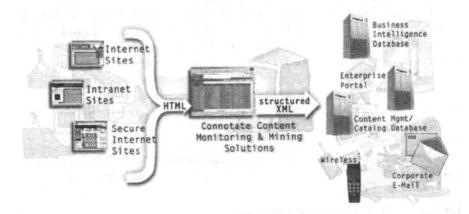

Fig. 16. vTag Web MiningServer Architecture. Figure taken from white paper on Web content and services mining [27]

unstructured data iinto structured form and store it in a database. Creation of such information agents requires no special skills and can be done easily using the graphical user interface provided. The content that is converted to a structured format like XML can be used for business intelligence, supply chain integrations etc. The converted content can also be sent as an e-mail or a message to a user in his mobile.

5 Research Directions

Even though we are going through an inevitable phase of "irrational despair" following a phase of "irrational exuberance" about the commercial potential of the Web, the adoption and usage of the Web continues to grow unabated [96]. As the Web and its usage grows, it will continue to generate evermore content, structure, and usage data, and the value of Web mining will keep increasing. Outlined here are some research directions that must be pursued to ensure that we continue to develop Web mining technologies that will enable this value to be realized.

5.1 Web Metrics and Measurements

From an experimental human behaviorist's viewpoint, the Web is the perfect experimental apparatus. Not only does it provides the ability of measuring human behavior at a micro level, it eliminates the bias of the subjects knowing that they are participating in an experiment, and allows the number of participants to be many orders of magnitude larger than conventional studies. However, we have not yet begun to appreciate the true impact of a revolutionary experimental apparatus for human behavior studies. The Web Lab of Amazon [3] is one of the early efforts in this direction. It is regularly used to measure the user impact of various proposed changes – on operational metrics such as site visits and visit/buy ratios, as well as on financial metrics such as revenue and profit   before a deployment decision is made. For example, during Spring 2000 a 48 hour long experiment on the live site was carried out, involving over one million user sessions, before the decision to change Amazon's logo was made. Research needs to be done in developing the right set of Web metrics. and their measurement procedures, so that various Web phenomena can be studied.

5.2 Process Mining

Mining of "market basket" data, collected at the point-of-sale in any store, has been one of the visible successes of data mining. However, this data provides only the end result of the process, and that too decisions that ended up in product purchase. Click-stream data provides the opportunity for a detailed

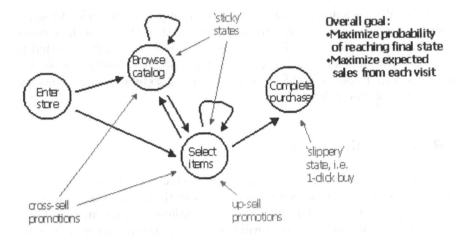

Fig. 17. Shopping Pipeline modeled as State Transition Diagram

look at the decision making process itself, and knowledge extracted from it can be used for optimizing the process, influencing the process, etc. [97]. Underhill [78] has conclusively proven the value of process information in understanding users' behavior in traditional shops. Research needs to be carried out in (i) extracting process models from usage data, (ii) understanding how different parts of the process model impact various Web metrics of interest, and (iii) how the process models change in response to various changes that are made, i.e. changing stimuli to the user. Figure 17 shows an approach of modeling online shopping as a state transition diagram.

5.3 Temporal Evolution of the Web

Society's interaction with the Web is changing the Web as well as the way people interact. While storing the history of all of this interaction in one place is clearly too staggering a task, at least the changes to the Web are being recorded by the pioneering Internet Archive project [44]. Research needs to be carried out in extracting temporal models of how Web content, Web structures, Web communities, authorities, hubs, etc. evolve over time. Large organizations generally archive (at least portions of) usage data from there Web sites. With these sources of data available, there is a large scope of research to develop techniques for analyzing of how the Web evolves over time. Figure 18 shows how content, structure and usage of Web information can evolve over time.

5.4 Web Services Optimization

As services over the Web continue to grow [50], there will be a continuing need to make them robust, scalable and efficient. Web mining can be applied to

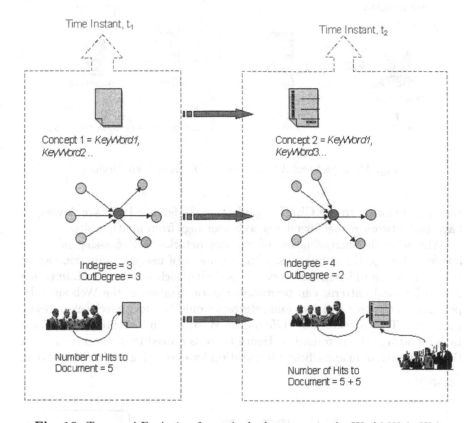

Fig. 18. Temporal Evolution for a single document in the World Wide Web

better understand the behavior of these services, and the knowledge extracted can be useful for various kinds of optimizations. The successful application of Web mining for predictive pre-fetching of pages by a browser has been demonstrated in [73]. Research is needed in developing Web mining techniques to improve various other aspects of Web services.

5.5 Distributed Web Mining

The data on the World Wide Web is huge and distributed across different web sites. To analyse such a data one can integrate all the data to one site and perform the required analysis. However, such an approach is time consuming and not scalable. A better approach would be to analyse the data locally at the different locations and build an overall model. This can be done in two diffrent ways: surreptious and co-operative. Surreptious approaches are those in which the user behavior across different web sites is tracked and integrated without the user explicitly having to submit any information. In the more co-operative approaches like the D-DOS attacks, such unusual behavior is

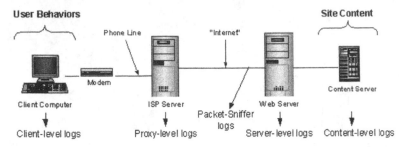

Fig. 19. High Level Architecture of Different Web Services

centrally reported to the CERT organisation. Reference [19] have developed bayesian network model for mining web user logs from multiple sites.

Also with the increasing use of wireless netorks and accessing of of the internet through the cell phones, a large amount of usage information can be collected across different web server logs. With such information interesting user behavioral patterns can be mined. Personalization of the Web sites depending on the user locations and interests would be more effective analysing such data. Thus a concept of "Life on the Web" for an individual can defined by integrating such information. Hence there is a need to develop models for the distributed data and efficient integration for extracting useful information.

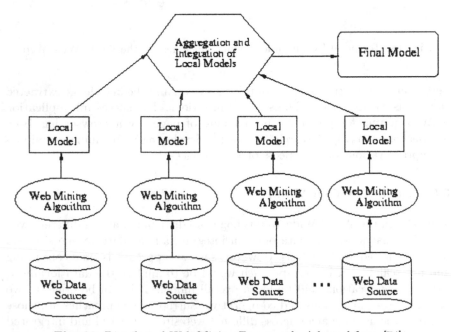

Fig. 20. Distributed Web Mining Framework. Adopted from [74]

The data distributed across different servers could have different nature and also the computing resources at different locations may vary.Hence, there is a need to develop different web mining algorithms to extract useful models [74].

5.6 Mining Information from E-mails-Discovering Evolving User Trends

E-mails have found to contain a huge amount information both in terms of its content, its usage and the evolving network built by sending e-mails. Target marketing using e-mails is one field that has been proved to be very effective according to a recent survey done by DoubleClick [33]. E-mails are a big source for multi-channel purchases. E-mails provide very useful information to track user interests and purchasing behavior and helps in increasing the level of personalization that can be offered. For example, according to the survey conducted by DoubleClick, women are more receptive to promotions and discounts and their idea of spam differs from that of men. And hence e-mail can serve as an excellent online source for marketing products related to women. However, e-mail faces the problem of spam that annoys users. Both consumers and companies providing free e-mail services are using tools to limit the spam to make the web life of the user more comfortable. Limiting spam helps in removing the removing the "noise" from the data provided by e-mails.

While marketing is one such area where E-mail provides an excellent source of information, this could be extended to other areas too. For example, there are E-mail groups that consists of prospective graduate students. Such groups would provide an excellent feedback of what the student interests are and what are the universities that are popular. Mining this kind of information would be useful for Universities that give admissions and also for porspective students in the later years. Mining information from such e-mail groups would help to uderstand the user needs and the underlying trends.

E-mails also form a network of directed graphs with the nodes being people or a work position (like operator@. webmaster@) and an e-mail sent from one one node to the other represented as a directed edge. This kind of a network is also dynamic and possesses a temporal dimension. The basic structure of the netwrok and its usage will be very useful in providing information about online communties. significant persons or abnormal user behavior.

5.7 Fraud and Threat Analysis

The anonymity provided by the Web has led to a significant increase in attempted fraud, from unauthorized use of individual credit cards to hacking into credit card databases for blackmail purposes [84]. Yet another example is auction fraud. which has been increasing on popular sites like eBay

[USDoJ2002]. Since all these frauds are being perpetrated through the Internet, Web mining is the perfect analysis technique for detecting and preventing them. Research issues include developing techniques to recognize known frauds, and characterize and then recognize unknown or novel frauds, etc. The issues in cyber threat analysis and intrusion detection are quite similar in nature [57].

5.8 Web Mining and Privacy

While there are many benefits to be gained from Web mining, a clear drawback is the potential for severe violations of privacy. Public attitude towards privacy seems to be almost schizophrenic, i.e. people say one thing and do quite the opposite. For example, famous case like [34] and [31] seem to indicate that people value their privacy, while experience at major e-commerce portals shows that over 97% of all people accept cookies with no problems – and most of them actually like the personalization features that can be provided based on it. Reference [92] have demonstrated that people were willing to provide fairly personal information about themselves, which was completely irrelevant to the task at hand, if provided the right stimulus to do so. Furthermore, explicitly bringing attention to information privacy policies had practically no effect. One explanation of this seemingly contradictory attitude towards privacy may be that we have a bi-modal view of privacy, namely that "I'd be willing to share information about myself as long as I get some (tangible or intangible) benefits from it, as long as there is an implicit guarantee that the information will not be abused". The research issue generated by this attitude is the need to develop approaches, methodologies and tools that can be used to verify and validate that a Web service is indeed using an end-user's information in a manner consistent with its stated policies.

6 Conclusions

As the Web and its usage continues to grow, so grows the opportunity to analyze Web data and extract all manner of useful knowledge from it. The past five years have seen the emergence of Web mining as a rapidly growing area, due to the efforts of the research community as well as various organizations that are practicing it. In this paper we have briefly described the key computer science contributions made by the field, a number of prominent applications, and outlined some promising areas of future research. Our hope is that this overview provides a starting point for fruitful discussion.

Acknowledgements

The ideas presented here have emerged in discussions with a number of people over the past few years – far too numerous to list. However, special mention must be made of Robert Cooley, Mukund Deshpande, Joydeep Ghosh, Ronny Kohavi, Ee-Peng Lim, Brij Masand, Bamshad Mobasher, Ajay Pandey, Myra Spiliopoulou, Pang-Ning Tan, Terry Woodfield, and Masaru Kitsuregawa discussions with all of whom have helped develop the ideas presented herein. This work was supported in part by the Army High Performance Computing Research Center contract number DAAD19-01-2-0014. The ideas and opinions expressed herein do not necessarily reflect the position or policy of the government (either stated or implied) and no official endorsement should be inferred. The AHPCRC and the Minnesota Super-computing Institute provided access to computing facilities.

References

1. ACM Portal. http://portal.acm.org/portal.cfm.
2. Alexa research. http://www.alexa.com.
3. Amazon.com. http://www.amazon.com.
4. America Online. http://www.aol.com, 2002.
5. Giuseppe Attardi, Antonio Gullí, and Fabrizio Sebastiani. Automatic Web page categorization by link and context analysis. In Chris Hutchison and Gaetano Lanzarone, editors, *Proceedings of THAI-99, European Symposium on Telematics, Hypermedia and Artificial Intelligence*, pp. 105–119, Varese, IT, 1999.
6. BEA Weblogic Server. http://www.bea.com/products/weblogic/server/index.shtml.
7. B. Berendt, A. Hotho, and G. Stumme. Towards semantic web mining, 2002.
8. Krishna Bharat and Monika R. Henzinger. Improved algorithms for topic distillation in a hyperlinked environment. In *Proceedings of SIGIR-98, 21st ACM International Conference on Research and Development in Information Retrieval*, pp. 104–111, Melbourne, AU, 1998.
9. B. Padmanabhan and A. Tuzhilin. A Belief-Driven Method for Discovering Unexpected Patterns. In *Knowledge Discovery and Data Mining*, pp. 94–100, 1998.
10. B. Prasetyo, I. Pramudiono, K. Takahashi, M. Toyoda, and M. Kitsuregawa. Naviz — user behavior visualization of dynamic page.
11. Broadvision 1-to-1 portal. http://www.bvportal.com/.
12. I.V. Cadez, D. Heckerman, C. Meek, P. Smyth, and S. White. Visualization of navigation patterns on a Web site using model-based clustering. In *Knowledge Discovery and Data Mining*, pp. 280–284, 2000.
13. S. Chakrabarti, B. Dom, D. Gibson, J. Kleinberg, P. Raghavan, and S. Rajagopalan. Automatic resource list compilation by analyzing hyperlink structure and associated text. In *Proceedings of the 7th International World Wide Web Conference*. 1998.
14. S. Chakrabarti, M. Joshi, K. Punera, and D. Pennock. The structure of broad topics on the web, 2002.
15. Soumen Chakrabarti. Integrating the document object model with hyperlinks for enhanced topic distillation and information extraction. In *World Wide Web*, pp. 211–220, 2001.

16. Soumen Chakrabarti, Mukul Joshi, and Vivek Tawde. Enhanced topic distillation using text, markup tags, and hyperlinks. In *Research and Development in Information Retrieval*, pp. 208–216, 2001.

17. Soumen Chakrabarti, Martin van den Berg, and Byron Dom. Distributed hypertext resource discovery through examples. In *The VLDB Journal*, pp. 375–386, 1999.

18. Soumen Chakrabarti, Martin van den Berg, and Byron Dom. Focused crawling: a new approach to topic-specific Web resource discovery. *Computer Networks (Amsterdam, Netherlands: 1999)*, 31(11–16):1623–1640, 1999.

19. R. Chen, K. Sivakumar, and H. Kargupta. Distributed web mining using Bayesian networks from multiple data streams. In *Proceedings of ICDM 2001*, pp. 75–82, 2001.

20. E.H. Chi, J.Pitkow, J.Mackinlay, P.Pirolli, R.Gossweiler, and S.K. Card. Visualizing the evolution of web ecologies. In *Proceedings of the Conference on Human Factors in Computing Systems CHI'98*, 1998.

21. E.H. Chi, P.Pirolli, K.Chen, and J.E. Pitkow. Using Information Scent to model user information needs and actions and the Web. In *Proceedings of CHI 2001*, pp. 490–497, 2001.

22. C.H. Moh, E.P. Lim, and W.K. Ng. DTD-Miner: A Tool for Mining DTD from XML Documents. WECWIS, 2000.

23. CiteSeer – Scientific Literature Digital Library. http://citeseer.nj.nec.com/cs.

24. The CLEVER Project. http://www.almaden.com/cs/k53/clever.html.

25. E. Colet. Using Data Mining to Detect Fraud in Auctions, 2002.

26. Connotate technologies. http://www.connotate.com/.

27. Web Services Content Mining: Extract, Monitor and Deliver. http://www.connotate.com/web_mining_white_paper.pdf.

28. R. Cooley. Web Usage Mining: Discovery and Application of Interesting Patterns from Web Data, 2000.

29. R. Cooley, J. Srivastava, and B. Mobasher. Web Mining: Information and Pattern Discovery on the World Wide Web, 1997.

30. DBLP Bibiliography. http://www.informatik.uni-trier.de/ ley/db/.

31. DoubleClick's DART Technology. http://www.doubleclick.com/dartinfo/, 2002.

32. DoubleClick's Lawsuit. http://www.wired.com/news/business/0,1367,36434,00.html, 2002.

33. DoubleClick 2003 Consumer Email Study. http://www.doubleclick.com/us/knowledge_central/documents/research/dc_consumeremailstudy_0310.pdf, 2003.

34. C. Dembeck and P. A. Greenberg. Amazon: Caught Between a Rock and a Hard Place. http://www.ecommercetimes.com/perl/story/2467.html, 2002.

35. D. Gibson, J.M. Kleinberg, and P. Raghavan. Inferring Web Communities from Link Topology. In *UK Conference on Hypertext*, pp. 225–234, 1998.

36. eBay Inc. http://www.ebay.com.

37. The FOCUS project. http://www.cs.berkeley.edu/ soumen/focus/.

38. G. Flake, S. Lawrence, and C.L. Giles. Efficient Identification of Web Communities. In *Sixth ACM SIGKDD International Conference on Knowledge Discovery and Data Mining*, pp. 150–160, Boston, MA, August 20–23 2000.

39. Google Recognized As Top Business-To-Business Media Property. http://www.google.com/press/pressrel/b2b.html.
40. Google News. http://news.google.com.
41. Google Inc. http://www.google.com.
42. T. Haveliwala. Topic-sensitive PageRank. In Proceedings of the Eleventh International World Wide Web Conference, Honolulu, Hawaii, May 2002., 2002.
43. I-mode. http://http://www.eurotechnology.com/imode/index.html.
44. The Internet Archive Project. http://www.archive.org/.
45. J. Borges and M. Levene. Mining Association Rules in Hypertext Databases. In *Knowledge Discovery and Data Mining*, pp. 149–153, 1998.
46. J. Ghosh and J. Srivastava. Proceedings of Workshop on Web Analytics. http://www.lans.ece.utexas.edu/workshop_index2.htm, 2001.
47. J. Ghosh and J. Srivastava. Proceedings of Workshop on Web Mining. http://www.lans.ece.utexas.edu/workshop_index.htm, 2001.
48. L.R. Ford Jr and D.R. Fulkerson. Maximal Flow through a network, 1956.
49. J. Srivastava, R. Cooley. M. Deshpande, and P.N. Tan. Web Usage Mining: Discovery and Applications of Usage Patterns from Web Data. *SIGKDD Explorations*, 1(2):12–23, 2000.
50. R.H. Katz. Pervasive Computing: It's All About Network Services, 2002.
51. K. Bollacker, S. Lawrence, and C.L. Giles. CiteSeer: An autonomous web agent for automatic retrieval and identification of interesting publications. In Katia P. Sycara and Michael Wooldridge, editors, *Proceedings of the Second International Conference on Autonomous Agents*, pp. 116–123, New York, 1998. ACM Press.
52. J.M. Kleinberg. Authoritative sources in a hyperlinked environment. *Journal of the ACM*, 46(5):604–632, 1999.
53. R. Kohavi, M. Spiliopoulou, and J. Srivastava. Proceedings of WebKDD2000 – Web Mining for E-Commerce   Challenges and Opportunities, 2001.
54. Ravi Kumar, Prabhakar Raghavan, Sridhar Rajagopalan, and Andrew Tomkins. Trawling the Web for emerging cyber-communities. *Computer Networks (Amsterdam. Netherlands: 1999)*, 31(11–16):1481–1493, 1999.
55. K. Wang and H. Liu. Discovering Typical Structures of Documents: A Road Map Approach. In *21st Annual International ACM SIGIR Conference on Research and Development in Information Retrieval*, pp. 146–154, 1998.
56. L. Adamic and E. Adar. Friends and Neighbors on the Web.
57. A. Lazarevic, P. Dokas. L. Ertoz, V. Kumar, J. Srivastava, and P.N. Tan. Data mining for network intrusion detection, 2002.
58. Tim Berners Lee. What the semantic web can represent. http://www.w3.org/DesignIssues/RDFnot.html, 1998.
59. R. Lempel and S. Moran. The stochastic approach for link-structure analysis (SALSA) and the TKC effect. *Computer Networks (Amsterdam, Netherlands: 1999)*. 33(1–6):387 401. 2000.
60. L. Getoor, E. Segal. B. Tasker. and D. Koller. Probabilistic models of text and link structure for hypertext classification. IJCAI Workshop on Text Learning:Beyond Supervision. Seattle, WA, 2001.
61. L. Page, S. Brin. R. Motwani, and T. Winograd. The pagerank citation ranking: Bringing order to the web. Technical report, Stanford Digital Library Technologies Project. 1998.

62. S.K. Madria, S.S. Bhowmick, W.K Ng, and E.P Lim. Research Issues in Web Data Mining. In *Data Warehousing and Knowledge Discovery*, pp. 303–312, 1999.
63. B. Masand and M. Spiliopoulou. Proceedings of WebKDD1999 – Workshop on Web Usage Analysis and User Profiling, 1999.
64. B. Masand, M. Spiliopoulou, J. Srivastava, and O. Zaiane. Proceedings of WebKDD2002 – Workshop on Web Usage Patterns and User Profiling, 2002.
65. A.O. Mendelzon and D. Rafiei. What do the Neighbours Think? Computing Web Page Reputations. *IEEE Data Engineering Bulletin*, 23(3):9–16, 2000.
66. MicroSoft.NET Passport. http://www.microsoft.com/netservices/passport/.
67. Top 50 US Web and Digital Properties. http://www.jmm.com/xp/jmm/press/mediaMetrixTop50.xml.
68. E. Morphy. Amazon pushes 'personalized store for every customer'. http://www.ecommercetimes.com/perl/story/13821.html, 2001.
69. M. Perkowitz and O. Etzioni. Adaptive Web Sites: Conceptual Cluster Mining. In *IJCAI*, pp. 264–269, 1999.
70. N. Imafuji and M. Kitsuregawa. Effects of maximum flow algorithm on identifying web community. In *Proceedings of the fourth international workshop on Web information and data management*, pp. 43–48. ACM Press, 2002.
71. O. Etzioni. The World-Wide Web: Quagmire or Gold Mine? *Communications of the ACM*, 39(11):65–68, 1996.
72. OWL. http://www.w3.org/TR/owl-features/.
73. A. Pandey, J. Srivastava, and S. Shekhar. A web intelligent prefetcher for dynamic pages using association rules – a summary of results, 2001.
74. Byung-Hoon Park and Hillol Kargupta. Distributed data mining: Algorithms, systems, and applications.
75. P. Desikan, J. Srivastava, V. Kumar, and P.N. Tan. Hyperlink Analysis-Techniques & Applications. Technical Report 2002-152, Army High Performance Computing Research Center, 2002.
76. Peter Pirolli, James Pitkow, and Ramana Rao. Silk from a sow's ear: Extracting usable structures from the web. In *Proc. ACM Conf. Human Factors in Computing Systems, CHI*. ACM Press, 1996.
77. P.K. Reddy and M. Kitsuregawa. An approach to build a cyber-community hierarchy. Workshop on Web Analytics,held in Conjunction with Second SIAM International Conference on Data Mining, 2002.
78. P. Underhill. Why we buy: The Science of shopping. Touchstone Books, 2000.
79. R. Cooley, B. Mobasher, and J. Srivastava. Data Preparation for Mining World Wide Web Browsing Patterns. *Knowledge and Information Systems*, 1(1):5–32, 1999.
80. R. Kohavi. Mining e-commerce data: The good, the bad, and the ugly. In Foster Provost and Ramakrishnan Srikant, editors, *Proceedings of the Seventh ACM SIGKDD International Conference on Knowledge Discovery and Data Mining*, pp. 8–13, 2001.
81. R. Kohavi, B. Masand, M. Spiliopoulou, and J. Srivastava. Proceedings of WebKDD2001 – Mining Log Data Across All Customer Touchpoints, 2001.
82. R. Kosala and H. Blockeel. Web mining research: A survey. *SIGKDD: SIGKDD Explorations: Newsletter of the Special Interest Group (SIG) on Knowledge Discovery& Data Mining, ACM*, 2, 2000.
83. S. Brin and L. Page. The anatomy of a large-scale hypertextual Web search engine. *Computer Networks and ISDN Systems*, 30(1–7):107–117, 1998.

84. D. Scarponi. Blackmailer Reveals Stolen Internet Credit Card Data. http://abcnews.go.com/sections/world/DailyNews/internet000110.html, 2000.

85. Science Citation Index. http://www.isinet.com/isi/products/citation/sci/.

86. S. Lawrence. Online or invisible? *Nature*, 411(6837):521, 2001.

87. S. Lawrence, C.L. Giles, and K. Bollacker. Digital Libraries and Autonomous Citation Indexing. *IEEE Computer*, 32(6):67–71, 1999.

88. J. F. Sowa. Conceptual Structures Information Processing in Mind and Machine. Addison Wesley, reading et al. 1984.

89. M. Spiliopoulou. Data Mining for the Web. Proceedings of the Symposium on Principles of Knowledge Discovery in Databases (PKDD), 1999.

90. T. Springer. Google LaunchesNews Service. http://www.computerworld.com/developmenttopics/websitemgmt/story/0,10801,74470,00.html, 2002.

91. J. Srivastava and B. Mobasher. Web Mining: Hype or Reality? . 9th IEEE International Conference on Tools With Artificial Intelligence (ICTAI '97), 1997.

92. S. Spiekermann. J. Grossklags, and B. Berendt. Privacy in 2nd generation E-Commerce: privacy preferences versus actual behavior. In *ACM Conference on Electronic Commerce*. pp. 14–17, 2001.

93. P. Tan and V. Kumar. Discovery of web robot sessions based on their navigational patterns, 2002.

94. Vignette StoryServer. http://www.cio.com/sponsors/110199_vignette_stor-y2.html.

95. V. Katz and W.S. Li. Topic Distillation in hierarchically categorized Web Documents. In *Proceedings of 1999 Workshop on Knowledge and Data Engineering Exchange*, IEEE, 1999.

96. Hosting Firm Reports Continued Growth. http://thewhir.com/market-watch/ser053102.cfm. 2002.

97. K.L. Ong and W. Keong. Mining Relationship Graphs for Eective Business Objectives.

98. W. Woods. What's in a link: Foundations for semantic networks. Academic Press. New York. 1975.

99. Yahoo!. Inc. http://www.yahoo.com.

100. Yodlee. Inc. http://www.yodlee.com.

Privacy-Preserving Data Mining

C. Clifton, M. Kantarcıoğlu, and J. Vaidya

Purdue University
{clifton, kanmurat, jsvaidya}@cs.purdue.edu

The growth of data mining has raised concerns among privacy advocates. Some of this is based on a misunderstanding of what data mining does. The previous chapters have shown how data mining concentrates on extraction of rules, patterns and other such summary knowledge from large data sets. This would not seem to inherently violate privacy, which is generally concerned with the release of individual data values rather than summaries.

To some extent, this has been recognized by privacy advocates. For example, the Data-Mining Moratorium Act proposed in the U.S. Senate in January 2003 would have stopped all data-mining activity by the Department of Defense [15]. A later version is more specific, defining "data-mining" as searches for individual information based on profiles [25]. While data mining may help in the *development* of such profiles, with the possible exception of outlier detection data mining would not be a forbidden activity under the later bill.

Although data mining *results* may have survived the scrutiny of privacy advocates, the data mining *process* still faces challenges. For example, the Terrorism Information Awareness (TIA) program, formerly known as Total Information Awareness, proposed government use of privately held databases (e.g., credit records) to aid in the discovery and prevention of terrorist activity [9]. This raised justifiable concerns, leading to a shutdown of the program [22] and proposals for restriction on government use of privately held databases [25]. The real problem is the potential for misuse of the information. The TIA program did not propose collection of new data, only access to existing collections. However, providing a single point of access to many collections, and linking individuals across these collections, provides a much more complete view of individuals than can be gleaned from any individual collection. While this could significantly improve capabilities to identify and track terrorists, it also makes it easier to misuse this information for unethical or illegal harassment of political dissidents, unpopular officials, or even neighbors of a rogue agent or hacker who manages to break into the system.

Information gives knowledge gives power, and many feel that the potential for misuse of this power exceeds the benefit.

Privacy-preserving data mining has emerged as an answer to this problem. The goal of privacy-preserving data mining is to develop data mining models without increasing the risk of misuse of the data used to generate those models. This is accomplished by ensuring that nobody but the original possessor of data sees individual data values. Since no real or virtual "data warehouse" providing integrated access to data is used, the potential for misuse of data is not increased by the data mining process. While the potential for misuse of the produced data mining models is not eliminated, these are considered to be sufficiently removed from individual data values (or perhaps so important) that the threat to individual privacy is not an issue.

Privacy-preserving data mining work is divided into two broad classes. One, first proposed in [3], is based on adding noise to the data before providing it to the data miner. Since real data values are not revealed (the noise can be added at the data source), individual privacy is preserved. The challenge in this class is developing algorithms that achieve good results in spite of the noise in the data. While these techniques have been shown effective, there is growing concern about the potential for noise reduction and thus compromise of individual privacy [1,10,21]. As a result, we will touch only briefly on this approach, giving a summary of how the method of [3] works in Sect. 2.

The second class of privacy-preserving data mining comes out of the cryptography community [24]. The idea is that the data sources collaborate to obtain data mining results without revealing *anything* except those results. This approach is based on the definitions and standards that have guided the cryptography community, in particular Secure Multiparty Computation [16, 38]. The disadvantage to this approach is that the algorithms are generally distributed, requiring active participation of the data sources. However, as this model limits any privacy breach to that inherent in the results, this chapter will emphasize this class of privacy-preserving data mining techniques.

1 Privacy-Preserving Distributed Data Mining

Privacy-preserving distributed data mining uses algorithms that require parties to collaborate to get results, while provably preventing disclosure of data except the data mining results. As a simple example, assume several supermarkets wish to collaborate to obtain global "market basket" association rules, without revealing individual purchases or even the rules that hold at individual stores. To simplify, assume we only wish to compute the global support count for a single itemset, e.g., "beer" and "diapers". Each market first computes the number of market baskets it has that contain both items. A designated starting market also generates a random number R. The starting party adds its support count S_1 to R, and sends $R + S_1$ to the second market. The second market adds its support count, sending $R + S_1 + S_2$ to the third market. This continues, with the nth market sending $R + \sum_{i=1}^{n} S_i$ to the first market. The first market subtracts R to obtain the desired result.

A crucial assumption for security is that markets $i+1$ and $i-1$ do not collude to learn S_i. However, if we can assume no collusion, all operations take place over a closed field, and the choice of R is uniformly distributed over the field, we can see that no market learns anything about the other market's support counts except what can be inferred from the result and one's own data. While this does not guarantee that individual values are not revealed (for example, if the global support count is 0, we know the count is 0 for every market), it does preserve as much privacy as possible assuming we must obtain the results (in this case, global support count.)

The concept of privacy in this approach is based on a solid body of theoretical work. We briefly discuss some of this work now, then describe several techniques for privacy-preserving distributed data mining to demonstrate how this theory can be applied in practice.

1.1 Privacy Definitions and Proof Techniques

Secure Multiparty Computation (SMC) originated with Yao's Millionaires' problem [38]. The basic problem is that two millionaires would like to know who is richer, but neither wants to reveal their net worth. Abstractly, the problem is simply comparing two numbers, each held by one party, without either party revealing its number to the other. Yao presented a solution for any efficiently computable function restricted to two parties and semi-honest adversaries.

What do we mean by *semi-honest*? The secure multiparty computation literature makes use of two models of what an adversary may do to try to obtain information. These are:

Semi-Honest: Semi-honest (or Honest but Curious) adversaries will follow the protocol faithfully, but are allowed to try to infer the secret information of the other parties from the data they see during the execution of the protocol.

Malicious: Malicious adversaries may do anything they like to try to infer secret information (within the bounds of polynomial computational power). They can abort the protocol at any time, send spurious messages, spoof messages, collude with other (malicious) parties, etc.

Reference [16] extended Yao's result to an arbitrary number of parties as well as malicious adversaries. The basic idea is based on circuit evaluation: The function is represented as a boolean circuit. Each party gives the other(s) a randomly determined share of their input, such that the exclusive or of the shares gives the actual value. The parties collaborate to compute a share of the output of each gate. Although the exclusive or of the shares gives the output of the gate, the individual shares are random in isolation. Since each party learns nothing but its share, nothing is revealed, and these shares can be used in the next gate in the circuit. At the end, the shares are combined to produce the final result. Since the intermediate shares were randomly determined, nothing is revealed except the final result.

Informally, the definition of privacy is based on equivalence to having a trusted third party perform the computation. Imagine that each of the data sources gives their input to a (hypothetical) trusted third party. This party, acting in complete isolation, computes the results and reveals them. After revealing the results, the trusted party forgets everything it has seen. A secure multiparty computation approximates this standard: no party learns more than with the (hypothetical) trusted third party approach.

One fact is immediately obvious: no matter how secure the computation, some information about the inputs may be revealed. If one's net worth is $100,000, and the other party is richer, one has a lower bound on their net worth. This is captured in the formal SMC definitions: any information that can be inferred from one's own data and the result can be revealed by the protocol. For example, assume one party attributes A and B for all individuals, and another party has attribute C. If mining for association rules, gives that $AB \Rightarrow C$ with 100% confidence, then if one knows that A and B hold for some individual it is okay to learn C for that individual during the data mining process. Since this could be inferred from the result anyway, privacy is not compromised by revealing it during the *process* of data mining. Thus, there are two kinds of information leaks; the information leak from the function computed irrespective of the process used to compute the function and the information leak from the specific process of computing the function. Whatever is leaked from the function itself is unavoidable as long as the function has to be computed. In secure computation the second kind of leak is provably prevented. There is *no* information leak whatsoever due to the process. Some algorithms improve efficiency by trading off some security (leak a small amount of information). Even if this is allowed, the SMC style of proof provides a tight bound on the information leaked; allowing one to determine if the algorithm satisfies a privacy policy.

This leads to the primary proof technique used to demonstrate the security of privacy-preserving distributed data mining: a simulation argument. Given only its own input and the result, a party must be able to simulate what it sees during execution of the protocol. Note that "what it sees" is not constant: for example, what each party sees during the secure summation described above is dependent on the first party's choice of R. So the view to be simulated is actually a distribution. The formal definition for secure multiparty computation captures this: the *distribution* of values produced by the simulator for a given input and result must be equivalent to the distribution of values seen during real executions on the same input. The key challenge is to simulate this view *without knowledge* of the other party's input (and based only on the given party's input and output). The ability to simulate shows that the view of the party in a real protocol execution could actually have been generated by itself (without any interaction and just been given the output). Therefore, anything that the party can learn, it can learn from its input and output (because just running the simulation locally is equivalent to participating in a real protocol).

This is captured in the following definition (based on that of [17], however for readability we present it from the point of view of one party.)

Definition 1. *Privacy with respect to semi-honest behavior.*

Let $f : \{0,1\}^* \times \{0,1\}^* \longmapsto \{0,1\}^* \times \{0,1\}^*$ *be a probabilistic, polynomial-time functionality. Let Π be a two-party protocol for computing f.*

The view *of a party during an execution of Π on (x,y), denoted $VIEW^{\Pi}$ (x,y) is $(x,r,m_1,\ldots,m_t)$, where r represents the outcome of the party's internal coin tosses, and m_i represents the ith message it has received. The final outputs of the parties during an execution are denoted $OUTPUT_1^{\Pi}(x,y)$ and $OUTPUT_2^{\Pi}(x,y)$.*

Π privately computes f if there exists a probabilistic polynomial time algorithm S such that

$$\{(S(x.f(x,y)),)\}_{x,y\in\{0,1\}^*}$$
$$\equiv^C \{(VIEW^{\Pi}(x,y).OUTPUT_2^{\Pi}(x,y))\}_{x,y\in\{0,1\}^*}$$

where $\equiv^C$ denotes computational indistinguishability. Note that a party's own output is implicit in its view.

The definition given above is restricted to two parties. The basic idea holds for extension to more than two parties. Reference [17] proved that this definition is essentially equivalent to the "trusted third party" definition, showing that any computation meeting this simulation argument in fact meets our intuitive expectations of security. A similar, but considerably more complex, definition exists for malicious adversaries. Because of the complexity, we will stick to the semi-honest definition. However, many applications require something stronger than semi-honest protocols. Intermediate definitions are possible (e.g., the secure association rules discussed at the beginning of this section is secure against malicious parties that do not collude), but formal frameworks for such definitions remain to be developed.

One key point is the restriction of the simulator to polynomial time algorithms. and that the views only need to be *computationally* indistinguishable. Algorithms meeting this definition need not be proof against an adversary capable of trying an exponential number of possibilities in a reasonable time frame. While some protocols (e.g.. the secure sum described above) do not require this restriction. most make use of cryptographic techniques that are only secure against polynomial time adversaries. This is adequate in practice (as with cryptography); security parameters can be set to ensure that the computing resources to break the protocol in any reasonable time do not exist.

A second key contribution is the composition theorem of [17], stated informally here:

Theorem 1. *Composition Theorem for the semi-honest model.*

Suppose that g is privately reducible to f and that there exists a protocol for privately computing f. Then there exists a protocol for privately computing g.

Informally, the theorem states that if a protocol is shown to be secure except for several invocations of sub-protocols, and if the sub-protocols themselves are proven to be secure, then the entire protocol is secure. The immediate consequence is that, with care, we can combine secure protocols to produce new secure protocols.

While the general circuit evaluation method has been proven secure by the above definition, it poses significant computational problems. Given the size and computational cost of data mining problems, representing algorithms as a boolean circuit results in unrealistically large circuits. The challenge of privacy-preserving distributed data mining is to develop algorithms that have reasonable computation and communication costs on real-world problems, and prove their security with respect to the above definition. While the secure circuit evaluation technique may be used within these algorithms, use must be limited to constrained sub-problems. For example, by adding secure comparison to the protocol at the beginning of this Section, the protocol can simply reveal if support and confidence exceed a threshold without revealing actual values. This mitigates the 100% confidence privacy compromise described above.

We now describe several techniques whose security properties have been evaluated using the standards described above. These examples demonstrate some key concepts that can be used to develop privacy-preserving distributed data mining algorithms, as well as demonstrating how algorithms are proven to be secure.

1.2 Association Rules

Association Rule mining is one of the most important data mining tools used in many real life applications. It is used to reveal unexpected relationships in the data. In this section, we will discuss the problem of computing association rules within a horizontally partitioned database framework. We assume homogeneous databases: All sites have the same schema, but each site has information on different entities. The goal is to produce association rules that hold globally, while limiting the information shared about each site.

The association rules mining problem can formally be defined as follows [2]: Let $I = \{i_1, i_2, \ldots, i_n\}$ be a set of items. Let DB be a set of transactions, where each transaction T is an itemset such that $T \subseteq I$. Given an itemset $X \subseteq I$, a transaction T *contains* X if and only if $X \subseteq T$. An association rule is an implication of the form $X \Rightarrow Y$ where $X \subseteq I, Y \subseteq I$ and $X \cap Y = \emptyset$. The rule $X \Rightarrow Y$ has *support* s in the transaction database DB if $s\%$ of transactions in DB contain $X \cup Y$. The association rule holds in the transaction database DB

with *confidence c* if *c*% of transactions in *DB* that contain *X* also contains *Y*. An itemset *X* with *k* items is called a *k*-itemset. The problem of mining association rules is to find all rules whose support and confidence are higher than certain user specified minimum support and confidence.

Clearly, computing association rules without disclosing individual transactions is straightforward. We can compute the global support and confidence of an association rule $AB \Rightarrow C$ knowing only the local supports of AB and ABC, and the size of each database:

$$support_{AB \Rightarrow C} = \frac{\sum_{i=1}^{\#sites} support\_count_{ABC}(i)}{\sum_{i=1}^{sites} database\_size(i)}$$

$$support_{AB} = \frac{\sum_{i=1}^{\#sites} support\_count_{AB}(i)}{\sum_{i=1}^{sites} database\_size(i)}$$

$$confidence_{AB \Rightarrow C} = \frac{support_{AB \Rightarrow C}}{support_{AB}}$$

Note that this doesn't require sharing any individual transactions. and protects individual data privacy, but it does require that each site disclose what rules it supports, and how much it supports each potential global rule. What if this information is sensitive? Clearly, such an approach will not be secure under SMC definitions.

A trivial way to convert the above simple distributed method to a secure method in SMC model is to use secure summation and comparison methods to check whether threshold are satisfied for every potential itemset. For example, for every possible candidate 1-itemset, we can use the secure summation and comparison protocol to check whether the threshold is satisfied.

Figure 1 gives an example of testing if itemset *ABC* is globally supported. Each site first computes its local support for *ABC*, or specifically the number of itemsets by which its support exceeds the minimum support threshold (which may be negative). The parties then use the previously described secure summation algorithm (the first site adds a random to its local excess support, then passes it to the next site to add its excess support, etc.) The only change is the final step: the last site performs a secure comparison with the first site to see if the *sum* $\geq R$. In the example, $R + 0$ is passed to the second site, which adds its excess support (-4) and passes it to site 3. Site 3 adds its excess support; the resulting value (18) is tested using secure comparison to see if it exceeds the Random value (17). It is, so itemset *ABC* is supported globally.

Due to huge number of potential candidate itemsets, we need to have a more efficient method. This can be done by observing the following lemma: If a rule has *support* > *k*% globally, it must have *support* > *k*% on at least one of the individual sites. A distributed algorithm for this would work as follows: Request that each site send all rules with support at least *k*. For each

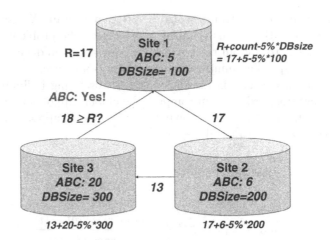

Fig. 1. Determining if itemset support exceeds 5% threshold

rule returned, request that all sites send the count of their transactions that support the rule, and the total count of all transactions at the site. From this, we can compute the global support of each rule, and (from the lemma) be certain that all rules with support at least k have been found. This has been shown to be an effective pruning technique [7]. In order to use the above lemma, we need to compute the union of locally large sets. We then use the secure summation and comparison only on the candidate itemsets contained in the union.

Revealing candidate itemsets means that the algorithm is no longer fully secure: itemsets that are large at one site, but not globally large, would not be disclosed by a fully secure algorithm. However, by computing the union *securely*, we prevent disclosure of which site, or even how many sites, support a particular itemset. This release of innocuous information (included in the final result) enables a completely secure algorithm that approaches the efficiency of insecure distributed association rule mining algorithms. The function now being computed reveals more information than the original association rule mining function. However, the key is that we have provable limits on what is disclosed. We now demonstrate how to securely compute a union.

Secure Union of Locally Large Itemsets

One way to securely compute a union is to directly apply secure circuit evaluation as follows: For each possible large k-itemset, each site can create a 0/1 vector such that if the ith itemset is locally supported at the site, it will set the ith bit of its vector to 1 otherwise it will set it to 0. Let's denote this vector as v_j for site j and let $v_j(i)$ be the ith bit of this vector. All the itemsets are arranged according to lexicographic order. Now for any given itemset, we can find its index i, and evaluate $\vee_{j=1}^{n} v_j(i)$ where $\vee$ is the or gate. Assuming that we use a secure generic circuit evaluation for or gate ($\vee$), the above protocol is secure and reveals nothing other than the set union result. However expensive circuit evaluation is needed for each potential large k-itemset. This secure method does not use the fact that local pruning eliminates some part of the large itemsets. We will now give a much more efficient method for this problem. Although the new method reveals a little more information than the above protocol, a precise description of what is revealed is given, and we prove that nothing else is revealed.

To obtain an efficient solution without revealing what each site supports, we instead exchange locally large itemsets in a way that obscures the source of each itemset. We assume a secure commutative encryption algorithm with negligible collision probability. Intuitively, under commutative encryption, the order of encryption does not matter. If a plaintext message is encrypted by two different keys in a different order, it will be mapped to the same ciphertext. Formally, commutativity ensures that $E_{k1}(E_{k2}(x)) = E_{k2}(E_{k1}(x))$. There are several examples of commutative encryption schemes, RSA is perhaps the best known.

The detailed algorithm is given in Algorithm 1. We now briefly explain the key phases of the algorithm. The main idea is that each site encrypts the locally supported itemsets, along with enough "fake" itemsets to hide the actual number supported. Each site then encrypts the itemsets from other sites. In Phases 2 and 3, the sets of encrypted itemsets are merged. Due to commutative encryption, duplicates in the locally supported itemsets will be duplicates in the encrypted itemsets, and can be deleted. The reason this occurs in two phases is that if a site knows which fully encrypted itemsets come from which sites, it can compute the size of the intersection between any set of sites. While generally innocuous, if it has this information for itself, it can guess at the itemsets supported by other sites. Permuting the order after encryption in Phase 1 prevents knowing exactly which itemsets match, however separately merging itemsets from odd and even sites in Phase 2 prevents any site from knowing the fully encrypted values of its own itemsets.

Phase 4 decrypts the merged frequent itemsets. Commutativity of encryption allows us to decrypt all itemsets in the same order regardless of the order they were encrypted in, preventing sites from tracking the source of each itemset.

The detailed algorithm assumes the following representations: F represents the data that can be used as fake itemsets. $|LLe_{i(k)}|$ represents the set of the encrypted k itemsets at site i. E_i is the encryption and D_i is the decryption by site i.

An illustration of the above protocol is given in Fig. 2. Using commutative encryption, each party encrypts its own frequent itemsets (e.g., Site 1 encrypts itemset C). The encrypted itemsets are then passed to other parties, until all parties have encrypted all itemsets. These are passed to a common party to eliminate duplicates, and to begin decryption. (In the figure, the full set of itemsets are shown to the left of Site 1, after Site 1 decrypts.) This set is then

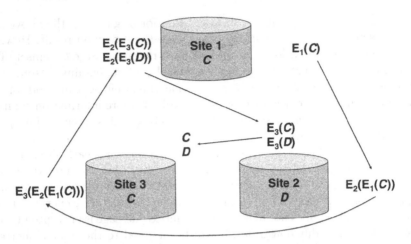

Fig. 2. Determining global candidate itemsets

passed to each party, and each party decrypts each itemset. The final result is the common itemsets (C and D in the figure).

Clearly, Algorithm 1 finds the union without revealing which itemset belongs to which site. It is not, however, secure under the definitions of secure multi-party computation. It reveals the number of itemsets having common support between sites, e.g., sites 3, 5, and 9 all support some itemset. It does not reveal *which* itemsets these are, but a truly secure computation (as good as giving all input to a "trusted party") could not reveal even this count. Allowing innocuous information leakage (the number of itemsets having common support) allows an algorithm that is sufficiently secure with much lower cost than a fully secure approach.

If we deem leakage of the number of commonly supported itemsets as acceptable, we can prove that this method is secure under the definitions of secure multi-party computation. The idea behind the proof is to show that given the result, the leaked information, and a site's own input, a site can simulate everything else seen during the protocol. Since the simulation generates everything seen during execution of the protocol, the site clearly learns nothing new from the protocol beyond the input provided to the simulator. One key is that the simulator does not need to generate exactly what is seen in any particular run of the protocol. The exact content of messages passed during the protocol is dependent on the random choice of keys; the simulator must generate an equivalent distribution, based on random choices made by the simulator, to the distribution of messages seen in real executions of the protocol. A formal proof that this proof technique shows that a protocol preserves privacy can be found in [17]. We use this approach to prove that Algorithm 1 reveals only the union of locally large itemsets and a clearly bounded set of innocuous information.

Algorithm 1 [20] Finding secure union of large itemsets of size k

Require: $N \geq 3$ sites numbered $0..N-1$, set F of non-itemsets.

Phase 0: Encryption of all the rules by all sites
for each site i **do**
    generate $LL_{i(k)}$ (Locally Large k-itemsets)
    $LLe_{i(k)} = \emptyset$
    for each $X \in LL_{i(k)}$ **do**
        $LLe_{i(k)} = LLe_{i(k)} \cup \{E_i(X)\}$
    end for
    for $j = |LLe_{i(k)}| + 1$ to $|CG_{(k)}|$ **do**
        $LLe_{i(k)} = LLe_{i(k)} \cup \{E_i(\text{random selection from } F)\}$
    end for
end for

Phase 1: Encryption by all sites
for Round $j = 0$ to $N - 1$ **do**
    if Round $j = 0$ **then**
        Each site i sends permuted $LLe_{i(k)}$ to site $(i+1) \bmod N$
    else
        Each site i encrypts all items in $LLe_{(i-j \bmod N)(k)}$ with E_i, permutes, and
        sends it to site $(i+1) \bmod N$
    end if
end for{At the end of Phase 1, site i has the itemsets of site $(i+1) \bmod N$
encrypted by every site}

Phase 2: Merge odd/even itemsets
Each site i sends $LLe_{i+1 \bmod N}$ to site $1 - ((i+1 \bmod N) \bmod 2)$
Site 0 sets $RuleSet_1 = \cup_{j=1}^{\lceil (N-1)/2 \rceil} LLe_{(2j-1)(k)}$
Site 1 sets $RuleSet_0 = \cup_{j=0}^{\lfloor (N-1)/2 \rfloor} LLe_{(2j)(k)}$

Phase 3: Merge all itemsets
Site 1 sends permuted $RuleSet_1$ to site 0
Site 0 sets $RuleSet = RuleSet_0 \cup RuleSet_1$

Phase 4: Decryption
for $i = 0$ to $N - 1$ **do**
    Site i decrypts items in $RuleSet$ using D_i
    Site i sends permuted $RuleSet$ to site $i + 1 \bmod N$
end for
Site $N - 1$ decrypts items in $RuleSet$ using D_{N-1}
$RuleSet_{(k)} = RuleSet - F$
Site $N - 1$ broadcasts $RuleSet_{(k)}$ to sites $0..N - 2$

Theorem 2. *[20] Algorithm 1 privately computes the union of the locally large itemsets assuming no collusion, revealing at most the result $\cup_{i=1}^{N} LL_{i(k)}$ and:*

1. *Size of intersection of locally supported itemsets between any subset of odd numbered sites,*
2. *Size of intersection of locally supported itemsets between any subset of even numbered sites, and*
3. *Number of itemsets supported by at least one odd and one even site.*

Proof. Phase 0: Since no communication occurs in Phase 0, each site can simulate its view by running the algorithm on its own input.

Phase 1: At the first step, each site sees $LLe_{i-1(k)}$. The size of this set is the size of the global candidate set $CG_{(k)}$, which is known to each site. Assuming the security of encryption, each item in this set is computationally indistinguishable from a number chosen from a uniform distribution. A site can therefore simulate the set using a uniform random number generator. This same argument holds for each subsequent round.

Phase 2: In Phase 2, site 0 gets the fully encrypted sets of itemsets from the other even sites. Assuming that each site knows the source of a received message, site 0 will know which fully encrypted set $LLe_{(k)}$ contains encrypted itemsets from which (odd) site. Equal itemsets will now be equal in encrypted form. Thus, site 0 learns if any odd sites had locally supported itemsets in common. We can still build a simulator for this view, using the information in point 2 above. If there are k itemsets known to be common among all $\lfloor N/2 \rfloor$ odd sites (from point 1), generate k random numbers and put them into the simulated $LLe_{i(k)}$. Repeat for each $\lfloor N/2 \rfloor - 1$ subset, etc., down to 2 subsets of the odd sites. Then fill each $LLe_{i(k)}$ with randomly chosen values until it reaches size $|CG_{i(k)}|$. The generated sets will have exactly the same combinations of common items as the real sets, and since the *values* of the items in the real sets are computationally indistinguishable from a uniform distribution, their simulation matches the real values.

The same argument holds for site 1, using information from point 2 to generate the simulator.

Phase 3: Site 1 eliminates duplicates from the $LLe_{i(k)}$ to generate $RuleSet_1$. We now demonstrate that Site 0 can simulate $RuleSet_1$. First, the size of $RuleSet_1$ can be simulated knowing point 2. There may be itemsets in common between $RuleSet_0$ and $RuleSet_1$. These can be simulated using point 3: If there are k items in common between even and odd sites, site 0 selects k random items from $RuleSet_0$ and inserts them into $RuleSet_1$. $RuleSet_1$ is then filled with randomly generated values. Since the encryption guarantees that the values are computationally indistinguishable from a uniform distribution, and the set sizes $|RuleSet_0|$, $|RuleSet_1|$, and $|RuleSet_0 \cap RuleSet_1|$ (and thus $|RuleSet|$) are identical in the simulation and real execution, this phase is secure.

Phase 4: Each site sees only the encrypted items after decryption by the preceding site. Some of these may be identical to items seen in Phase 2, but since all items must be in the union, this reveals nothing. The simulator for site i is built as follows: take the values generated in Phase 2 step $N-1-i$, and place them in the *RuleSet*. Then insert random values in *RuleSet* up to the proper size (calculated as in the simulator for Phase 3). The values we have not seen before are computationally indistinguishable from data from a uniform distribution, and the simulator includes the values we have seen (and knew would be there), so the simulated view is computationally indistinguishable from the real values.

The simulator for site $N-1$ is different, since it learns $RuleSet_{(k)}$. To simulate what it sees in Phase 4, site $N-1$ takes each item in $RuleSet_{(k)}$, the final result, and encrypts it with E_{N-1}. These are placed in *RuleSet*. *RuleSet* is then filled with items chosen from F, also encrypted with E_{N-1}. Since the choice of items from F is random in both the real and simulated execution, and the real items exactly match in the real and simulation, the *RuleSet* site $N-1$ receives in Phase 4 is computationally indistinguishable from the real execution.

Therefore, we can conclude that above protocol is privacy-preserving in the semi-honest model with the stated assumptions.   ∎

The information disclosed by points 1–3 could be relaxed to the number of itemsets support by 1 site, 2 sites,..., N sites if we assume anonymous message transmission. The number of jointly supported itemsets can also be masked by allowing sites to inject itemsets that are not really supported locally. These fake itemsets will simply fail to be globally supported, and will be filtered from the final result when global support is calculated as shown in the next section. The jointly supported itemsets "leak" then becomes an upper bound rather than exact, at an increased cost in the number of candidates that must be checked for global support. While not truly zero-knowledge, it reduces the confidence (and usefulness) of the leaked knowledge of the number of jointly supported itemsets. In practical terms, revealing the size (but not content) of intersections between sites is likely to be of little concern.

A complimentary problem of mining association rules over vertically partitioned data is addressed in [34,37]. While we do not describe these techniques here, we would like to emphasize that the different model of distribution requires very different solution techniques.

1.3 Decision Trees

The first paper discussing the use of Secure Multiparty Computation for data mining gave a procedure for constructing decision trees [24], specifically running ID3 [31] between two parties, each containing a subset of the training entities. Of particular interest is the ability to maintain "perfect" security in the SMC sense, while trading off efficiency for the quality of the resulting decision tree.

Building an ID3 decision tree is a recursive process, operating on the decision attributes R, class attribute C, and training entities T. At each stage, one of three things can happen:

1. R may be empty; i.e., the algorithm has no attributes on which to make a choice. In this case a leaf node is created with the class of the leaf being the majority class of the transactions in T.
2. All the transactions in T may have the same class c, in which case a leaf is created with class c.
3. Otherwise, we recurse:
   (a) Find the attribute A that is the most effective classifier for transactions in T, specifically the attribute that gives the highest information gain.
   (b) Partition T based on the values a_i of A.
   (c) Return a tree with root labeled A and edges a_i, with the node at the end of edge a_i constructed from calling ID3 with $R - \{A\}, C, T(A_i)$.

In step 3a, *information gain* is defined as the change in the entropy relative to the class attribute. Specifically, the entropy

$$H_C(T) = \sum_{c \in C} -\frac{|T(c)|}{|T|} \log \frac{|T(c)|}{|T|}.$$

Analogously, the entropy after classifying with A is

$$H_C(T|A) = \sum_{a \in A} -\frac{|T(a)|}{|T|} H_C(T(a)).$$

Information gain is

$$Gain(A) \stackrel{def}{=} H_C(T) - H_C(T|A).$$

The goal, then, is to find A that maximizes $Gain(A)$, or minimizes $H_C(T|A)$. Expanding, we get:

$$
\begin{aligned}
H_C(T|A) &= \sum_{a \in A} \frac{|T(a)|}{|T|} H_C(T(A)) \\
&= \frac{1}{|T|} \sum_{a \in A} |T(a)| \sum_{c \in C} -\frac{|T(a,c)|}{|T(a)|} \log \left(\frac{|T(a,c)|}{|T(A)|} \right) \\
&= \frac{1}{|T|} \left(-\sum_{a \in A} \sum_{c \in C} |T(a,c)| \log(|T(a,c)|) + \sum_{a \in A} |T(a)| \log(|T(a)|) \right)
\end{aligned}
$$

(1)

Looking at this from the point of view of privacy preservation, we can assume that R and C are known to both parties. T is divided. In Step 1 we need only determine the class value of the majority of the transactions in

T. This can be done using circuit evaluation 1. Since each party is able to compute the count of local items in each class, the input size of the circuit is fixed by the number of classes, rather than growing with the (much larger) training data set size.

Step 2 requires only that we determine if all of the items are of the same class. This can again be done with circuit evaluation, here testing for equality. Each party gives as input either the single class c_i of all of its remaining items, or the special symbol $\perp$ if its items are of multiple classes. The circuit returns the input if the input values are equal, else it returns $\perp$.[1]

It is easy to prove that these two steps preserve privacy: Knowing the tree, we know the majority class for Step 1. As for Step 2, if we see a tree that has a "pruned" branch, we know that all items must be of the same class, or else the branch would have continued. Interestingly, if we test if all items are in the same class before testing if there are no more attributes (reversing steps 1 and 2, as the original ID3 algorithm was written), the algorithm would not be private. The problem is that Step 2 reveals if all of the items are of the same class. The decision tree doesn't contain this information. However, if a branch is "pruned" (the tree outputs the class without looking at all the attributes), we know that all the training data at that point are of the same class - otherwise the tree would have another split/level. Thus Step 2 doesn't reveal any knowledge that can't be inferred from the tree *when the tree is pruned* - the given order ensures that this step will only be taken if pruning is possible.

This leaves Step 3. Note that once A is known, steps 3b and 3c can be computed locally -- no information exchange is required, so no privacy breach can occur. Since A can be determined by looking at the result tree, revealing A is not a problem, provided nothing but the proper choice for A is revealed. The hard part is Step 3a: computing the attribute that gives the highest information gain. This comes down to finding the A that minimizes (1).

Note that since the database is horizontally partitioned, $|T(a)|$ is really $|T_1(a)| + |T_2(a)|$, where T_1 and T_2 are the two databases. The idea is that the parties will compute (random) shares of each $(|T_1(a,c)| + |T_2(a,c)|) \log(|T_1(a,c)| + |T_2(a,c)|)$, and $(|T_1(a)| + |T_2(a)|) \log(|T_1(a)| + |T_2(a)|)$. The parties can then locally add their shares to give each a random share of $H_C(T|A)$. This is repeated for each attribute A, and a (small) circuit, of size linear in the number of attributes, is constructed to select the A that gives the largest value.

The problem, then is to efficiently compute $(x+y)\log(x+y)$. Lindell and Pinkas actually give a protocol for computing $(x+y)\ln(x+y)$, giving shares of $H_C(T|A) \cdot |T| \cdot \ln 2$. However, the constant factors are immaterial since the goal is simply to find the A that minimizes the equation. In [24] three protocols are

---

[1]The paper by Lindell and Pinkas gives other methods for computing this step, however circuit evaluation is sufficient - the readers are encouraged to read [24] for the details.

given: Computing shares of $\ln(x+y)$, computing shares of $x \cdot y$, and the protocol for computing the final result. The last is straightforward: Given shares u_1 and u_2 of $\ln(x+y)$, the parties call the multiplication protocol twice to give shares of $u_1 \cdot y$ and $u_2 \cdot x$. Each party then sums three multiplications: the two secure multiplications, and the result of multiplying its input (x or y) with its share of the logarithm. This gives each shares of $u_1 y u_2 x + u_1 x + u_2 y = (x+y)(u_1 + u_2) = (x+y)\ln(x+y)$.

The logarithm and multiplication protocols are based on oblivious polynomial evaluation [27]. The idea of oblivious polynomial evaluation is that one party has a polynomial P, the other has a value for x, and the party holding x obtains $P(x)$ without learning P or revealing x. Given this, the multiplication protocol is simple: The first party chooses a random r and generates the polynomial $P(y) = xy - r$. The resulting of evaluating this on y is the second party's share: $xy - r$. The first party's share is simply r.

The challenge is computing shares of $\ln(x+y)$. The trick is to approximate $\ln(x+y)$ with a polynomial, specifically the Taylor series:

$$\ln(1+\epsilon) = \sum_{i=1}^{k} \frac{(-1)^{i-1}\epsilon^i}{i}$$

Let 2^n be the closest power of 2 to $(x+y)$. Then $(x+y) = 2^n(1+\epsilon)$ for some $-1/2 \le \epsilon \le 1/2$. Now

$$\ln(x) = \ln(2^n(1+\epsilon)) = n\ln 2 + \epsilon - \frac{\epsilon^2}{2} + \frac{\epsilon^3}{3} - \cdots$$

We determine shares of $2^N n \ln 2$ and $2^N \epsilon$ (where N is an upper bound on n) using circuit evaluation. This is a simple circuit. $\epsilon \cdot 2^n = (x+y) - 2^n$, and n is obtained by inspecting the two most significant bits of $(x+y)$. There are a small (logarithmic in the database size) number of possibilities for $2^N n \ln 2$, and $\epsilon \cdot 2^N$ is obtained by left shifting $\epsilon \cdot 2^n$.

Assume the parties share of $2^N n \ln 2$ are α_1 and α_2, and the shares of $2^N \epsilon$ are β_1 and β_2. The first party defines

$$P(x) = \sum_{i=1}^{k} \frac{(-1)^{i-1}}{2^{N(i-1)}} \frac{(\alpha_1 + x)^i}{i} - r$$

and defines it's share $u_1 = \beta_1 + r$. The second party defines its share as $\beta_2 + P(\alpha_2)$. Note that $P(\alpha_2)$ computes the Taylor series approximation times 2^N, minus the random r. Since 2_N is public, it is easily divided out later, so the parties do get random shares of an approximation of $\ln(x+y)$.

As discussed in Sect. 1.2, all arithmetic is really done over a sufficiently large field, so that the random values (e.g., shares) can be chosen from a uniform distribution. In addition, the values in the Taylor series are multiplied by the least common multiple of $2, \ldots, k$ to eliminate fractions.

The key points to remember are the use of oblivious polynomial evaluation, and the use of an efficiently computable (bounded) approximation when efficiently and privately computing the real value is difficult.

There has also been a solution for constructing ID3 on vertically partitioned data [12]. This work assumes the class of the training data is shared, but some the attributes are private. Thus most steps can be evaluated locally. The main problem is computing which site has the best attribute to split on – each can compute the gain of their own attributes without reference to the other site.

1.4 Third Party Based Solutions

The use of an outside party often enables more efficient solutions to secure computations. The key issues are what level of trust is placed in this third, outside party; and what level of effort is required of the third party. Generally the trust issue is rooted in collusion: What happens if parties collude to violate privacy? This gives us a hierarchy of types of protocols:

No trusted third party. The most general type of protocol meets the strong statements of Definition 1: No party learns anything beyond what it can infer from the results. If parties collude, they are treated as one from the point of view of the definition: What can they infer from their combined inputs and results?

Non-colluding untrusted third party protocol. These protocols allow all parties to utilize an untrusted third party to do part of the computation. The third party learns nothing by itself (it need not even see the results). Provided this third party does not collude with one or more of the other parties. this method preserves privacy as well as a fully secure protocol.

Typically data is sent to this party in some "encrypted" form such that it cannot make any sense of the data by itself. This party performs some computations and replies to the local parties, which then remove the effect of the encryption to get back the final result. The key is that the untrusted third party does not see any "cleartext" data and is assumed to not collude with any of the other parties.

Commodity server protocol. Commodity server protocols also requires a non-colluding third party. They differ from non-colluding untrusted third party protocols in that only one way communication is allowed from the commodity server to the other parties. Because of this, the commodity server clearly learns nothing (absent collusion). The general approach is to use the commodity server to generate complementary data (e.g., public-private encryption key pairs). each part of which is given to a different party.

The commodity server model has been proven to be powerful enough to do all secure computation [5]. Thus, in terms of scope and power commodity server protocols are equivalent to protocols with an untrusted third party.

They are simpler to prove secure though typically more complicated than untrusted third party protocols.

Trusted third party protocol. The gold standard for security is the assumption that we have a trusted third party to whom we can give all data. The third party performs the computation and delivers only the results – except for the third party, it is clear that nobody learns anything not inferable from its own input and the results. The goal of secure protocols is to reach this same level of privacy preservation, without the (potentially insoluble) problem of finding a third party that everyone trusts.

The complexity of fully secure protocols generally increases with an increasing number of parties. Simple (completely secure) solutions for two parties do not extend easily to more than two parties. In such scenarios, it is often worthwhile to reduce the single complex problem to a series of two-party sub-problems. One approach is to make use of untrusted non-colluding third party protocols, using some of the participating parties to serve as "untrusted" parties for other parties in the protocol. If the target function consists of additive sub-blocks, or the target function can be reduced to a combination of associative functions, such an approach is possible. The key is to find an untrusted third party solution to the two-party problem, then securely combine the two-party results in a way that gives the desired final result.

We now give a couple of examples typifying these cases, and show solutions that illuminate the basic concept. First consider the following geometric function: Consider an n-dimensional space split between r different parties, $P_1, \ldots, P_r$. P_i owns a variable number n_i of dimensions/axes such that $\sum_{i=1}^{r} n_i = n$. A point X in this n-dimensional space would have its n co-ordinates split between the r parties. Thus, party i would know n_i of the co-ordinates of the point X. We assume that there is some way of linking the co-ordinates of the same point together across all the parties (i.e., a join key). Now, assume there are k points $Y_1, \ldots, Y_k$ split between the r parties. The target function is to jointly compute the index i of the point Y_i that is the "closest" to point X according to some distance metric $\mathcal{D}$.

Why is this problem interesting? K-means clustering over vertically partitioned data can be easily reduced to this problem. K-means clustering is an iterative procedure that starts with K arbitrary cluster means. In each iteration, all of the points are assigned to the current closest cluster (based on distance from mean). Once all the points are assigned, the cluster means are recomputed based on the points assigned to each cluster. This procedure is repeated until the clusters converge. One easy convergence condition is to stop when the difference between the old means and the new means is sufficiently small. The key step, assigning a point to a cluster, is done by finding the closest cluster to the point. This is solved by our earlier "abstract" geometrical problem.[2]

---

[2]A solution for clustering in horizontally partitioned data has also been developed [23], this relies heavily on secure summation.

One possible distance metric is the Minkowski distance, d_M. The Minkowski distance d_M between two points X and Y is defined as

$$d_M = \left\{ \sum_{i=1}^{n} (x_i - y_i)^m \right\}^{\frac{1}{m}}$$

Note that $m = 2$ gives the Euclidean distance, while $m = 1$ gives the Manhattan distance. Instead of comparing two distances, we get the same result by comparing the mth power of the distances. Note that if we do not take the mth root, the target function is additive. We can exploit this additiveness to get an efficient protocol.

The problem is formally defined as follows. Consider r parties $P_1, \ldots, P_r$, each with their own k-element vector $\mathbf{X_i}$:

$$P_1 \text{ has } \mathbf{X_1} = \begin{bmatrix} x_{11} \\ x_{21} \\ \vdots \\ x_{k1} \end{bmatrix}, P_2 \text{ has } \begin{bmatrix} x_{12} \\ x_{22} \\ \vdots \\ x_{k2} \end{bmatrix}, \ldots, P_r \text{ has } \begin{bmatrix} x_{1r} \\ x_{2r} \\ \vdots \\ x_{kr} \end{bmatrix}.$$

The goal is to compute the index l that represents the row with the minimum sum. Formally, find

$$\underset{i=1..k}{argmin} \left(\sum_{j=1..r} x_{ij} \right)$$

For use in k-means clustering, $x_{ij} = |\mu_{ij} - point_j|$, or site P_j's component of the distance between a point and the cluster i with mean μ_i.

The security of the algorithm is based on three key ideas.

1. Disguise the site components of the distance with random values that cancel out when combined.
2. Compare distances so only the comparison result is learned; no party knows the distances being compared.
3. Permute the order of clusters so the real meaning of the comparison results is unknown.

The algorithm also requires three non-colluding sites. These parties may be among the parties holding data, but could be external as well. They need only know the number of sites r and the number of clusters k. Assuming they do not collude with each other, they learn nothing from the algorithm. For simplicity of presentation, we will assume the non-colluding sites are P_1, P_2, and P_r among the data holders.

The algorithm proceeds as follows. Site P_1 generates a length k random vector $\mathbf{V_i}$ for each site i. such that $\sum_{i=1}^{r} \mathbf{V_i} = \mathbf{0}$. P_1 also chooses a permutation π of $1..k$. P_1 then engages each site P_i in the permutation algorithm (see Sect. 1.4) to generate the sum of the vector $\mathbf{V_i}$ and P_i's distances $\mathbf{X_i}$. The resulting vector is known only to P_i, and is permuted by π known only to P_1, i.e., P_i has $\pi(\mathbf{V_i} + \mathbf{X_i})$, but does not know π or $\mathbf{V_i}$. P_1 and $P_3 \ldots P_{r-1}$ send their vectors to P_r.

Sites P_2 and P_r now engage in a series of secure addition/comparisons to find the (permuted) index of the minimum distance. Specifically, they want to find if $\sum_{i=1}^{r} x_{li} + v_{li} < \sum_{i=1}^{r} x_{mi} + v_{mi}$. Since $\forall l, \sum_{i=1}^{r} v_{li} = 0$, the result is $\sum_{i=1}^{r} x_{li} < \sum_{i=1}^{r} x_{mi}$, showing which cluster (l or m) is closest to the point. P_r has all components of the sum except $\mathbf{X_2} + \mathbf{V_2}$. For each comparison, we use a secure circuit evaluation that calculates $a_2 + a_r < b_2 + b_r$, without disclosing anything but the comparison result. After $k - 1$ such comparisons, keeping the minimum each time, the minimum cluster is known.

P_2 and P_r now know the minimum cluster in the permutation π. They do not know the real cluster it corresponds to (or the cluster that corresponds to any of the others items in the comparisons.) For this, they send the minimum i back to site P_1. P_1 broadcasts the result $\pi^{-1}(i)$, the proper cluster for the point.

Algorithm 2 reproduces the full algorithm from [36]. We now describe the two key building blocks borrowed from the Secure Multiparty Computation literature. The secure addition and comparison consists of a circuit that has two inputs from each party, sums the first input of both parties and the second input of both parties, and returns the result of comparing the two sums. This (simple) circuit is evaluated securely using the generic algorithm. Though the generic algorithm is impractical for large inputs and many parties, it is quite efficient for a limited number invocations of the *secure_add_and_compare* function. For two parties, the message cost is $O(circuit\_size)$, and the number of rounds is constant. We can add and compare numbers with $O(m = \log(number\_of\_entities))$ bits using an $O(m)$ size circuit. A graphical depiction of stages 1 and 2 is given in Figs. 3(a) and 3(b).

We now give the permutation algorithm of [11], which simultaneously computes a vector sum and permutes the order of the elements in the vector.

The permutation problem is an asymmetric two party algorithm, formally defined as follows. There exist 2 parties, A and B. B has an n-dimensional vector $\mathbf{X} = (x_1, \ldots, x_n)$, and A has an n-dimensional vector $\mathbf{V} = (v_1, \ldots, v_n)$. A also has a permutation π of the n numbers. The goal is to give B the result $\pi(\mathbf{X} + \mathbf{V})$, without disclosing anything else. In particular, neither A nor B can learn the other's vector, and B does not learn π. For our purposes, the $\mathbf{V}$ is a vector of random numbers from a uniform random distribution, used to hide the permutation of the other vector.

The solution makes use of a tool known as *Homomorphic Encryption*. An encryption function $\mathcal{H} : \mathcal{R} \to \mathcal{S}$ is called *additively homomorphic* if there is an efficient algorithm *Plus* to compute $H(x + y)$ from $H(x)$ and $H(y)$ that does not reveal x or y. Many such systems exist; examples include systems by [6, 26, 28], and [30]. This allows us to perform addition of encrypted data without decrypting it.

Algorithm 2 *closest_cluster* [36]: Find minimum distance cluster

---

Require: r parties, each with a length k vector $\mathbf{X}$ of distances. Three of these parties (trusted not to collude) are labeled P_1, P_2, and P_r.

1: {Stage 1: Between P_1 and all other parties}
2: P_1 generates r random vectors $\mathbf{V_i}$ summing to $\mathbf{0}$ (see Algorithm 3).
3: P_1 generates a random permutation π over k elements
4: **for all** $i = 2 \ldots r$ **do**
5:     $\mathbf{T_i}$ (at P_i) $= add\_and\_permute(\mathbf{V_i}, \pi(\text{at } P_1), \mathbf{X_i}(\text{at } P_i))$ {This is the permutation algorithm described in Sect. 1.4}
6: **end for**
7: P_1 computes $\mathbf{T_1} = \pi(\mathbf{X_1} + \mathbf{V_1})$
8:
9: {Stage 2: Between all but P_2 and P_r}
10: **for all** $i = 1, 3 \ldots r - 1$ **do**
11:     P_i sends $\mathbf{T_i}$ to P_r
12: **end for**
13: P_r computes $\mathbf{Y} = \mathbf{T_1} + \sum_{i=3}^{r} \mathbf{T_i}$
14:
15: {Stage 3: Involves only P_2 and P_r}
16: $minimal \leftarrow 1$
17: **for** j=2..k **do**
18:     **if** $secure\_add\_and\_compare(Y_j + T_{2j} < Y_{minimal} + T_{2minimal})$ **then**
19:         $minimal \leftarrow j$
20:     **end if**
21: **end for**
22:
23: {Stage 4: Between P_r (or P_2) and P_1}
24: Party P_r sends $minimal$ to P_1
25: P_1 broadcasts the result $\pi^{-1}(minimal)$

Algorithm 3 genRandom [36]: Generates a (somewhat) random matrix $V_{k \times r}$

Require: Random number generator *rand* producing values uniformly distributed over $0..n - 1$ spanning (at least) the domain of the distance function $-_D$.
Ensure: The sum of the resulting vectors is $\mathbf{0}$.

1: **for all** i = 1 \ldots k **do**
2:     $PartSum_i \leftarrow 0$
3:     **for** j = 2 \ldots r **do**
4:         $V_{ij} \leftarrow rand()$
5:         $PartSum_i \leftarrow PartSum_i + V_{ij} \pmod{n}$
6:     **end for**
7:     $V_{i1} \leftarrow -PartSum_i \pmod{n}$
8: **end for**

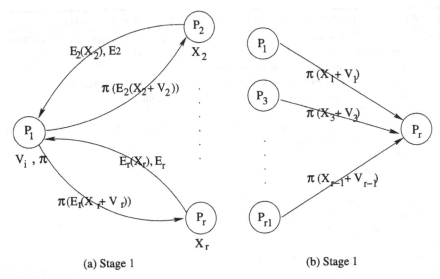

(a) Stage 1   (b) Stage 1

Fig. 3. Closest Cluster Computation

The permutation algorithm consists of the following steps:

1. B generates a public-private keypair (E_k, D_k) for a homomorphic encryption scheme.
2. B encrypts its vector $\mathbf{X}$ to generate the encrypted vector $\mathbf{X'} = (x'_1, \ldots, x'_n)$, $x'_i = E_k(x_i)$.
3. B sends $\mathbf{X'}$ and the public key E_k to A.
4. A encrypts its vector $\mathbf{V}$ generating the encrypted vector $\mathbf{V'} = (v'_1, \ldots, v'_n)$, $v'_i = E_k(v_i)$.
5. A now multiplies the components of the vectors $\mathbf{X'}$ and $\mathbf{V'}$ to get $\mathbf{T'} = (t'_1, \ldots, t'_n)$, $t'_i = x'_i * v'_i$.
   Due to the homomorphic property of the encryption,

$$x'_i * v'_i = E_k(x_i) * E_k(v_i) = E_k(x_i + v_i)$$

   so $\mathbf{T'} = (t'_1, \ldots, t'_n), t'_i = E_k(x_i + v_i)$.
6. A applies the permutation π to the vector $\mathbf{T'}$ to get $\mathbf{T'_p} = \pi(\mathbf{T'})$, and sends $\mathbf{T'_p}$ to B.
7. B decrypts the components of $\mathbf{T'_p}$ giving the final result $\mathbf{T_p} = (t_{p_1}, \ldots, t_{p_n})$, $t_{p_i} = x_{p_i} + v_{p_i}$.

Intersection

We now give an algorithm that demonstrates another way of using untrusted, non-colluding third parties. The specific algorithm is for computing the size of set intersection. This is useful for finding association rules in vertically partitioned data.

Assume the database is a boolean matrix where a 1 in cell (i, j) represents that the ith transaction contains feature j. The TID-list representation of the database has a transaction identifier list associated with every feature/attribute. This TID-list contains the identifiers of transactions that contain the attribute. Now, to count a frequency of an itemset $<AB>$ (where A and B are at different sites). it is necessary to count the number of transactions having both A and B. If we intersect the TID-lists for both A and B, we get the transactions containing both A and B. The size of this intersection set gives the (in)frequency of the itemset. Thus, the frequency of an itemset can be computed by securely computing the size of the intersection set of the TID-lists.

In addition to being useful for association rule mining, the algorithm illuminates a general technique that we can use to extend untrusted third party solutions for two parties to multiple parties. This works whenever the target function is associative. As in Sect. 1.2, the approach leaks some innocuous information, but can be proven to leak no more than this information. We start with the general technique. then discuss the specific application to set intersection.

Given k parties, the goal is to compute a function $y \in F_g$, where $y = f(x_1, \ldots, x_k)$, where $x_1, \ldots, x_k$ are the local inputs of the k parties. If the function can be decomposed into smaller invocations of an associative function, we can rewrite $y = x_1 \otimes x_2 \otimes \cdots \otimes x_k$. If we have a protocol f_s to securely compute the two-input function $\otimes$, we can construct a protocol to compute y as follows.

The key idea is to create two partitions P_0 and P_1. Split the k parties equally into the two partitions. We can now use the parties in partition P_i as untrusted third parties to evaluate partition P_{1-i}. To visualize this, construct a binary tree on the partition P_i with the leaves being the parties in P_i (Fig. 4).[3] There can be at most $|P_i| - 1$ interior nodes in the binary tree. Due to the (almost) equi-partitioning, the following invariant always holds: $|P_{1-i}| \geq |P_i| - 1$, for both values of i. Thus, there are sufficient parties in the other partition to act as interior nodes. The role of the parties in partition P_{1-i} is to act as the commodity server or untrusted third party for the parties in partition P_j.

In the first round, the $k/4$ of the parties from the other partition act as third parties for the $k/2$ parties in the first partition. For the remaining $\log k/2 - 1$ rounds the other $k/4$ parties of the 2nd partition act as third parties upwards along the tree. Each third party receives some form of the intermediate result, and utilizes it in the next round. It is important to analyze the amount of data revealed to the third party at this point and modify the protocol if necessary to limit the information disclosure. The entire process is illustrated in Fig. 4. where we show the process for partition P_0 consisting

---

[3]While the example assumes k is a power of 2, a proper assignment of parties to partitions is also possible if the tree is not complete. This is described in [37].

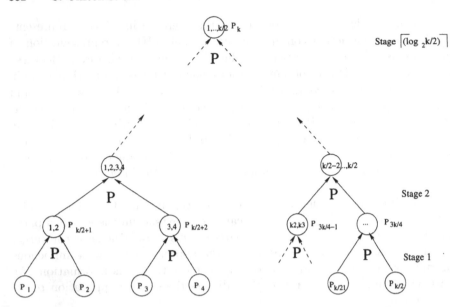

Fig. 4. The general protocol process applied on partition P_0

of the first $k/2$ parties. Thus, all of the parties $P_{k/2+1},\ldots,P_k$ act as third parties/commodity servers in a single call to the protocol f_s when applied to the parties at the leaf nodes. There are a total of $\log k/2$ rounds in which several calls to the protocol f_s are made in parallel.

Once a similar process is done for the other partition P_1, the two topmost representatives of the two parties use a secure two party protocol f' to compute the final result. Every party possibly acquires some information about a few of the other parties, which goes against the precept of secure multi-party computation. But as long as the information revealed is held within strict (and provable) bounds, it is often worthwhile to trade this limited information disclosure for efficiency and practicality.

We summarize the use of this approach to solve secure set intersection [35, 37]. The problem is defined as follows. There are k parties, $P_1,\ldots,P_k$, each with a local set S_k drawn from a common (global) universe U. They wish to compute $|\cap_{j=1}^k S_j|$, i.e., the cardinality of the common intersection set. This is useful for several applications for example data mining association rules (see [37] for details.)

We now outline a two party protocol $f_\cap$ using a third untrusted party to compute $|S_a \cap S_b|$ for two parties A and B. The key idea behind protocol $f_\cap$ is to use commutative encryption (as described in Sect. 1.2) to allow comparing items without revealing them. Parties A and B generate encryption keys E_a and E_b respectively. A encrypts the items in its set S_a with E_a and sends them to B. Similarly, B encrypts the items in S_b with E_b and sends them to A. Each site now encrypts the received items with its own key, and sends the doubly-encrypted sets S_a' and S_b' to U. U now finds the intersection of these two sets. Because of commutativity of the encryption, an item $x \in S_a \cap S_b$ will correspond to an item $E_a(E_b(x)) = E_b(E_a(x))$ that appears in both S_a' and S_b'. Therefore, the size of the intersection $|S_a' \cap S_b'| = |S_a \cap S_b|$. Thus U learns the size of the intersection, but learns nothing about the items *in* the intersection.

Extending this to more than two parties is simple. We use the tree based evaluation for each partition. The lowest layer (consisting of leaves) proceeds as above. At the higher layers, the parties encrypt with the keys of their sibling's children. Since a party never sees any of the values from the sibling's children (even after encryption), knowing the keys gives no information. More details are given in [37].

2 Privacy Preservation through Noise Addition

The other approach to privacy-preserving data mining is based on adding random noise to the data, then providing the noisy dataset as input to the data mining algorithm. The privacy-preserving properties are a result of the noise: Data values for individual entities are distorted, and thus individually identifiable (private) values are not revealed. An example would be a survey: A company wishes to mine data from a survey of private data values. While the respondents may be unwilling to provide those data values directly, they would be willing to provide randomized/distorted results.

What makes this work interesting is how the mining of the noisy data set is done. Naïvely running a data mining algorithm on the data may work – for example, adding noise from a gaussian distribution centered at 0 will preserve averages – but does not always give good results. However, using knowledge of how the noise was generated enables us to do better. In particular, what is used is knowledge of the distribution that the noise came from (e.g., uniform or gaussian and the appropriate parameters). Knowing the distribution the random values came from does not reveal the specific values used to mask each entity, so privacy is still preserved. However, as we shall see the knowledge of the distribution of the noise does enable us to improve data mining results.

The problem addressed in [3] was building decision trees. If we return to the description of ID3 in Sect. 1.3, we see that Steps 1 and 3c do not reference the (noisy) data. Step 2 references only the class data, which is assumed to be known (for example, the survey may be demographics of existing customers – the company already knows which are high-value customers, and wants to know what demographics correspond to high-value customers.)

This leaves Steps 3a and 3b: Finding the attribute with the maximum information gain and partitioning the tree based on that attribute. Looking

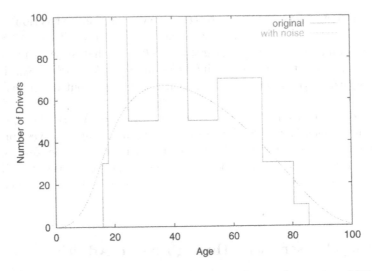

Fig. 5. Original distribution vs. distribution after random noise addition

at (1), the only thing needed is $|T(a,c)|$ and $|T(a)|$.[4] $|T(a)|$ requires partitioning the entities based on the attribute value, exactly what is needed for Step 3b. The problem is that the attribute values are modified, so we don't know which entity really belongs in which partition.

Figure 5 demonstrates this problem graphically. There are clearly peaks in the number of drivers under 25 and in the 25–35 age range, but this doesn't hold in the noisy data. The ID3 partitioning should reflect the peaks in the data.

A second problem comes from the fact that the data is assumed to be ordered (otherwise "adding" noise makes no sense.) As a result, where to divide partitions is not obvious (as opposed to categorical data). Again, reconstructing the distribution can help. We can see that in Fig. 5 partitioning the data at ages 30 and 50 would make sense – there is a natural "break" in the data at those points anyway. However, we can only see this from the actual distribution. The split points are not obvious in the noisy data.

Both these problems can be solved if we know the distribution of the original data, even if we do not know the original values. The problem remains that we may not get the *right* entities in each partition, but we are likely to get enough that the statistics on the class of each partition will still hold (In [3] experimental results are given demonstrating this fact.)

---

[4]Reference [3] actually uses the gini coefficient rather than information gain. While this may affect the quality of the decision tree, it has no impact on the discussion here. We stay with information gain for simplicity.

What remains is the problem of estimating the distribution of the real data (X) given the noisy data (w) and the distribution of the noise (Y). This is accomplished through Bayes' rule:

$$
\begin{aligned}
F_X'(a) &\equiv \int_{-\infty}^{a} f_X(z|X+Y=w)dz \\
&= \int_{-\infty}^{a} \frac{f_{X+Y}(w|X=z)f_X(z)}{f_{X+Y}(w)}dz \\
&= \int_{-\infty}^{a} \frac{f_{X+Y}(w|X=z)f_X(z)}{\int_{-\infty}^{\infty} f_{X+Y}(w|X=z')f_X(z')dz'}dz \\
&= \frac{\int_{-\infty}^{a} f_{X+Y}(w|X=z)f_X(z)dz}{\int_{-\infty}^{\infty} f_{X+Y}(w|X=z)f_X(z)dz} \\
&= \frac{\int_{-\infty}^{a} f_Y(w-z)f_X(z)dz}{\int_{-\infty}^{\infty} f_Y(w-z)f_X(z)dz}
\end{aligned}
$$

Given the actual data values $w_i = x_i + y_i$, we use this to estimate the distribution function as follows:

$$
F_X'(a) = \frac{1}{n}\sum_{i=1}^{n} F_{X_i}' = \frac{1}{n}\sum_{i=1}^{n} \frac{\int_{-\infty}^{a} f_Y(w_i-z)f_X(z)dz}{\int_{-\infty}^{\infty} f_Y(w_i-z)f_X(z)dz}
$$

Differentiating gives us the posterior density function:

$$
f_X'(a) = \frac{1}{n}\sum_{i=1}^{n} \frac{f_Y(w_i-a)f_X(a)}{\int_{-\infty}^{\infty} f_Y(w_i-z)f_X(z)dz} \tag{2}
$$

The only problem is, we don't know the real density function f_X. However, starting with an assumption of a uniform distribution, we can use (2) to iteratively refine the density function estimate, converging on an estimate of the real distribution for X.

In [3] several optimizations are given, for example partitioning the data to convert the integration into sums. They also discuss tradeoffs in *when* to compute distributions: Once for each attribute? Separately for each class? For only the data that makes it to each split point? They found that reconstructing each attribute separately for each class gave the best performance/accuracy tradeoff, with classification accuracy substantially better than naïvely running on the noisy data, and approaching that of building a classifier directly on the real data.

One question with this approach is how much privacy is given? With the secure multiparty computation based approaches, the definition of privacy is clear. However, given a value that is based on the real value, how do we know how much noise is enough? Agrawal and Srikant proposed a metric based the confidence in estimating a value within a specified width: If it can be estimated with $c\%$ confidence that a value x lies in the interval $[x_l, x_h]$, then the privacy at the $c\%$ confidence level is $|x_h - x_l|$. The quantify this in terms of a percentage: The privacy metric for noise from a uniform distribution is the confidence times twice the interval width of the noise: 100% privacy corresponds to a 50% confidence that the values is within two distribution widths of the real value, or nearly 100% confidence that it is within one width. They have an equivalent definition for noise from a gaussian distribution.

Agrawal and Aggarwal (not the same Agrawal) pointed out problems with this definition of privacy [1]. The very ability to reconstruct distributions may give us less privacy than expected. Figure 5 demonstrates this. Assume the noise is known to come from a uniform distribution over $[-15, 15]$, and the actual/reconstructed distribution is as shown by the bars. Since there are no drivers under age 16 (as determined from the reconstructed distribution), a driver whose age is given as 1 in the "privacy-preserving" dataset is known to be 16 years old – all privacy for this individual is lost. They instead give a definition based on entropy (discussed in Sect. 1.3). Specifically, if a random variable Y has entropy $H(Y)$, the privacy is $2^{H(Y)}$. This has the nice property that for a uniform distribution, the privacy is equivalent to the width of the interval from which the random value is chosen. This gives a meaningful way to compare different sources of noise.

They also provide a solution to the loss of privacy obtained through reconstructing the original data distribution. The idea is based on conditional entropy. Given the reconstructed distribution X, the privacy is now $2^{H}(Y|X)$. This naturally captures the expected privacy in terms of the interval width description: a reconstruction distribution that eliminates part of an interval (or makes it highly unlikely) gives a corresponding decrease in privacy.

There has been additional work in this area, such as techniques for association rules [14, 32]. Techniques from signal processing have also been applied to distribution reconstruction [21], generalizing much of this work. One problem is the gap between known abilities to reconstruct distributions and lower bounds on ability to reconstruct actual data values: the jury is still out on how effective these techniques really are at preserving privacy.

3 Conclusions and Recommendations

While privacy-preserving data mining does have the potential to reconcile the concerns of data mining proponents and privacy advocates, it has not reached the level of an effective panacea. Two issues remain.

First, the rigor required of the cryptography and security protocols communities must be brought to this field. While some of the work in this field does approach this level of rigor, much of the work does not. For some work, particularly with the noise addition approach, it is not clear if a determined adversary could compromise privacy (and in some cases, it is clear that they can [10, 21].) The distributed approach has a clear set of standards borrowed from the cryptography community, it is important that work be judged against these standards. In particular, work must move beyond the semi-honest model. This could mean developing efficient solutions secure against malicious adversaries, or possibly new definitions such as "proof against collusion" that meet practical needs and are defined with the rigor of the semi-honest and malicious definitions.

The second problem may be more challenging. Privacy-preserving data mining has operated under the assumption that data mining results do not of themselves compromise privacy. This is not necessarily true. While there has been some work on restricting data mining results [4, 18, 19, 29, 33], this has emphasized protection against revealing specific results. This work does not address connections between the results and compromise of source data items. While work on limiting classification strength may address this issue [8], the proposed method also prevents the data from being useful for data mining in any form. Achieving a reasonable connection between individual privacy and data mining results is still an open problem. Until this is solved, the full concerns of privacy advocates will not have been addressed.

That said, privacy-preserving data mining in its current form does have practical applications. In some circumstances, the value of the data mining results may exceed the potential cost to individual privacy. This is most likely true where the individual data items reflect commercial data (e.g., intellectual property) rather than personal information. For example, U.S. antitrust law frowns on the general sharing of information between competitors, however if the shared information is limited to that absolutely necessary to achieve some consumer benefit the sharing is likely to pass legal muster. The concept that use of information is allowed when necessary also shows up in the European Community privacy recommendations [13], reconciling these laws with the potential privacy breach of data mining results will be easier if privacy-preserving data mining techniques are used to ensure that *only* the results are disclosed.

In summary, while privacy-preserving data mining has achieved significant success, many challenges remain.

References

1. D. Agrawal and C. C. Aggarwal. "On the design and quantification of privacy preserving data mining algorithms," in *Proceedings of the Twentieth ACM SIGACT-SIGMOD-SIGART Symposium on Principles of Database Systems*. Santa Barbara, California, USA: ACM, May 21–23 2001, pp. 247–255. [Online]. Available: http://doi.acm.org/10.1145/375551.375602
2. R. Agrawal and R. Srikant, "Fast algorithms for mining association rules," in *Proceedings of the 20th International Conference on Very Large Data Bases*. Santiago. Chile: VLDB. Sept. 12 15 1994, pp. 487–499. [Online]. Available: http://www.vldb.org/dblp/db/conf/vldb/vldb94-487.html

3. ——, "Privacy-preserving data mining," in *Proceedings of the 2000 ACM SIG-MOD Conference on Management of Data*. Dallas, TX: ACM, May 14–19 2000, pp. 439–450. [Online]. Available: http://doi.acm.org/10.1145/342009. 335438

4. M. Atallah, E. Bertino, A. Elmagarmid, M. Ibrahim, and V. Verykios, "Disclosure limitation of sensitive rules," in *Knowledge and Data Engineering Exchange Workshop (KDEX'99)*, Chicago, Illinois, Nov. 8 1999, pp. 25–32. [Online]. Available: http://ieeexplore.ieee.org/iel5/6764/18077/00836532.pdf? isNumber=18077&prod=CNF&arnumber=00836532

5. D. Beaver, "Commodity-based cryptography (extended abstract)," in *Proceedings of the twenty-ninth annual ACM symposium on Theory of computing*. El Paso, Texas, United States: ACM Press, 1997, pp. 446–455.

6. J. Benaloh, "Dense probabilistic encryption," in *Proceedings of the Workshop on Selected Areas of Cryptography*, Kingston, Ontario, May 1994, pp. 120–128. [Online]. Available: http://research.microsoft.com/crypto/papers/dpe.ps

7. D. W.-L. Cheung, V. Ng, A. W.-C. Fu, and Y. Fu, "Efficient mining of association rules in distributed databases," *IEEE Transactions on Knowledge and Data Engineering*, vol. 8, no. 6, pp. 911–922, Dec. 1996.

8. C. Clifton, "Using sample size to limit exposure to data mining," *Journal of Computer Security*, vol. 8, no. 4, pp. 281–307, Nov. 2000. [Online]. Available: http://iospress.metapress.com/openurl.asp?genre=article&issn=0926-227X& volume=8&issue=4&spage=281

9. "Total information awareness (TIA) system," Defense Advanced Research Projects Agency. [Online]. Available: http://www.darpa.mil/iao/TIASystems. htm

10. I. Dinur and K. Nissim, "Revealing information while preserving privacy," in *Proceedings of the twenty-second ACM SIGMOD-SIGACT-SIGART symposium on Principles of database systems*. ACM Press, 2003, pp. 202–210. [Online]. Available: http://doi.acm.org/10.1145/773153.773173

11. W. Du and M. J. Atallah, "Privacy-preserving statistical analysis," in *Proceeding of the 17th Annual Computer Security Applications Conference*, New Orleans, Louisiana, USA, December 10–14 2001. [Online]. Available: http://www.cerias.purdue.edu/homes/duw/research/paper/acsac2001.ps

12. W. Du and Z. Zhan, "Building decision tree classifier on private data," in *IEEE International Conference on Data Mining Workshop on Privacy, Security, and Data Mining*, C. Clifton and V. Estivill-Castro, Eds., vol. 14. Maebashi City, Japan: Australian Computer Society, Dec. 9 2002, pp. 1–8. [Online]. Available: http://crpit.com/Vol14.html

13. "Directive 95/46/EC of the european parliament and of the council of 24 october 1995 on the protection of individuals with regard to the processing of personal data and on the free movement of such data," *Official Journal of the European Communities*, vol. No I., no. 281, pp. 31–50, Oct. 24 1995. [Online]. Available: http://europa.eu.int/comm/internal_market/privacy/

14. A. Evfimievski, R. Srikant, R. Agrawal, and J. Gehrke, "Privacy preserving mining of association rules," in *The Eighth ACM SIGKDD International Conference on Knowledge Discovery and Data Mining*, Edmonton, Alberta, Canada, July 23–26 2002, pp. 217–228. [Online]. Available: http://doi.acm.org/ 10.1145/775047.775080

15. M. Feingold, M. Corzine, M. Wyden, and M. Nelson, "Data-mining moratorium act of 2003," U.S. Senate Bill (proposed), Jan. 16 2003. [Online]. Available: http://thomas.loc.gov/cgi-bin/query/z?c108:S.188:

16. O. Goldreich, S. Micali, and A. Wigderson, "How to play any mental game – a completeness theorem for protocols with honest majority," in *19th ACM Symposium on the Theory of Computing*, 1987, pp. 218–229. [Online]. Available: http://doi.acm.org/10.1145/28395.28420

17. O. Goldreich, "Secure multi-party computation," Sept. 1998, (working draft). [Online]. Available: http://www.wisdom.weizmann.ac.il/ oded/pp.html

18. T. Johnsten and V. Raghavan, "Impact of decision-region based classification algorithms on database security," in *Proceedings of the Thirteenth Annual IFIP WG 11.3 Working Conference on Database Security*, Seattle, Washington, July 26–28 1999. [Online]. Available: http://www.cacs.usl.edu/Publications/Raghavan/JR99.ps.Z

19. ——, "A methodology for hiding knowledge in databases," in *Proceedings of the IEEE ICDM Workshop on Privacy, Security and Data Mining*, ser. Conferences in Research and Practice in Information Technology, vol. 14. Maebashi City, Japan: Australian Computer Society, Dec. 9 2002, pp. 9–17. [Online]. Available: http://crpit.com/confpapers/CRPITV14Johnsten.pdf

20. M. Kantarcıoğlu and C. Clifton, "Privacy-preserving distributed mining of association rules on horizontally partitioned data," *IEEE Transactions on Knowledge and Data Engineering*, vol. 16. no. 4, July 2004.

21. H. Kargupta, S. Datta, Q. Wang, and K. Sivakumar, "On the privacy preserving properties of random data perturbation techniques," in *Proceedings of the Third IEEE International Conference on Data Mining (ICDM'03)*, Melbourne, Florida, Nov. 19–22 2003.

22. M. Lewis, "Department of defense appropriations act, 2004," July 17 2003, title VIII section 8120. Enacted as Public Law 108–87. [Online]. Available: http://thomas.loc.gov/cgi-bin/bdquery/z?d108:h.r.02658:

23. X. Lin, C. Clifton. and M. Zhu, "Privacy preserving clustering with distributed EM mixture modeling." *Knowledge and Information Systems*, to appear 2004.

24. Y. Lindell and B. Pinkas, "Privacy preserving data mining," in *Advances in Cryptology – CRYPTO 2000*. Springer-Verlag, Aug. 20–24 2000, pp. 36–54. [Online]. Available: http://link.springer.de/link/service/series/0558/bibs/1880/18800036.htm

25. M. Murkowski and M. Wyden, "Protecting the rights of individuals act," U.S. Senate Bill (proposed), July 31 2003. [Online]. Available: http://thomas.loc.gov/cgi-bin/bdquery/z?d108:s.01552:

26. D. Naccache and J. Stern, "A new public key cryptosystem based on higher residues," in *Proceedings of the 5th ACM conference on Computer and communications security*. San Francisco. California. United States: ACM Press, 1998, pp. 59–66.

27. M. Naor and B. Pinkas. "Oblivious transfer and polynomial evaluation," in *Proceedings of the thirty-first annual ACM symposium on Theory of computing*. Atlanta. Georgia. United States: ACM Press, 1999, pp. 245–254.

28. T. Okamoto and S. Uchiyama, "A new public-key cryptosystem as secure as factoring." in *Advances in Cryptology – Eurocrypt '98. LNCS 1403*. Springer-Verlag, 1998, pp. 308–318.

29. S. R. M. Oliveira and O. R. Zaïane, "Protecting sensitive knowledge by data sanitization," in *Proceedings of the Third IEEE International Conference on Data Mining (ICDM'03)*. Melbourne, Florida, Nov. 19–22 2003.

30. P. Paillier, "Public key cryptosystems based on composite degree residuosity classes," in *Advances in Cryptology – Eurocrypt '99 Proceedings, LNCS 1592.* Springer-Verlag, 1999, pp. 223–238.

31. J. R. Quinlan, "Induction of decision trees," *Machine Learning*, vol. 1, no. 1, pp. 81–106, 1986. [Online]. Available: http://ipsapp008.kluweronline.com/content/getfile/4984/119/4/abstract.htm

32. S. J. Rizvi and J. R. Haritsa, "Maintaining data privacy in association rule mining," in *Proceedings of 28th International Conference on Very Large Data Bases.* Hong Kong: VLDB, Aug. 20–23 2002, pp. 682–693. [Online]. Available: http://www.vldb.org/conf/2002/S19P03.pdf

33. Y. Saygin, V. S. Verykios, and C. Clifton, "Using unknowns to prevent discovery of association rules," *SIGMOD Record*, vol. 30, no. 4, pp. 45–54, Dec. 2001. [Online]. Available: http://www.acm.org/sigmod/record/issues/0112/SPECIAL/5.pdf

34. J. Vaidya and C. Clifton, "Privacy preserving association rule mining in vertically partitioned data," in *The Eighth ACM SIGKDD International Conference on Knowledge Discovery and Data Mining*, Edmonton, Alberta, Canada, July 23–26 2002, pp. 639–644. [Online]. Available: http://doi.acm.org/10.1145/775047.775142

35. ——, "Leveraging the "multi" in secure multi-party computation," in *Workshop on Privacy in the Electronic Society held in association with the 10th ACM Conference on Computer and Communications Security*, Washington, DC, Oct. 30 2003.

36. ——, "Privacy-preserving *k*-means clustering over vertically partitioned data," in *The Ninth ACM SIGKDD International Conference on Knowledge Discovery and Data Mining*, Washington, DC, Aug. 24–27 2003.

37. ——, "Secure set intersection cardinality with application to association rule mining," *Journal of Computer Security*, submitted.

38. A. C. Yao, "How to generate and exchange secrets," in *Proceedings of the 27th IEEE Symposium on Foundations of Computer Science.* IEEE, 1986, pp. 162–167.

Printing: Strauss GmbH, Mörlenbach
Binding: Schäffer, Grünstadt